Next Generation Sequencing Based Clinical Molecular Diagnosis of Human Genetic Disorders

Lee-Jun C. Wong
Editor

Next Generation Sequencing Based Clinical Molecular Diagnosis of Human Genetic Disorders

Springer

Editor
Lee-Jun C. Wong
Molecular Diagnostic Laboratory
Baylor College of Medicine
Houston, Texas, USA

ISBN 978-3-319-56416-6 ISBN 978-3-319-56418-0 (eBook)
DOI 10.1007/978-3-319-56418-0

Library of Congress Control Number: 2017939356

Printed on acid-free paper

This Springer imprint is published by Springer Nature
The registered company is Springer International Publishing AG
The registered company address is: Gewerbestrasse 11, 6330 Cham, Switzerland

Preface

Recent development of high-throughput next generation sequencing (NGS) technology has transformed the way DNA-based molecular diagnostic testing is performed in clinical laboratories. In the past few years, clinically validated NGS has been applied to routine molecular diagnosis of human genetic diseases. In this book, we review the outcome of NGS in clinical practice.

Pitfalls of traditional PCR-based Sanger sequencing can be overcome by NGS. This book first reviews the technologies of NGS and their advantages over traditional Sanger sequencing and why NGS has become the new gold standard for clinical molecular diagnosis. In addition, clinically validated deep NGS can accurately detect not only single nucleotide variants (SNVs) but also copy number variants (CNVs). The ability to simultaneously detect and quantify SNVs and CNVs in multiple genes makes NGS an ideal comprehensive approach for molecular diagnosis. Applications of NGS to genetic analysis of various disease areas, such as metabolic disorders, retinal disease, hearing impairment, primary immunodeficiency, bone disorders, hereditary cancer, RASopathies, complex neuromuscular disorders, diabetes, cardiovascular genetic diseases, mitochondrial DNA-related disorders, whole exome sequencing in clinical setting, family-based studies, and carrier screening, are described in individual chapters. The complete coverage of clinical utility of NGS is of course beyond the scope of this book. We hope that through the examples described in this book, readers will have an overview of how NGS is utilized in clinical diagnostic laboratories.

I am indebted to contributing authors who have made this book possible.

Houston, Texas, USA　　　　Lee-Jun C. Wong

Contents

Overview of the Clinical Utility of Next Generation Sequencing in Molecular Diagnoses of Human Genetic Disorders

Lee-Jun C. Wong

Abstract Massively Parallel Sequencing, currently embodied as Next Generation Sequencing, has now been widely applied to clinical molecular diagnoses of human genetic disorders. This chapter describes the clinical utility of the Next Generation Sequencing (NGS), the scope of its application, its power in detecting DNA changes that were not previously possible by conventional methods, and its evolution into the new gold standard of molecular diagnoses.

Keywords Next generation sequencing • NGS • Clinical utility of NGS • Comprehensive molecular diagnosis by deep NGS

1 Introduction

The ability to massively parallel sequence a large number of genes by next generation sequencing (NGS) technologies has changed the paradigm of molecular diagnosis of human genetic disorders [1–5]. The NGS approach employs a broad spectrum of advanced chemistry, technological, computational and bioinformatics tools that exceed the performance of traditional gold standard Sanger sequencing for mutation detection at different levels to resolve clinically relevant questions in a more timely and cost effective manner. When the NGS technologies were first developed, they were primarily used for research purposes to facilitate mutation and gene discoveries. This was due to the complex sequencing technologies and novel analytical bioinformatics involved, in addition to the challenges of reliable clinical interpretation of a large number of variants [4, 5]. These difficult issues have been gradually resolved through vigorous clinical validation, meanwhile, the NGS technologies have been constantly improved and reliably applied to clinical diagnoses

L.-J.C. Wong, Ph.D. (✉)
Department of Molecular and Human Genetics, Baylor College of Medicine, Houston, TX, USA
e-mail: ljwong@bcm.edu

L.-J.C. Wong (ed.), *Next Generation Sequencing Based Clinical Molecular Diagnosis of Human Genetic Disorders*, DOI 10.1007/978-3-319-56418-0_1

during the past few years [1–3, 6–12]. This chapter provides a brief overview of the general application and results of NGS clinical practice in different disease areas, as well as the advances that NGS has brought to the molecular diagnostic paradigm.

2 Traditional Step-Wise Molecular Diagnostic Approaches Are No longer Adequate

The whole purpose of molecular diagnosis of human genetic disorders is to definitively identify and characterize the molecular etiology of the diseases such that accurate and prompt patient management can be administered. Furthermore, appropriate genetic counseling for families and prenatal diagnosis may be provided as an option to reduce the risk of having subsequently affected babies. Traditionally, prenatal risk assessment utilized a variety of approaches, including linkage markers, restriction fragment length polymorphism (RFLP), allele-specific oligonucleotide (ASO), allele refractory mutation system (ARMS), oligonucleotide ligation assay (OLA), pyrosequencing, and Sanger sequence analyses [13]. The methods used to search for unknown point mutations, such as single-strand conformation polymorphism (SSCP), denaturing gradient gel electrophoresis (DGGE), and temperature gradient gel electrophoresis (TGGE), temporal temperature gradient gel electrophoresis (TTGE), heteroduplex analyses (HDA), denaturing high-performance liquid chromatography (dHPLC), and protein truncation test (PTT) [13], require sequence analysis to confirm the identified nucleotide changes. Due to the tedious confirmation steps following the detection of changes, these mutation scanning methods have been replaced by direct Sanger sequencing. Thus, Sanger analysis became the gold standard for the analysis of single nucleotide variants (SNVs) before the NGS era.

The luster of gold standard Sanger sequencing has faded as other genetic and genomic tools have been developed to supplement genetic analyses. For example, Sanger sequencing does not detect large deletions or duplications unless one is fortunate enough to sequence the breakpoint. The traditional means of looking for deletion or duplication mutations has been Southern blot analysis. More recently, targeted oligonucleotide array comparative genome hybridization (aCGH) or single nucleotide polymorphism (SNP) array analyses have been developed for the detection of large chromosomal deletions/duplications [11, 14–16]. With the advent of the rapid discovery of new genes, genetic conditions, and phenotypic ranges, the number of recognized polygenic conditions has escalated. Thus, one-by-one analysis of single genes or candidate genes is not a cost and time efficient way to identify the molecular etiologies of diseases. Furthermore, PCR-based Sanger sequencing is notorious for allele dropout [2]. The need to address these drawbacks has spurred the birth of a comprehensive, effective next generation approach.

3 Next Generation Sequencing

Pitfalls of traditional PCR-based Sanger sequencing analyses are resolved by capture-based next generation sequencing methods that not only allow the simultaneous sequencing of many genes, but also allow the simultaneous detection of Copy Number Variations (CNVs) and even low level SNVs [17, 18]. The application of NGS to clinical diagnosis comes in several different scales. Depending on the size of the targeted region and the application purpose, the coverage depth varies. While complexity increases as the focus shifts from a single large gene, to a group of genes, to the whole exome (~20,000 genes), and ultimately, to the whole genome, the read depth coverage and the ability to call CNVs decrease.

The typical workflow from sample receiving to NGS analysis on an Illumina sequencing platform contains several steps: the DNA extraction, DNA fragmentation and target sequence enrichment, followed by library construction and sample indexing before loading onto the sequencer for cluster formation and sequence generation. The sequence images are then converted to base calls followed by filtering for high quality base calls, sequence alignment, analysis and variant annotation, and finally reporting and interpretation. Due to the batching of a large number of samples, the complexity of each step, the reagent costs, and the complex hardware and software utilized, quality control procedures are required to ensure that each step works properly. QC measures are also needed to ensure that and the final results are accurately and appropriately interpreted according to each patient's clinical presentation.

The sequence analyses consist of three major steps. The primary analysis involves the image capture, the conversion of image to base calls, and the assignment of quality scores to base calls. The secondary analysis is the filtering of reads based on quality followed by alignment and assembly of the reads. Finally, the tertiary analysis scores variant calls based on the NCBI's reference sequence, annotates the variants, and then interprets the results for reporting. The quality control procedures should be incorporated to monitor the performance of each step.

4 Target Sequence Enrichment and NGS Platforms

The whole human genome encodes about 20,000 genes. The targeted whole exome is only about 1–2% of the entire genome. For NGS panels, the total target sequence is usually no more than 1 Mb, and the number of targeted genes is about 5–500. Therefore, the genomic regions of interest in NGS panels are only a small fraction of the entire genome. This small target fraction needs to be enriched before NGS analysis. Two major methods are usually used for target sequence enrichment; the PCR-based gene-specific amplification and the capture-based hybridization to a library of probes within the target regions [17]. Regardless of the enrichment method, the DNA templates are fragmented to an optimal size and ligated with adaptors that contain common sequencing primers and sample indexes for subsequent sequence analysis [1, 12, 17, 19].

Which enrichment method is optimal depends on the number of genes and the characteristics of the target gene sequence structure. Amplicon-based enrichment has the pitfalls of allele drop out, limitation to scale up, and it cannot determine the *cis* or *trans* configuration when two heterozygous alleles are close enough. In addition, it is difficult to optimize the PCR conditions when the number of amplicons is large. Capture-based enrichment is less biased and is particularly efficient for large genes with a large number of exons. Uniform hybridization conditions are used for target sequence capture. Allele drop-out is not an issue. Since there is no bias of PCR efficiency, the variation of sequence read depth for a given individual exon is very small among different samples or different runs, thus, allowing CNV analysis [11].

Two major principles for the second generation sequencing (aka NGS); sequence by synthesis (Illumina, Roche 454, and Ion Torrent) and by ligation (SOLiD by Applied Biosystem). The chemistries for the detection of the three synthesis based sequencing methods are different. Roche 454 is based on the conversion of the released inorganic PPi from DNA synthesis to ATP, which in turn is used by luciferase to emit light [20]. The chemistry of Ion Torrent is similar to that of Roche 454, except that it measures minor pH changes caused by the nucleotide incorporation event(s) [21]. Since the release of PPi or proton is independent of which nucleotide is incorporated, sequential reactions with each of the four different nucleotidesare necessary. The detection of homopolymer stretches is by the quantification of cumulative signal intensity. Thus, it is difficult to accurately determine the number of nucleotides in a stretch of homopolymer, particularly when the number is large. For example, more than 3 or 4 of the same nucleotides in a row [22]. On the other hand, Illumina's platform detects different color of fluorescent emission released by each specific nucleotide one at a time. Thus, there is no ambiguity in detecting the exact number of homopolymers, since only one nucleotide is incorporated each time between signal measurements [17, 22]. The SOLiD platform uses numerous primers specific to target sequences for ligation. The read length is short and the reagent cost is extremely high. Consequently the SOLiD methodis not suitable for sequencing large targets. Constant improvements to the resolution, speed, capacity, and accuracy, have allowed the Illumina platforms to dominate the sequencing market. A comprehensive comparison of various NGS platforms can be found in reference [17].

5 Clinical Validation

The entire procedure of all NGS-based analysis for clinical molecular diagnosis must be validated. First, the performance parameters, including test specificity, sensitivity, positive predictive value, reproducibility, and accuracy need to be validated on fully characterized controls. Each performance parameter must be defined clearly [1, 19]. For example, if all target sequences are not analyzed, what percentage and which part of the target regions is not sufficiently covered? Are the insufficiently covered regions consistent from sample to sample and run to run? If so, are there strategies to completely cover these regions? The second step of validation is to determine if the analysis adequately detects all types of mutations including SNVs, CNVs, complex chromosomal rearrangement such as inversions, translocations, insertion of *Alu* repeat sequences, and conversion between active and pseudo genes, etc., using known mutation positive samples [2, 9, 11]. The step 3 of validation is to blindly analyze samples from patients with a specific clinical phenotype to determine the diagnostic yield using specifically designed panels.

For quantitative analysis of mosaicism and/or mitochondrial DNA (mtDNA) mutation heteroplasmy, the experimental errors and limit of detection must be defined [12]. Finally the protocols for variant classification should be documented and interpreted accordingly based on patient's clinical phenotype in the report [23].

6 Panel Design

Genes included in an NGS test panel are usually selected based on a common pathway, a defined phenotype, shared or similar phenotypes, known complex disorders, or the whole exome. Examples of pathway-based panels include: urea cycle disorders (UCD), glycogen storage diseases (GSD), congenital deficiency of glycosylation (CDG), fatty acid oxidation (FAO), cobalamin metabolism/elevated methylmalonic acid (cbl/MMA), peroxisomal disorders, pyruvate dehydrogenase complex deficiency, metabolic myopathy and rhabdomyolysis. Examples of phenotype-based panels include: Usher syndrome, retinal disease related [7, 24], hearing impairment and ciliopathies. Complex disorders include diabetes, epilepsies, hypoglycemia, Leigh disease, hereditary cancers [9], neuromuscular disorders [8], primary immunedeficiency or severe combined immunodeficiency (SCID), and mitochondrial disorders [1, 6, 12, 19, 25–28]. The individual NGS-based tests are described in individual chapters throughout this book. Many of these can be divided into sub-panels. For example, epilepsies may include infantile and metabolic epilepsies, while hereditary cancers may include specific cancer sites, such as breast, colon/GI, lung, brain and renal/pheochromocytoma/paraganglioma. Neuromuscular disorders can be divided into at least 15 sub-categories including congenital myopathy, metabolic myopathy, congenital muscular dystrophy, congenital myasthenic syndrome, motor neuron disorders, arthrogryposis, etc. [8] Perhaps the most complex are the mitochondrial disorders [1, 12, 26] which may involve 1500 nuclear encoded genes that are associated with mitochondrial structure and function, in addition to the mitochondrial genome, which encodes for 37 genes that are either integral components of the mitochondrial electron transport chain or directly involved in their synthesis [25]. Among the 1500 mitochondrial related genes, currently the following groups are best known: the protein components of each of the oxidative phosphorylation complexes I-V, complexes assembly genes, genes involved in mitochondrial genome biosynthesis and the maintenance of mitochondrial DNA (mtDNA) integrity, mitochondrial aminoacyl tRNA synthetases (ARS2's).

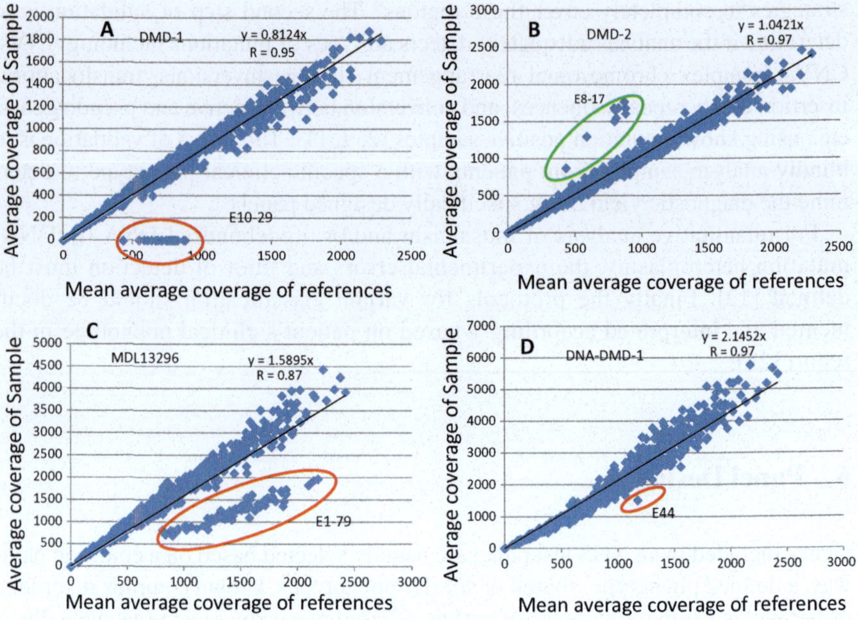

Fig. 1 One-step comprehensive NGS analysis of *DMD*. Each *dot* represents the normalized coverage depth of an exon. The *X*-axis is the coverage depth of the reference sample, and the *Y*-axis is the coverage depth of the test sample. Normal coverage of each exon places the *dots* on the 45 degree line. However, duplication in female (increase in coverage depth by 1.5 fold, from 2 copies to 3 copies) will place the *dots* above the 45° line, while deletion (from 2 copies to one copy) will place the *dot* below the *line*. (**a**) affected male with hemizygous deletion of exons 10–29, (**b**) female carrier with a heterozygous deletion of the entire *DMD* gene, (**c**) female carrier with a duplication of exons 8–17, and (**d**) a female carrier of one single exon deletion

7 Examples of Clinical Application

7.1 *The Advantage of NGS-Based Analysis of Large Genes*

Amplicon-based Sanger analyses for large genes or genes containing a large number of exons are time consuming and costly. Examples include *DMD* (79 exons for dystrophin), *TTN* (363 exons for titin), *RYR1* (106 exons), *NEB* (149 exons), all collagen genes (each with >50 small exons), and *BRCA1/2* genes (each contains a large exon >10 kb); these are more efficiently analyzed by capture-based NGS. A comprehensive capture-based NGS analysis for *DMD* is particularly important and practical. The majority (about 2/3) of molecular defects in *DMD* are large intragenic deletion or duplication, which can be detected by deep NGS, in an affected male (Fig. 1a) or a female carrier (Fig. 1b, d for deletion or Fig. 1c for duplication). Traditionally, the deletion/duplications were detected by Southern blot analysis. MLPA analysis has largely supplanted Southern blot analysis. Typically, if deletion/duplication analysis was negative, follow-up consisted of PCR amplification and Sanger sequencing of all 79 DMD exons. The procedures are tedious and expensive. Deep NGS simplifies the process by simultaneously detecting both t CNVs and SNVs.

Other examples of genes well suited to NGS analysis include the *BRCA1/2* genes. Not only are these large genes, but each contains an exon that is greater than 10 kb. Regardless of size, capture-based NGS analysis allows quick detection of SNVs and CNVs in the clinical laboratory settings with a fast turnaround time of 2 weeks and at a cost which is a small fraction of the combined cost of sequencing and array CGH.

7.2 Pathway Driven Panel Analysis

Defects in genes involved in the same metabolic pathway often result in similar clinical phenotypes. Depending on the affected step(s), the enzyme activities, accumulated metabolites and/or absent metabolites may be detected by enzyme assays or biochemical methods. However, enzymatic activity assays usually depend on invasive tissue biopsies and affected metabolites may not be easily detected. Thus, for definitive diagnosis, the identification of genetic defects is essential for appropriate patient management, accurate genetic counseling and family planning. Since a metabolic pathway may involve a number of genes with similar phenotype, analysis of genes one-by-one is not cost and time efficient. Therefore, massively parallel analysis of all the genes involved in the same pathway is the solution. The NGS-based metabolic panels are discussed in Chap. 3. Clinically available pathway-based gene panels evaluated by NGS may be found in GeneTests (https://www.genetests.org/) or individual laboratory websites (e.g. https://www.bcm.edu/research/medical-genetics-labs/).

8 Caveats of Capture-Based NGS

In general, regardless of the target size, increasing the amount of capture probes and/or total amount of DNA template input will increase the average coverage depth of a sample. However, this is not always true for all targeted regions (or exons). This is because some target regions have high GC content, short repeats, off-target homologous regions, pseudogenes, or complicated secondary structure preventing efficient sequencing. These regions are difficult to cover sufficiently, despite the increase in capture probes or input DNA. The number of insufficiently covered exons in a given panel depends on the target size and the DNA structural characteristics of the target sequences. For a 16 gene GSD panel of 294 exons, at an average coverage depth of 300× and above, all coding regions are sufficiently covered. However, if the average coverage depth is reduced to 200×, there will be 2–3 exons that are not sufficiently covered [3]. On the other hand, an Usher panel containing 9 large genes with a total of 362 exons always had 3 poorly covered exons even when the average coverage depth reached 500× and beyond. Due to the intrinsic DNA structure, the same 3 exons remain insufficiently covered even when the average coverage depth increases to 1500× or 2000×. The number of insufficiently covered exons increases to 28 for the 200 gene Mitome panel that contains 1790 exons. Thus, these insufficiently covered exons must be sequenced separately using a different strategy. Insufficiently covered exons are usually sequenced by designing specific or long range PCR primers for amplification followed by sequencing [2, 6, 9, 29]. A patient diagnosed with Leigh syndrome, supported by electron transport chain analysis showing isolated respiratory chain complex IV deficiency provides a case in point The DNA sample from this patient was analyzed by capture-based NGS for 200 mitochondrial related genes. The analysis identified a heterozygous deleterious frameshift mutation (c.817_826dup10) in exon 8 of the *SURF1* gene, which encodes a complex IV assembly factor. While SURF1 deficiency would be consistent with the clinical and biochemical phenotype, a second mutant allele could not be detected. Due to a known lack of coverage by capture/NGS analysis, exon1 of the *SURF1* gene was backfilled by PCR-based Sanger sequencing. The backfill revealed a heterozygous c.22C > T(p.Q8*) truncation mutation in exon 1. Consequently, this patient had a confirmed molecular diagnosis of SURF1 deficiency. This example underscores the importance of thorough validation, not only knowing what percentage of the target sequences are not sufficiently covered, but also knowing which regions are insufficiently covered so that they can be properly investigated with a second method.

9 Allele Dropout: Pitfalls of Amplicon-Based Analysis

Two common reasons for allele dropout in amplicon-based sequence analysis are SNPs at the PCR primer sites and large deletions including one or both of the primer sites. The variant (either SNV or CNV) in this region derived from the parent carrying the SNPs at the primer site(s) or the deletion will not be detected, while the variant in this region derived from the carrier parent will be detected as apparently homozygous [2, 3, 11]. Thus, amplicon-based sequencing may miss a pathogenic variant or an exonic deletion due to allele dropout.

10 Factors Affecting Diagnostic Yields of NGS-Based Panels

Diagnostic yields of target sequencing vary from panel to panel. Single gene analysis of *POLG*, a highly polymorphic gene, yields a low diagnostic rate (~5%) due to the extremely diverse clinical phenotype [30], whereas the diagnostic yield for the *OTC* gene is relatively high (>76%, 61% due to SNVs and 15% due to CNVs) [31]. This is because ornithine transcarbamoylase deficiency has a defined biochemical phenotype (highly elevated plasma ammonia) that can be easily detected and because mutations in the *OTC* gene are the most common cause of proximal urea cycle defects. OTC deficiency is an X-linked disease usually presenting in males, although carrier females may manifest as a consequence of skewed X-inactivation or metabolic crisis. Panels such as GSD [31], bone disease (osteoporosis and osteopetrosis) [32], non-syndromic retinitis pigmentosa (RP) [7] and Usher syndrome all have relatively defined clinical phenotype; consequently, the diagnostic yields are reasonably high (from 76% to 94%). Conversely, mitochondrial disorders are by far the most complex disorders. Mitochondrial production and function involves the entire mitochondrial genome and an estimated 1500 nuclear genes. A diagnostic yield of only about 25% is produced by analyzing 200 of the most commonly known genes associated with mitochondrial disease. Similarly, the diagnostic yield of a clinically undiagnosed genetic disorder analyzed by the whole exome, containing all 20,000 genes, is only about 26% [33].

Factors affecting yields of NGS-based target gene panels may include enrichment methods, the average coverage depth, the performance parameters of the NGS-based tests, number and sequence characteristics of the genes, the percentage of problematic exons, allele dropout, and strategic CNV analysis.

11 Conclusion

Thorough understanding of the principles, utilities, challenges and limitations of NGS will help with interpretation of the results and improvement of the technologies for future clinical application.

References

1. Zhang, W., Cui, H., Wong, L.J.: Comprehensive one-step molecular analyses of mitochondrial genome by massively parallel sequencing. Clin. Chem. **58**(9), 1322–1331 (2012)
2. Wang, J., et al.: Capture-based high-coverage NGS: a powerful tool to uncover a wide spectrum of mutation types. Genet. Med. **18**(5), 513–521 (2016)
3. Wang, J., et al.: Clinical application of massively parallel sequencing in the molecular diagnosis of glycogen storage diseases of genetically heterogeneous origin. Genet. Med. Off. J Am. Coll. Med. Genet. **15**(2), 106–114 (2013)
4. Metzker, M.L.: Sequencing technologies - the next generation. Nat. Rev. Genet. **11**(1), 31–46 (2010)
5. Bamshad, M.J., et al.: Exome sequencing as a tool for mendelian disease gene discovery. Nat. Rev. Genet. **12**(11), 745–755 (2011)
6. Yu, H., et al.: Rapid molecular diagnostics of severe primary immunodeficiency determined by using targeted next-generation sequencing. J. Allergy Clin. Immunol. **138**(4), 1142–1151 e2 (2016)
7. Wang, J., et al.: Dependable and efficient clinical utility of target capture-based deep sequencing in molecular diagnosis of retinitis pigmentosa. Invest. Ophthalmol. Vis. Sci. **55**(10), 6213–6223 (2014)
8. Tian, X., et al.: Expanding genotype/phenotype of neuromuscular diseases by comprehensive target capture/NGS. Neurol. Genet. **1**, e14 (2015). doi:10.1212/NXG.0000000000000015

9. Li, J., et al.: A comprehensive strategy for accurate mutation detection of the highly homologous PMS2. J. Mol. Diagn. **17**(5), 545–553 (2015)
10. Jones, M.A., et al.: Molecular diagnostic testing for congenital disorders of glycosylation (CDG): detection rate for single gene testing and next generation sequencing panel testing. Mol. Genet. Metab. **110**(1–2), 78–85 (2013)
11. Feng, Y., et al.: Improved molecular diagnosis by the detection of exonic deletions with target gene capture and deep sequencing. Genet. Med. Off. J Am. Coll. Med. Genet. **17**(2), 99–107 (2015)
12. Cui, H., et al.: Comprehensive next-generation sequence analyses of the entire mitochondrial genome reveal new insights into the molecular diagnosis of mitochondrial DNA disorders. Genet. Med. Off. J Am. Coll. Med. Genet. **15**(5), 388–394 (2013)
13. Landsverk, M.L., Wong, L.J.C.: Chapter 2. Clinical molecular diagnostic techniques: a brief review. In: Next Generation Sequencing: Translation to Clinical Diagnosis. Springer, New York. ISBN: 978-1-4614-7000-7 (2013)
14. Wang, J., et al.: Targeted array CGH as a valuable molecular diagnostic approach: experience in the diagnosis of mitochondrial and metabolic disorders. Mol. Genet. Metab. **106**(2), 221–230 (2012)
15. Stankiewicz, P., Lupski, J.R.: Structural variation in the human genome and its role in disease. Annu. Rev. Med. **61**, 437–455 (2010)
16. Chinault, A.C., et al.: Application of dual-genome oligonucleotide array-based comparative genomic hybridization to the molecular diagnosis of mitochondrial DNA deletion and depletion syndromes. Genet. Med. Off. J Am. Coll. Med. Genet. **11**(7), 518–526 (2009)
17. Hong, C., Wong, L.J.C.: Chapter 3. Methods of gene enrichment and massively parallel sequencing technologies. In: Next Generation Sequencing: Translation to Clinical Diagnostics. Springer, New York. ISBN: 978-1-4614-7000-7 (2013)
18. Qin, L., et al.: Detection and quantification of mosaic mutations in disease genes by next-generation sequencing. J. Mol. Diagn. **18**(3), 446–453 (2016)
19. Zhang, V.W., Wong, L.J.C.: Chapter 15. Validation of NGS-based DNA testing and implementation of quality control procedures. In: Next Generation Sequencing: Translation to Clinical Diagnosis. Springer, New York. ISBN: 978-1-4614-7000-7 (2013)
20. Margulies, M., et al.: Genome sequencing in microfabricated high-density picolitre reactors. Nature. **437**(7057), 376–380 (2005)
21. Rothberg, J.M., et al.: An integrated semiconductor device enabling non-optical genome sequencing. Nature. **475**(7356), 348–352 (2011)
22. Wicker, T., Schlagenhauf, E., et al.: 454 sequencing put to the test using the complex genome of barley. BMC Genomics. **7**, 275 (2006)
23. Wang, J., et al.: Chapter 6. Algorithms and guidelines for interpretation of DNA variants. In: Next Generation Sequencing: Translation to Clinical Diagnosis. Springer, New York. ISBN: 978-1-4614-7000-7 (2013)
24. Rui, C., et al.: Chapter 10. NGS analysis of heterogeneous retinitis pigmentosa. In: Next Generation Sequencing: Translation to Clinical Diagnosis. Springer, New York. ISBN: 978-1-4614-7000-7 (2013)
25. LJC, W.: Next generation molecular diagnosis of mitochondrial disorders. Mitochondrion. **13**(4), 379–387 (2013)
26. Wong, L.J.C.: Chapter 11. Next-generation sequencing analyses of the whole mitochondrial genome. In: Next Generation Sequencing: Translation to Clinical Diagnosis. Springer, New York. ISBN: 978-1-4614-7000-7 (2013)
27. Wong, L. (ed.): Next Generation Sequencing: Translation to Clinical Diagnostics. Chapter 11. Springer, New York. ISBN: 978-1-4614-7000-7E (2013)
28. Valeria, V., et al.: Chapter 12. Application of next-generation sequencing of nuclear genes for mitochondrial disorders. In: Next Generation Sequencing: Translation to Clinical Diagnosis. Springer, New York. ISBN: 978-1-4614-7000-7 (2013)
29. Yu, H., et al.: Detection of a novel intragenic rearrangement in the creatine transporter gene by next generation sequencing. Mol. Genet. Metab. **110**(4), 465–471 (2013)

30. Tang, S., et al.: Mitochondrial DNA polymerase gamma mutations: an ever expanding molecular and clinical spectrum. J. Med. Genet. **48**(10), 669–681 (2011)
31. Shchelochkov, O.A., et al.: High-frequency detection of deletions and variable rearrangements at the ornithine transcarbamylase (OTC) locus by oligonucleotide array CGH. Mol. Genet. Metab. **96**(3), 97–105 (2009)
32. Sule, G., et al.: Next-generation sequencing for disorders of low and high bone mineral density. Osteoporos. Int. J. Established. Result. Coop. Between. Eur. Found. Osteoporos. Natl. Osteoporos. Found. USA. **24**(8), 2253–2259 (2013)
33. Yang, Y., et al.: Clinical whole-exome sequencing for the diagnosis of mendelian disorders. N. Engl. J. Med. **369**(16), 1502–1511 (2013)

Detection of Copy Number Variations (CNVs) Based on the Coverage Depth from the Next Generation Sequencing Data

Yanming Feng, David Chen, and Lee-Jun C. Wong

Abstract Intragenic copy number variations (CNVs) in the human genome are significant contributors to the inherited genetic disorders. Currently the most established methods to detect CNVs are array comparative genomic hybridization (aCGH) and MPLA. With the fast adaption of next generation sequencing (NGS) in the clinical sequencing, increasing interest has been attributed to the detection of CNV from NGS data. In this chapter, we describe an easy-to-implement strategy to detect and visualize exonic CNVs from captured NGS data, as well as the confirmation. We also discuss the specificity and sensitivity of this strategy.

Keywords Exonic deletion • aCGH • Molecular diagnosis • Copy number variation • CNV • Next generation sequencing • NGS

1 Introduction

Intragenic copy number variations (CNVs) in the human genome are significant contributors to the inherited genetic disorders [10, 14]. It has been reported that approximately 12% of the human genome has CNV [11]. The pathogenicity of CNVs is variable, and the role of some pathogenic CNVs is still unknown. Intragenic CNVs involving genes matching the clinical phenotype are most likely pathogenic due to the change in gene dosage (whole gene deletion/duplication) or the disruption of the gene (out-of-frame exonic deletion/duplication). In clinical settings, if only a heterozygous pathogenic variant is identified in the candidate

Y. Feng, Ph. D. (✉) • D. Chen
Baylor Genetics, 2450 Holcombe BVLD, Houston, TX 77021, USA
e-mail: yanmingf@bcm.edu

L.-J.C. Wong
Baylor Genetics, 2450 Holcombe BVLD, Houston, TX 77021, USA

Department of Molecular and Human Genetics, Baylor College of Medicine, One Baylor Plaza, Houston, TX 77030, USA

L.-J.C. Wong (ed.), *Next Generation Sequencing Based Clinical Molecular Diagnosis of Human Genetic Disorders*, DOI 10.1007/978-3-319-56418-0_2

gene for an autosomal recessive disorder by sequencing, and the phenotype is consistent with the disease gene, search for the second mutant allele shifts to the identification of intragenic deletions or duplications.

Exon targeted array comparative genomic hybridization (aCGH) is currently the most commonly used approach for the detection of exonic CNVs [17, 18]. Since backbone probes throughout the genome are included in the exon targeted array, resolutions are ranging from a few hundred bases to kilobases (Kb), to megabases (Mb), and even the entire chromosome [17, 18]. Multiplex ligation-dependent probe amplification (MLPA) is another commonly used method for CNV detection. However, specially designed probes for individual exons are required, thus, it is difficult to use MLPA for large scale CNV analysis. Various methodologies for CNV detection may not be readily available to some clinical laboratories or the assays developed by individual laboratories may not include a complete set of genes or exons of interest for technical or commercial reasons.

In recent years, next generation sequencing (NGS) technologies have been widely used in the clinical practice of molecular diagnosis of human genetic diseases [5, 15, 16, 23]. Since 85% of all known mutations are located in the coding regions and the intron/exon junctions [4], capture-based target gene enrichment followed by NGS analysis has been a cost effective way to identify point mutations and small indels that are less than 20 bp in the target genes. NGS with consistently deep coverage of individual target exons can potentially provide an opportunity for concurrent detection of copy number changes and point mutations in patients with inherited disorders.

2 Strategies for NGS Based CNV Detection

NGS based CNV detection strategies can be divided into four categories based on sequence reads and coverage depth: (1) Paired-end mapping [2, 7, 8]; (2) Split-read [20]; (3) Depth of coverage [1, 3, 12, 19, 21]; and (4) Assembly based [9, 13, 22].

2.1 Paired-End Mapping Method

Paired-end mapping (PEM) methods require paired-end reads. The distance of paired-end reads is predetermined. If the distance of a pair of mapped reads is significantly larger than the distribution of the predetermined distance, a possible insertion may be identified. If shorter, a possible deletion can be identified. Some programs have been developed using paired-end mapping method, such as PEMer, BreakDancer, and Variation Hunber [2, 7, 8].

2.2 Split-Read

Split-read method also need paired-end reads. Unlike paired-end mapping method, in which the break points are not in the reads, the split-read method need one perfectly matched read and one read contains the breakpoint so that this read cannot be perfectly mapped to the reference genomic sequence. This unmatched read is then split into several fragments, and the first and the last fragments are mapped to the reference genomic sequence. The unmatched reads are split into several short fragments too short to be mapped to the genomic reference sequence. This split-read method usually requires long reads. Pindel is a split-read based program [20].

2.3 Depth of Coverage Based

The depth of coverage information is embedded in all NGS data, thus, depth of coverage based methods have become the main method for CNV detection. NGS results from both paired-end and single-end reads can be used for coverage depth based methods. Many programs have been developed using the depth of coverage information, such as SegSeq, CNVseq, Rdxplorer, CNVnator, and ExomeCNV [1, 3, 12, 19, 21]. The fundamental hypothesis of the depth of coverage based method is that the coverage is related to the copy number.

2.4 Assembly Based

In paired-end mapping, split-read, and depth of coverage methods, the reads need to bemapped to a reference genomic sequence. In contrast, the assembly-based method does not need a reference genome to map the reads. Instead, the reads are assembled without a reference genomic sequence. The assembled sequence is then compared to the genome sequence. The difference usually contains the structural variation information, including CNV. Velvet, ABySS and SOAPdenovo are all assembly based method [9, 13, 22].

Each of these strategies has its own strength and weakness, and maybe adopted for different purposes. The paired-end mapping based methods and the split-read methods can indicate the location of the CNV so it is easier to find the breakpoint. However, they cannot determine the exact copy number. These two methods also require paired-end reads. The depth of coverage (DOC) based method does not need additional specific algorithm because DOC information is already embedded in all NGS data. This is an important advantageous point because in clinical settings, the major NGS approach is captured based, either target panel or whole exome sequencing (WES). Thus, DOC strategy is readily applicable. The assembly based method is different from the other three in that it does not need reference genome sequence for mapping. However, it does need long reads with continuous coverage, thus, both the data collection and processing are time and cost consuming. The method is the least commonly used.

3 Procedures to Detect CNVs Based on Depth of Coverage

3.1 Reference Samples

Most DOC based CNV detection methods share the similar principle that is to compare the average coverage depth of a test sample to DOC of a reference. DOC of a reference is usually the mean or medium DOC of a group of samples that are analyzed in the same batch. An ideal reference is with the lowest coefficients of variations in the coverage depth. A few factors may contribute to variations. One is that there are intrinsic CNVs in the reference samples. These CNVs could be present in any samples depending on their allele frequencies. These are most likely benign. The others are rare, clinically significant CNVs that may be associated with disease phenotypes. We can select reference samples that do not contain CNVs in the genes of interest. These samples may be available publicly or in the individual laboratories that have validated the reference samples by a second method, such as aCGH. Still, variations in coverage depth maybe due to batch effects, including sample quality, sample processing, sample or exon specific differences, as well as instrumentation, technical, and other experimental variations. These types of variations are usually characterized and minimized during validation steps, although they cannot be completely removed. The reference DOC file can be generated by averaging DOC from a group of samples that do not contain CNVs in regions of interests. Since pathogenic CNVs are rare, to further minimize variations from batch effects, in routine practice, NGS results of at least 20 samples performed under the same conditions as the testing samples are grouped to generate the reference file.

3.2 DOC Based CNV Detection Using Exon as Sliding Window

Unlike whole genome sequencing, in which sequence data are continuous, the fundamental elements of capture based NGS are exons. Capture probes are designed for individual exons as regions of interests. NGS reads are grouped by exons and are not continuous due to the interruption of introns that are not captured and sequenced. Therefore, it is most reasonable to use exon as the sliding window.

3.3 Normalization of the Depth of Sequence Read to the Total Amount of DNA Loaded to Sequencing Machine

The amount of DNA template loaded to the sequencing machine naturally determines the total sequence reads generated, thus, it also affects the depth of coverage of individual exons. Although the loading amount of DNA template is carefully controlled for each sample, variation among different samples is inevitable. For CNV detection, accurate quantification of the number of sequence read is critical because the read depth is what CNV detection based upon. The DOC in the NGS data is not only determined by the copy number, but also by the amount of total target DNA loaded unto the flow cell and sequenced. Thus, before the DOC of testing sample and reference sample is compared, the total coverage of each individual sample is normalized for equal loading of total DNA template, which is determined by the total mapped reads.

3.4 Generation of Reference File

The reference file of DOC of exons is essentially the average DOC of a group of selected samples performed in the same NGS batch. There are two important values in the reference file that is used for exon based CNV detection algorithm. One is the mean value (μ) of the first normalized DOC of an exon, which is later used for the testing sample normalization/comparison. The other is the standard deviation (σ) of this mean value, from which the coefficient of variation (CV) is obtained. CV is an indicator of the quality of the reference file.

3.5 Normalization of DOC of the Testing Sample (Second Normalization)

Unlike reference samples, in order to detect CNVs, the DOC of the testing sample is normalized twice. First, it is similar to reference samples, the DOC of each exon in the testing sample is normalized against the total mapped reads. The normalized DOC of an exon is then normalized again to the mean DOC (μ) of the corresponding exon. The mean DOC (μ) is the average DOC of a specific individual exon in the reference file.

3.6 Detection and Visualization of CNVs

Ideally, the final normalized DOC of an exon with normal copy number is 1 or around 1. The secondary normalized DOC is 0.5 for exons with heterozygous deletion, and 0 for homozygous deletion. Duplication with a total of 3copies, the

Index	Gene	CDS	Average Reads	Normalized	Reference	Norm/Ref	copies	Standard Deviation	CV	CNV Call
948	PHKB	1	185.19	1.67E-05	1.46E-05	1.1474937	2.294987	2.91E-06	0.199643	-
949	PHKB	1b	1024.4	9.25E-05	0.0001054	0.8779857	1.755971	1.21E-05	0.115065	-
950	PHKB	2	589.23	5.32E-05	0.0001076	0.4947187	1.040827	1.86E-05	0.172962	del
951	PHKB	3	533.32	4.82E-05	0.000107	0.450048	0.946845	1.71E-05	0.159316	del
952	PHKB	4	640.32	5.78E-05	0.0001245	0.4644381	0.977121	1.83E-05	0.1471	del
953	PHKB	5	488.39	4.41E-05	9.67E-05	0.4564205	0.960252	1.78E-05	0.184325	del
954	PHKB	6	520.8	4.70E-05	8.73E-05	0.5387064	1.133372	1.90E-05	0.217116	del
955	PHKB	7	231.45	2.09E-05	5.31E-05	0.3934672	0.827807	1.10E-05	0.206516	del
956	PHKB	8	566.67	5.12E-05	0.0001069	0.4786702	1.007063	2.34E-05	0.218864	del
957	PHKB	9	455.19	4.11E-05	9.07E-05	0.4535737	0.954263	1.66E-05	0.1836	del
958	PHKB	10	565.74	5.11E-05	0.0001095	0.466699	0.981877	1.77E-05	0.161279	del
959	PHKB	11	782.62	7.07E-05	6.94E-05	1.0190121	2.038024	1.60E-05	0.229942	-
960	PHKB	12	845.94	7.64E-05	8.03E-05	0.9520423	1.904085	1.83E-05	0.228602	-
961	PHKB	13	1113.9	0.0001006	0.0001156	0.8705247	1.741049	1.72E-05	0.148717	-
962	PHKB	14	1179.4	0.0001065	0.0001188	0.8965683	1.793137	1.58E-05	0.133099	-
963	PHKB	15	829.32	7.49E-05	7.53E-05	0.9951613	1.990323	1.00E-05	0.132983	-
...	...	...	...	...	...	...	...	...	...	...

Fig. 1 Example of DOC normalization and CNV call from NGS data. *Norm/Ref* is the final normalized coverage. CV is the coefficient of variation of the reference. The values in the CNV Call column are the automatic CNV calls based on the Norm/Ref value and the CV value

normalized DOC is 1.5. However due to the technical variation and various genomic properties, the final normalized DOC is in a range. Different exons have different variations. We developed a combo CNV detection and visual checking algorithm, which includes automatic CNV detection from the statistical aspect, and a visualization method for visual checking. To balance the sensitivity and specificity and avoid false negatives, we have these settings: (1) if the normalized value is less than 1–1.5CV, it is scored as a deletion; (2) if the normalized value is greater than 1 + 1.5CV, it is scored as duplication; (3) if the normalized value is in between, then, it is considered normal. An example of heterozygous deletion of E2-E10 of *PHKB* is shown in Fig. 1, in which each exon captured and sequenced is normalized and CNV is scored as described above. In this figure, column Norm/Ref is the final normalized DOC. Column CV is the coefficient of variation of the reference. The values in the CNV Call column are automatic CNV calls based on the Norm/Ref value and the CV value.

We have also generated a custom UCSC track file from the normalized DOC and the genomic coordinates. This file can be uploaded to UCSC genome browser to visualize the results. One advantage of the customized track file is that multiple samples can be visually simultaneously and compared. An example is shown in Fig. 2, in which four custom tracks in the figure represent the final normalized DOC of *PHKB* exons of four different samples, including one positive sample in the blue box, which has *PHKB* E2-E10 heterozygous deletion. Each vertical bar represents an exon. The height of the bar is the copy number of this exon. Exons in the red box are exons with only one copy (heterozygous deletion).

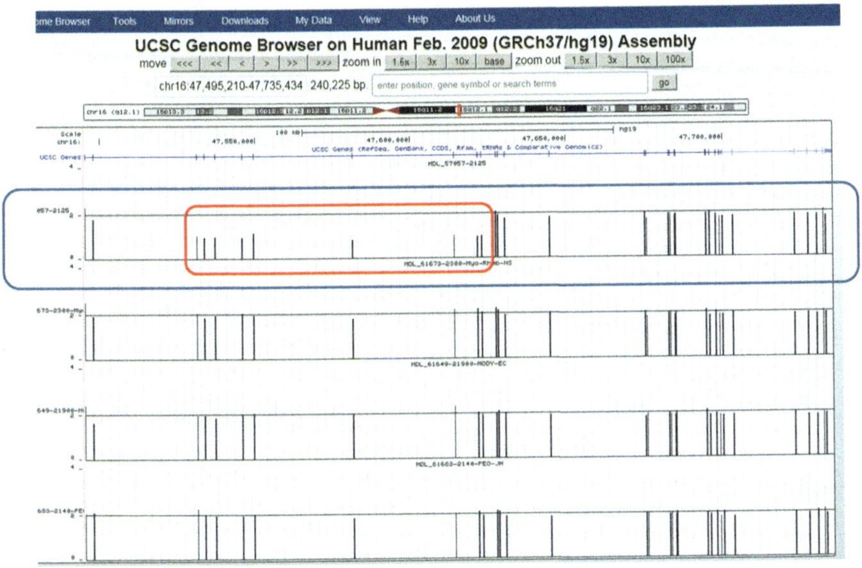

Fig. 2 Example of visualization of CNV call using UCSC genome browser with custom track. Four samples including one positive sample *circled in blue* and three negative samples are displayed together, along with the UCSC genes track which shows the genes and exons information. Nine exons with heterozygous deletions *circled in red* have half of the bar height of other normal exons

4 Confirmation of CNV

CNVs detected by DOC from NGS data can be confirmed by a second method, such as MPLA, aCGH or long range PCR (LR-PCR). High density aCGH is often used because it has the ability to reveal the boundary of the CNVs, if the breakpoints are not in the targeted exons. MLPA and LR-PCR are fast and cost effective ways to confirm exonic small CNVs and concurrent CNVs. Figure 3 is an example of the aCGH confirmation of a heterozygous deletion of E2-E10 of *PHKB* identified by using coverage based NGS data.

5 Sensitivity and Specificity

The sensitivity and specificity of NGS coverage based CNV detection was described previously [6]. In this paper, 12 validation samples were performed both NGS and aCGH, and the CNV detection results were compared. The total number of exons included in the comparison is 25,608. The sensitivity for the detection of deletion is 100% (9/9), but only 66.7% (2/3) for duplication. The specificity for the

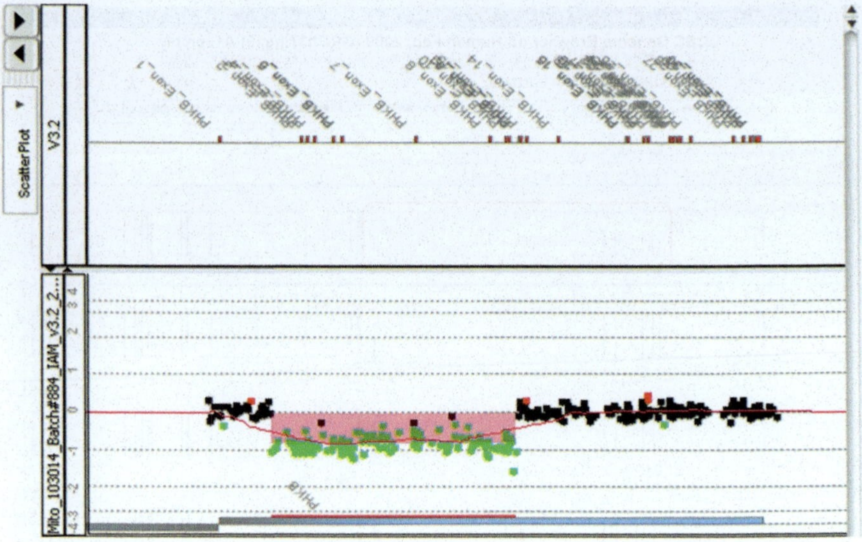

Fig. 3 aCGH confirmation of heterozygous deletion of PHKB E2-E10. Probes in *green* shows 1 copy of E2 to E10. Since there are probes in the intron region, aCGH usually can provide more information on the boundary of the deletion

detection of deletion and duplication is 99.92% and 99.86%, respectively. NGS coverage based CNV analysis is able to detect all deletions confirmed by aCGH at the single exon level without any false negative. The false positive rate of NGS based method is much higher for duplications (94.7%) than deletions (68.9%). The positive predicative value of duplication detection is only 5.3% (2/38). Even though all copy number losses detected by aCGH have been detected by NGS based analysis, the positive predicative rate is only 31% (9/29). This implies that all deletions detected by NGS based method require further confirmation with a second method, if the approach is to be used for clinical diagnostic purpose. In contrast, the negative predicative values for both deletions and duplications are 100%. This would suggest that a testing sample can be considered negative if the NGS based CNV analysis is negative.

6 Challenges and Issues

The most decisive step in the captured based NGS is the hybridization during the library preparation, which is affected by technical conditions and DNA properties. One outstanding factor is the GC content of the DNA. High GC content DNA is usually captured not as consistently as DNA with normal GC content. Some algorithms have been developed to correct the effect of GC content. However, our experience indicates that DNA with high GC content is more sensitive to subtle changes in experimental conditions during the hybridization step than DNA with normal GC content. So far, no good algorithms are able to take this intoaccount effectively. Exons with high GC content often show high coefficient of variations (CV). Fortunately, overall only less than 2% of all exons have high CV that CNVs cannot be determined reliably.

Another issue is the effect of homologous regions and pseudogenes on capture and sequencing coverage depth. Due to the presence of off-target high homologous sequences, the NGS data alignment software (aligner) sometimes cannot differentiate them to map sequences correctly. Therefore, the DOC may be distorted.

7 Future

Currently, the NGS based CNV detection algorithms have made great progress in the clinical utility in the diagnosis of inherited Mendelian diseases, in which the copy number of DNA is an integer, for example, 0, 1, 2 or 3. However, clinical utility of NGS has beengradually expanded to the detection of somatic mutations, in which the fraction of pathogenic variants is not present at 0%, 50% or 100%, as that is generally true for Mendelian mutations. It will be challenging and meaningful to investigate the performance of CNV detection in this situation.

Availability A script for the detection of CNVs used in this chapter was developed in Ruby and is available at https://sourceforge.net/projects/cnvanalysis.

References

1. Abyzov, A., Urban, A.E., Snyder, M., Gerstein, M.: CNVnator: an approach to discover, genotype, and characterize typical and atypical CNVs from family and population genome sequencing. Genome Res. **21**(6), 974–984 (2011)
2. Chen, K., Wallis, J.W., McLellan, M.D., Larson, D.E., Kalicki, J.M., Pohl, C.S., McGrath, S.D., Wendl, M.C., Zhang, Q., Locke, D.P., et al.: BreakDancer: an algorithm for high-resolution mapping of genomic structural variation. Nat. Methods. **6**(9), 677–681 (2009)
3. Chiang, D.Y., Getz, G., Jaffe, D.B., O'Kelly, M.J., Zhao, X., Carter, S.L., Russ, C., Nusbaum, C., Meyerson, M., Lander, E.S.: High-resolution mapping of copy-number alterations with massively parallel sequencing. Nat. Methods. **6**(1), 99–103 (2009)
4. Choi, M., Scholl, U.I., Ji, W., Liu, T., Tikhonova, I.R., Zumbo, P., Nayir, A., Bakkaloglu, A., Ozen, S., Sanjad, S., et al.: Genetic diagnosis by whole exome capture and massively parallel DNA sequencing. Proc. Natl. Acad. Sci. U. S. A. **106**(45), 19096–19101 (2009)
5. Cui, H., Li, F., Chen, D., Wang, G., Truong, C.K., Enns, G.M., Graham, B., Milone, M., Landsverk, M.L., Wang, J., et al.: Comprehensive next-generation sequence analyses of the entire mitochondrial genome reveal new insights into the molecular diagnosis of mitochondrial DNA disorders. Genet. Med. **15**(5), 388–394 (2013)
6. Feng, Y., Chen, D., Wang, G.L., Zhang, V.W., Wong, L.J.: Improved molecular diagnosis by the detection of exonic deletions with target gene capture and deep sequencing. Genet. Med. **17**(2), 99–107 (2015)

7. Hormozdiari, F., Alkan, C., Eichler, E.E., Sahinalp, S.C.: Combinatorial algorithms for structural variation detection in high-throughput sequenced genomes. Genome Res. **19**(7), 1270–1278 (2009)
8. Korbel, J.O., Abyzov, A., Mu, X.J., Carriero, N., Cayting, P., Zhang, Z., Snyder, M., Gerstein, M.B.: PEMer: a computational framework with simulation-based error models for inferring genomic structural variants from massive paired-end sequencing data. Genome Biol. **10**(2), R23 (2009)
9. Li, R., Zhu, H., Ruan, J., Qian, W., Fang, X., Shi, Z., Li, Y., Li, S., Shan, G., Kristiansen, K., et al.: De novo assembly of human genomes with massively parallel short read sequencing. Genome Res. **20**(2), 265–272 (2010)
10. Mills, R.E., Walter, K., Stewart, C., Handsaker, R.E., Chen, K., Alkan, C., Abyzov, A., Yoon, S.C., Ye, K., Cheetham, R.K., et al.: Mapping copy number variation by population-scale genome sequencing. Nature. **470**(7332), 59–65 (2011)
11. Redon, R., Ishikawa, S., Fitch, K.R., Feuk, L., Perry, G.H., Andrews, T.D., Fiegler, H., Shapero, M.H., Carson, A.R., Chen, W., et al.: Global variation in copy number in the human genome. Nature. **444**(7118), 444–454 (2006)
12. Sathirapongsasuti, J.F., Lee, H., Horst, B.A., Brunner, G., Cochran, A.J., Binder, S., Quackenbush, J., Nelson, S.F.: Exome sequencing-based copy-number variation and loss of heterozygosity detection: exomeCNV. Bioinformatics. **27**(19), 2648–2654 (2011)
13. Simpson, J.T., Wong, K., Jackman, S.D., Schein, J.E., Jones, S.J., Birol, I.: ABySS: a parallel assembler for short read sequence data. Genome Res. **19**(6), 1117–1123 (2009)
14. Stankiewicz, P., Lupski, J.R.: Structural variation in the human genome and its role in disease. Annu. Rev. Med. **61**(1), 437–455 (2010)
15. Tang, S., Wang, J., Zhang, V.W., Li, F.Y., Landsverk, M., Cui, H., Truong, C.K., Wang, G., Chen, L.C., Graham, B., et al.: Transition to next generation analysis of the whole mitochondrial genome: a summary of molecular defects. Hum. Mutat. **34**(6), 882–893 (2013)
16. Wang, J., Cui, H., Lee, N.C., Hwu, W.L., Chien, Y.H., Craigen, W.J., Wong, L.J., Zhang, V.W.: Clinical application of massively parallel sequencing in the molecular diagnosis of glycogen storage diseases of genetically heterogeneous origin. Genet. Med. **15**(2), 106–114 (2013)
17. Wang, J., Zhan, H., Li, F.Y., Pursley, A.N., Schmitt, E.S., Wong, L.J.: Targeted array CGH as a valuable molecular diagnostic approach: experience in the diagnosis of mitochondrial and metabolic disorders. Mol. Genet. Metab. **106**(2), 221–230 (2012)
18. Wong, L.J., Dimmock, D., Geraghty, M.T., Quan, R., Lichter-Konecki, U., Wang, J., Brundage, E.K., Scaglia, F., Chinault, A.C.: Utility of oligonucleotide array-based comparative genomic hybridization for detection of target gene deletions. Clin. Chem. **54**(7), 1141–1148 (2008)
19. Xie, C., Tammi, M.T.: CNV-seq, a new method to detect copy number variation using high-throughput sequencing. BMC Bioinf. **10**, 80 (2009)
20. Ye, K., Schulz, M.H., Long, Q., Apweiler, R., Ning, Z.: Pindel: a pattern growth approach to detect break points of large deletions and medium sized insertions from paired-end short reads. Bioinformatics. **25**(21), 2865–2871 (2009)
21. Yoon, S., Xuan, Z., Makarov, V., Ye, K., Sebat, J.: Sensitive and accurate detection of copy number variants using read depth of coverage. Genome Res. **19**(9), 1586–1592 (2009)
22. Zerbino, D.R., Birney, E.: Velvet: algorithms for de novo short read assembly using de Bruijn graphs. Genome Res. **18**(5), 821–829 (2008)
23. Zhang, W., Cui, H., Wong, L.J.: Application of next generation sequencing to molecular diagnosis of inherited diseases. Top. Curr. Chem. **336**, 19–45 (2012)

Next Generation Sequencing (NGS) Based Panel Analysis of Metabolic Pathways

Elizabeth B. Gorman and Lee-Jun C. Wong

Abstract Metabolic disorders are usually indicated by the presence of abnormal levels or absence of certain metabolite(s) due to defects in genes involved in common metabolic pathways that may consist of a series of reactions and other related reactions. One example is methylmalonic aciduria (MMA) due to defects in the conversion of methylmalonylCoA to succinylCoA, which requires a mutase and a cofactor; the adenosylcobalamin (adoCbl). Vitamin B12 is an inactive form of cobalamin that goes through at least eight steps of the cobalamin metabolic pathway to convert to its active form of AdoCbl. Thus, an NGS based comprehensive panel for MMA should include all these genes. There are hundreds of known disorders of metabolic pathways. In this chapter, we will describe the general approach to NGS based panel analysis, the design and gene selection of panels. We will provide a few examples. The advantages of deep coverage NGS panels over whole exome sequencing are also discussed.

Keywords Next Generation Sequencing • NGS • Metabolic pathways • Targeted NGS panels • Capture NGS • CNV • MSUD • UCD • GSD • Cobalamin • MODY • Hyperinsulinism • Hypoglycemia

1 Introduction

Massively parallel sequencing has been known as an effective approach to identify molecular defects in genetically heterogeneous disorders, particularly, if these disorders share the same pathway.

Metabolic disorders are well-suited for targeted parallel sequencing because they can be grouped into various metabolic pathways, each of which involves a series of biochemical reactions [1–6]. Defects in genes involving the same or related metabolic

E.B. Gorman (✉)
Baylor Genetics, Houston, TX, USA
e-mail: egorman@bcm.edu

L.-J.C. Wong
Baylor Genetics, Houston, TX, USA

Molecular and Human Genetics, Baylor College of Medicine, Houston, TX 77030, USA

L.-J.C. Wong (ed.), *Next Generation Sequencing Based Clinical Molecular Diagnosis of Human Genetic Disorders*, DOI 10.1007/978-3-319-56418-0_3

pathways result in the blockage of a specific step, thus, disrupt the pathway leading to a recognizable clinical or biochemical phenotype due to the accumulation or absence of certain metabolites that may be readily detectable by a simple biochemical analysis. However, for a definitive diagnosis of an inherited disease to be established, mutations in the causative gene(s) must be identified. Traditionally, the suspected defective genes are sequenced, one-by-one, by PCR based Sanger sequencing, until causative pathogenic mutations are identified. However, this approach is tedious, time consuming, and costly.

Metabolic Disorders are largely single gene disorders inherited in an autosomal recessive fashion, with some (see Table 2) X-linked. Examples include deficiencies of ornithine transcarbamylase, pyruvate dehydrogenase complex, and GSD Type IXa1 (pathogenic variants in *PHKA2*) (Table 1). Metabolic disorders are individually rare, but collectively common [7, 8]. According to a study in British Columbia conducted between 1969 and 1996, the overall incidence of inborn errors of metabolism was estimated to be 1 in 2500 births [7, 8]. A study in the Eastern Province of Saudi Arabia from 1983–2008 reported a slightly higher overall incidence of 3 in 2000 [8].

Table 1 Examples of metabolic NGS panels

Panel name	Number of genes	Total # exons	Target size (bp)	At 500-1000×, # exons <20×, homologous, special
Cobalamin Metabolism Disorders	20	239	43,567	0
Congenital Disorders of Glycosylation (CDG)	36	359	72,328	***ALG1* E6-E13,** *ALG9* E1, *COG1* E1, *MOGS* E1, *SRD5A3* E1, *TMEM165* E1
CoQ10 Deficiency	5	50	8665	*COQ2* E1, *PDSS1* E1, E2 & E6
Fatty Acid Oxidation Deficiency	22	272	44,788	***GLUD1* E1-E13, *HADHA* E20**
Glycogen Storage Disease (GSD) Comprehensive	23	355	61,878	***GAA* E18 del PCR, *GBE1* I15,** *GYG1* E1, ***PYGM*I5**
Hyperinsulinism	8	107	21,337	***GLUD1* E1-E13,** *INSR* E1
Hypoglycemia	87	1069	197,864	***ALG1* E6-E13,** *ALG9* E1, *COG1* E1, ***GLUD1* E1-E13,** *GYG1* E1, ***HADHA* E20,** *INSR* E1, *MOGS* E1, *SRD5A3* E1, *TMEM165* E1
Maple Syrup Urine Disease (MSUD)	4	44	7300	0
Maturity Onset Diabetes of the Young (MODY)	25	230	50,752	*CISD2* E3, *EIF2AK3* E1, *GATA6* E2, *KLF11* E1, *PDX1* E2, *PTF1A* E1
Peroxisomal Disorders	22	253	44,411	*PEX6* E1, *PEX7* E1, *PEX10* E1
Progressive Familial Intrahepatic Cholestasis (PFIC)	7	138	25,143	*JAG1* E1
UCD and Hyperammonemia	8	117	18,507	*NAGS* E1, ***SLC25A15*E2, E6 & E7**

Table 2 Clinically Validated Panels

Gene	NM#	HUGO name	Exons	Inheritence
Cobalamin Metabolism Panel: 21 genes, 250 CDS				
ABCD4	NM_005050.3	ATP Binding Cassette Subfamily D Member 4	19	AR
ACSF3	NM_174917.3	Acyl-CoA Synthetase Family Member 3	9	AR
CBS	NM_000071.2	Cystathionine-Beta-Synthase	15	AR
CD320	NM_016579.3	CD320 Molecule	5	AR
GIF	NM_005142.2	Gastric Intrinsic Factor (Vitamin B Synthesis)	9	AR
HCFC1	NM_005334.2	Host Cell Factor C1 (VP16-Accessory Protein)	26	XL
IVD	NM_002225.3	Isovaleryl-CoA Dehydrogenase	12	AR
LMBRD1	NM_018368.3	LMBR1 Domain Containing 1	16	AR
MCEE	NM_032601.3	Methylmalonyl-CoA Epimerase	3	AR
MMAA	NM_172250.2	Methylmalonic Aciduria (Cobalamin Deficiency) cblA Type	6	AR
MMAB	NM_052845.3	Methylmalonic Aciduria (Cobalamin Deficiency) cblB Type	9	AR
MMACHC	NM_015506.2	Methylmalonic Aciduria (Cobalamin Deficiency) cblC Type, with Homocystinuria	4	AR
MMADHC	NM_015702.2	Methylmalonic Aciduria and Homocystinuria, cblD Type	7	AR
MTHFR	NM_005957.4	Methylenetetrahydrofolate Reductase	11	AR
MTR	NM_000254.2	5-Methyltetrahydrofolate-Homocysteine Methyltransferase	33	AR
MTRR	NM_002454.2	5-Methyltetrahydrofolate-Homocysteine Methyltransferase Reductase	14	AR
MUT	NM_000255.3	Methylmalonyl-CoA Mutase	12	AR
SUCLA2	NM_003850.2	Succinate-CoA Ligase ADP-forming Beta Subunit	11	AR
SUCLG1	NM_003849.3	Succinate-CoA Ligase Alpha Subunit	9	AR

(continued)

Table 2 (continued)

Gene	NM#	HUGO name	Exons	Inheritence
SUCLG2	NM_003848.3	Succinate-CoA Ligase GDP-Forming Beta Subunit	11	No OMIM disease
TCN2	NM_000355.3	Transcobalamin 2	9	AR
Congenital Disorders of Glycosylation (CDG): 36 genes, 359 CDS				
ALG1	NM_019109.4	Chitobiosyldiphosphodolichol Beta-Mannosyltransferase	13	AR
ALG2	NM_033087.3	ALG2, Alpha-1,3/1,6-Mannosyltransferase	2	AR
ALG3	NM_005787.5	ALG3, Alpha-1,3- Mannosyltransferase	9	AR
ALG6	NM_013339.3	ALG6, Alpha-1,3-Glucosyltransferase	14	AR
ALG8	NM_024079.4	ALG8, Alpha-1,3-Glucosyltransferase	13	AR?
ALG9	NM_024740.2	ALG9, Alpha-1,2-Mannosyltransferase	16	AR
ALG11	NM_001004127.2	ALG11, Alpha-1,2-Mannosyltransferase	4	AR
ALG12	NM_024105.3	ALG12, Alpha-1,6-Mannosyltransferase	9	AR?
ALG13	NM_001099922.2	ALG13, UDP-N-Acetylglucosaminyltransferase Subunit	27	XL
ATP6V0A2	NM_012463.3	ATPase, H+ Transporting, Lysosomal V0 Subunit a2	20	AR
B4GALT1	NM_001497.3	UDP-Gal:BetaGlcNAc Beta 1,4- Galactosyltransferase, Polypeptide 1	6	AR
COG1	NM_018714.2	Component of Oligomeric Golgi Complex 1	14	AR?
COG4	NM_015386.2	Component of Oligomeric Golgi Complex 4	19	AR
COG5	NM_006348.3	Component of Oligomeric Golgi Complex 5	22	AR
COG6	NM_020751.2	Component of Oligomeric Golgi Complex 6	19	AR
COG7	NM_153603.3	Component of Oligomeric Golgi Complex 7	17	AR?
COG8	NM_032382.4	Component of Oligomeric Golgi Complex 8	5	AR?
DDOST	NM_005216.4	Dolichyl-Diphosphooligosaccharide--Protein Glycosyltransferase Subunit (non-catalytic)	11	AR
DOLK	NM_014908.3	Dolichol Kinase	1	AR

Gene	NM#	HUGO name	Exons	Inheritence
DPAGT1	NM_001382.3	Dolichyl-Phosphate (UDP-N-Acetylglucosamine) N-Acetylglucosaminephosphotransferase 1 (GlcNAc-1-P Transferase)	9	AR
DPM1	NM_003859.1	Dolichyl-Phosphate Mannosyltransferase Subunit 1, Catalytic	9	AR
DPM3	NM_153741.1	Dolichyl-Phosphate Mannosyltransferase Subunit 3	1	AR?
GNE	NM_005476.5	Glucosamine (UDP-N-Acetyl)-2-Epimerase/N-Acetylmannosamine Kinase	11	AR, AD
MGAT2	NM_002408.3	Mannosyl (Alpha-1,6-)-Glycoprotein Beta-1,2-N-Acetylglucosaminyltransferase	1	AR
MOGS	NM_006302.2	Mannosyl-Oligosaccharide Glucosidase	4	AR
MPDU1	NM_004870.3	Mannose-P-Dolichol Utilization Defect 1	7	AR
MPI	NM_002435.1	Mannose Phosphate Isomerase	8	AR
NGLY1	NM_018297.3	N-Glycanase 1	12	AR
PMM2	NM_000303.2	Phosphomannomutase 2	8	AR
RFT1	NM_052859.3	RFT1 Homolog	13	AR
SLC35A1	NM_006416.4	Solute Carrier Family 35 (CMP-Sialic Acid Transporter), Member A1	8	AR
SLC35A2	NM_001042498.2	Solute Carrier Family 35 (UDP-Galactose Transporter), Member A2	4	XL
SLC35C1	NM_018389.4	Solute Carrier Family 35 (GDP-Fucose Transporter), Member C1	2	AR
SRD5A3	NM_024592.4	Steroid 5 Alpha-Reductase 3	5	AR
TMEM165	NM_018475.4	Transmembrane Protein 165	6	AR
TUSC3	NM_006765.3	Tumor Suppressor Candidate 3	10	AR?

(continued)

Table 2 (continued)

Gene	NM#	HUGO name	Exons	Inheritence
CoQ10 Deficiency: 5 genes, 50 CDS				
ADCK3	NM_020247.4	Coenzyme Q8A	14	AR
COQ2	NM_015697.7	Coenzyme Q2, Polyprenyltransferase	7	AR
COQ9	NM_020312.3	Coenzyme Q9	9	AR
PDSS1	NM_014317.3	Prenyl (Decaprenyl) Diphosphate Synthase, Subunit 1	12	AR
PDSS2	NM_020381.3	Prenyl (Decaprenyl) Diphosphate Synthase, Subunit 2	8	AR
Fatty Acid Oxidation Deficiency: 22 genes, 272 CDS				
ACAD9	NM_014049.4	Acyl-CoA Dehydrogenase Family Member 9	18	AR
ACADL	NM_001608.3	Acyl-CoA Dehydrogenase, Long Chain	11	No OMIM disease
ACADM	NM_000016.4	Acyl-CoA Dehydrogenase, C-4 to C-12 Straight Chain	12	AR
ACADS	NM_000017.2	Acyl-CoA Dehydrogenase, C-2 to C-3 Short Chain	10	AR
ACADVL	NM_000018.2	Acyl-CoA Dehydrogenase, Very Long Chain	20	AR
CPT1A	NM_001876.3	Carnitine Palmitoyltransferase 1A	18	AR
CPT1B	NM_004377.2	Carnitine Palmitoyltransferase 1B	18	No OMIM disease
CPT2	NM_000098.2	Carnitine Palmitoyltransferase 2	5	AR
ETFA	NM_000126.3	Electron Transfer Flavoprotein Alpha Subunit	12	AR
ETFB	NM_001985.2	Electron Transfer Flavoprotein Beta Subunit	6	AR
ETFDH	NM_004453.2	Electron Transfer Flavoprotein Dehydrogenase	13	AR
GLUD1	NM_005271.3	Glutamate Dehydrogenase 1	13	AD
HADH	NM_005327.4	Hydroxyacyl-CoA Dehydrogenase	8	AR
HADHA	NM_000182.4	Hydroxyacyl-CoA Dehydrogenase/3-Ketoacyl-CoA Thiolase/Enoyl-CoA Hydratase (Trifunctional Protein), Alpha Subunit	20	AR
HADHB	NM_000183.2	Hydroxyacyl-CoA Dehydrogenase/3-Ketoacyl-CoA Thiolase/Enoyl-CoA Hydratase (Trifunctional Protein), Beta Subunit	15	AR

Gene	NM#	HUGO name	Exons	Inheritence
HMGCL	NM_000191.2	3-Hydroxymethyl-3-Methylglutaryl-CoA Lyase	9	AR
HMGCS2	NM_005518.3	3-Hydroxy-3-Methylglutaryl-CoA Synthase 2 (mitochondrial)	9	AR
HSD17B10	NM_004493.2	Hydroxysteroid 17-Beta Dehydrogenase 10	6	XL
LPIN1	NM_145693.2	Lipin 1	19	AR
SLC22A5	NM_003060.2	Solute Carrier Family 22 Member 5	10	AR
SLC25A20	NM_000387.5	Solute Carrier Family 25 Member 20	9	AR
TAZ	NM_000116.3	Tafazzin	11	XL
Glycogen Storage Disease (GSD): 23 genes, 355 CDS				
AGL	NM_000642.2	Amylo-Alpha-1, 6-Glucosidase, 4-Alpha-Glucanotransferase	33	AR
FBP1	NM_000507.3	Fructose-Bisphosphatase 1	7	AR
G6PC	NM_000151.2	Glucose-6-Phosphatase Catalytic Subunit	5	AR
GAA	NM_00152.3	Glucosidase Alpha, Acid	19	AR
GBE1	NM_000158.3	1,4-Alpha-Glucan Branching Enzyme 1	16	AR
GYG1	NM_004130.3	Glycogenin 1	8	AR
GYS1	NM_002103.4	Glycogen Synthase 1	16	AR
GYS2	NM_021957.3	Glycogen Synthase 2	16	AR
LAMP2	NM_002294.2	Lysosomal-Associated Membrane Protein 2	9	XL
LDHA	NM_005566.3	Lactate Dehydrogenase A	7	AR
PCK1	NM_002591.3	Phosphoenolpyruvate Carboxykinase 1 (soluble)	9	AR
PCK2	NM_004563.2	Phosphoenolpyruvate Carboxykinase 2 (mitochondrial)	10	AR
PFKM	NM_000289.5	Phosphofructokinase, Muscle	22	AR
PGAM2	NM_000290.3	Phosphoglycerate Mutase 2	3	AR
PGM1	NM_002633.2	Phosphoglucomutase 1	11	AR

(continued)

Table 2 (continued)

Gene	NM#	HUGO name	Exons	Inheritence
PHKA1	NM_002637.3	Phosphorylase Kinase Regulatory Subunit Alpha 1	32	XL
PHKA2	NM_000292.2	Phosphorylase Kinase Regulatory Subunit Alpha 2	33	XL
PHKB	NM_000293.2	Phosphorylase Kinase Regulatory Subunit Beta	31	AR
PHKG2	NM_000294.2	Phosphorylase Kinase Catalytic Subunit Gamma 2	9	AR
PYGL	NM_002863.4	Phosphorylase, Glycogen, Liver	20	AR
PYGM	NM_005609.2	Phosphorylase, Glycogen, Muscle	20	AR
SLC2A2	NM_000340.1	Solute Carrier Family 2 (Facilitated Glucose Transporter), Member 2	11	AR
SLC37A4	NM_001467.5	Solute Carrier Family 37 Member 4	8	AR
Hyperinsulinism: 8 genes, 107 CDS				
ABCC8	NM_000352.4	ATP-Binding Cassette, Sub-Family C (CFTR/MRP), Member 8	39	AR, AD
GCK	NM_000162.3	Glucokinase (Hexokinase 4)	10	AD
GLUD1	NM_005271.3	Glutamate Dehydrogenase 1	13	AD
HADH	NM_005327.4	Hydroxyacyl-CoA Dehydrogenase	8	AR
HNF4A	NM_175914.4	Hepatocyte Nuclear Factor 4, Alpha	10	AD
INSR	NM_000208.2	Insulin Receptor	22	AD
KCNJ11	NM_000525.3	Potassium Inwardly-Rectifying Channel, Subfamily J, Member 11	1	AD
SLC16A1	NM_003051.3	Solute Carrier Family 16 (Monocarboxylate Transporter), Member 1	4	AD
Maple Syrup Urine Disease (MSUD): 4 genes, 44 CDS				
BCKDHA	NM_000709.3	Branched Chain Keto Acid Dehydrogenase E1, Alpha Polypeptide	9	AR
BCKDHB	NM_183050.2	Branched Chain Keto Acid Dehydrogenase E1 Subunit Beta	10	AR

Gene	NM#	HUGO name	Exons	Inheritence
DBT	NM_001918.3	Dihydrolipoamide Branched Chain Transacylase E2	11	AR
DLD	NM_000108.3	Dihydrolipoamide Dehydrogenase	14	AR
Maturity Onset Diabetes of the Young (MODY): 25 genes, 230 CDS				
ABCC8	NM_000352.4	ATP-Binding Cassette, Sub-Family C (CFTR/MRP), Member 8	39	AD/AR
AKT2	NM_001626.5	V-Akt Murine Thymoma Viral Oncogene Homolog 2	13	AD
BLK	NM_001715.2	B Lymphoid Tyrosine Kinase	12	AD
CISD2	NM_001008388.4	CDGSH Iron Sulfur Domain 2	3	AR
CP	NM_000096.3	Ceruloplasmin (Ferroxidase)	19	AR
EIF2AK3	NM_004836.5	Eukaryotic Translation Initiation Factor 2-Alpha Kinase 3	17	AR
FOXP3	NM_014009.3	Forkhead Box P3	11	X-Linked
GATA6	NM_005257.4	GATA Binding Protein 6	6	AD
GCK	NM_000162.3	Glucokinase (Hexokinase 4)	10	AD
GLIS3	NM_152629.3	GLIS family zinc finger 3	9	AR
HNF1A	NM_000545.5	HNF1 homeobox A	10	AD
HNF1B	NM_000458.2	HNF1 homeobox B	9	AD
HNF4A	NM_175914.4	Hepatocyte Nuclear Factor 4, Alpha	10	AD
IER3IP1	NM_016097.4	Immediate Early Response 3 Interacting Protein 1	3	AR
INS	NM_000207.2	Insulin	2	AD
KCNJ11	NM_000525.3	Potassium Inwardly-Rectifying Channel, Subfamily J, Member 11	1	AD/AR
KLF11	NM_003597.4	Kruppel-Like Factor 11	4	AD
NEUROD1	NM_002500.4	Neuronal Differentiation 1	1	AD
NEUROG3	NM_020999.3	Neurogenin 3	1	AR
PAX4	NM_006193.2	Paired Box 4	9	AD

(continued)

Table 2 (continued)

Gene	NM#	HUGO name	Exons	Inheritence
PDX1	NM_000209.3	Pancreatic and Duodenal Homeobox 1	2	AR
PTF1A	NM_178161.2	Pancreas Specific Transcription Factor, 1a	2	AR
RFX6	NM_173560.3	Regulatory Factor X, 6	19	AR
SLC2A2	NM_000340.1	Solute Carrier Family 2 (Facilitated Glucose Transporter), Member 2	11	AR
WFS1	NM_006005.3	Wolfram Syndrome 1 (Wolframin)	7	AD/AR
Peroxisomal Disorders: 22 genes, 253 CDS				
ACOX1	NM_004035.6	Acyl-CoA Oxidase 1, Palmitoyl	14	AR
AGPS	NM_003659.3	Alkylglycerone Phosphate Synthase	20	AR
AMACR	NM_014324.5	Alpha-Methylacyl-CoA Racemase	5	AR
DNM1L	NM_012062.3	Dynamin 1-Like	20	AD
GNPAT	NM_014236.3	Glyceronephosphate O-Acyltransferase	16	AR
HSD17B4	NM_000414.3	Hydroxysteroid (17-Beta) Dehydrogenase 4	24	AR
PEX1	NM_000466.2	Peroxisomal Biogenesis Factor 1	24	AR
PEX2	NM_153818.1	Peroxisomal Biogenesis Factor 2	6	AR
PEX3	NM_003846.2	Peroxisomal Biogenesis Factor 3	4	AR
PEX5	NM_000286.2	Peroxisomal Biogenesis Factor 5	3	AR
PEX6	NM_002618.3	Peroxisomal Biogenesis Factor 6	4	AR
PEX7	NM_004565.2	Peroxisomal Biogenesis Factor 7	9	AR
PEX10	NM_004813.2	Peroxisomal Biogenesis Factor 10	11	AR
PEX11B	NM_002857.3	Peroxisomal Biogenesis Factor 11 beta	8	AR
PEX12	NM_000318.2	Peroxisomal Biogenesis Factor 12	1	AR
PEX13	NM_017929.5	Peroxisomal Biogenesis Factor 13	5	AR
PEX14	NM_003630.2	Peroxisomal Biogenesis Factor 14	12	AR
PEX16	NM_001131025.1	Peroxisomal Biogenesis Factor 16	15	AR
PEX19	NM_000287.3	Peroxisomal Biogenesis Factor 19	17	AR

Gene	NM#	HUGO name	Exons	Inheritence
PEX26	NM_000288.3	Peroxisomal Biogenesis Factor 26	10	AR
PHYH	NM_006214.3	Phytanoyl-CoA 2-Hydroxylase	9	AR
SCP2	NM_002979.4	Sterol Carrier Protein 2	16	AR
Progressive Familial Intrahepatic Cholestasis (PFIC): 7 genes, 138 CDS				
ABCB4	NM_000443.3	ATP Binding Cassette Subfamily B Member 4	27	AR, AD
ABCB11	NM_003742.2	ATP Binding Cassette Subfamily B Member 11	27	AR
AKR1D1	NM_005603.4	Aldo-Keto Reductase Family 1, Member D1	27	AR
ATP8B1	NM_000214.2	ATPase Phospholipid Transporting 8B1	26	AR, AD
JAG1	NM_005989.3	Jagged 1	9	AD
SLC25A13	NM_000295.4	Solute Carrier Family 25 Member 13	4	AR
SERPINA1	NM_014251.2	Serpin Peptidase Inhibitor, Clade A (Alpha-1 Antiproteinase, Antitrypsin), Member 1	18	AR
Urea Cycle Disorders and Hyperammonemia: 8 genes, 117 CDS				
ARG1	NM_000045.3	Arginase 1	8	AR
ASL	NM_000048.3	Argininosuccinate Lyase	16	AR
ASS1	NM_000050.4	Argininosuccinate Synthase 1	14	AR
CPS1	NM_001875.4	Carbamoyl-Phosphate Synthase 1	38	AR
NAGS	NM_153006.2	N-Acetylglutamate Synthase	7	AR
OTC	NM_000531.5	Ornithine Carbamoyltransferase	10	XL
SLC25A13	NM_014251.2	Solute Carrier Family 25 Member 13	18	AR
SLC25A15	NM_014252.3	Solute Carrier Family 25 Member 15	6	AR

Often, biochemical testing provides the first suspicion of a metabolic disorder. The abnormal increase or absence of certain metabolites in newborn screening may be suggestive of a specific disorder; for example, elevated phenylalanine for PKU, hyperammonemia for a urea cycle disorder; or a disorder of a specific pathway such as glycogen storage disease sharing the common features of hypoglycemia, hepatomegaly and/or myopathy, and even more complex metabolic disorders such as mitochondrial diseases or Leigh syndrome that the common metabolic indication may be elevated lactic acid. These complex metabolic disorders are genetically heterogeneous and can be caused by pathogenic variants in many different genes. The Next Generation Sequencing (NGS) using the probe hybridization capture based enrichment approach to analyze a group of target genes is an efficient way to provide a definitive genetic diagnosis and can be tailored based on the clinical indication. As NGS technology improves, turnaround time for NGS panels can be greatly reduced to potentially less time than currently necessary for a single Sanger sequencing test. This allows for prompt patient management or treatment, which is particularly important for the critically ill patient.

In this chapter, we will describe our experience in the clinical diagnosis of several metabolic disorders using the capture based NGS approach. We will discuss our approach to panel design as well as the pros and cons compared to other NGS platforms.

2 NGS Based Panel Analysis: General Approach

Molecular diagnosis is becoming easily accessible to most patients and clinicians. With the goal of precision medicine, accurate genetic diagnoses are critical for appropriate treatment options. In order to deliver a genetic test that provides a definitive diagnosis in a cost effective and timely manner, the traditional method based on sequential single-gene stepwise PCR-Sanger sequencing is no longer sufficient. The new comprehensive approach should allow sequencing all possible candidate disease causing genes simultaneously in one single step. In addition, single nucleotide variants in the coding exons do not account for all of the molecular etiologies of inherited diseases. The detection of large chromosomal and exonic copy number variations (CNVs) usually requires a second method, such as chromosomal array comparative genomic hybridization (aCGH) or exon targeted oligonucleotide aCGH [9]. Furthermore, coding regions with high GC content, repetitive sequence, and the homologous sequences of copy genes or pseudogenes require special approaches in order to ensure 100% sequence coverage of the regions of interest. Figure 1 depicts the flowchart of the general approach to NGS based panel analysis. Details are explained in the following sections.

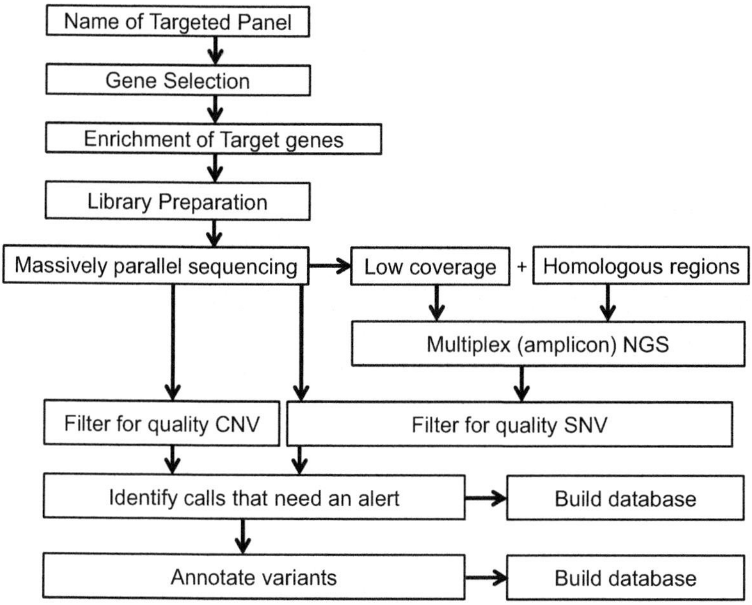

Fig. 1 Workflow of targeted gene panel design and implementation

2.1 Design of NGS-Based Panels

The first step to NGS panel (Fig. 1) analysis is to decide the purpose of the panel and which target genes to be included. NGS panels can range in scope from targeting as simple as a specific metabolic pathway or disease, to as complex as mitochondrial disorders that can include all related pathways and hundreds of genes. When designing a targeted NGS panel, several points need to be considered, depending on which pathways or diseases are to be analyzed. For example, a hyperammonemia due to urea cycle disorders (UCD) panel typically would include carbamoyl-phosphate synthetase I (CPS1), N-acetylglutamate Synthase (NAGS, NAG is a cofactor used by CPS1), Ornithine Carbamoyltransferase (OTC), Argininosuccinate Synthase 1 (ASS1), Argininosuccinate Lyase (ASL), and Arginase 1 (ARG1). UCDs can be further divided into proximal or distal UCDs by plasma citrulline levels. Proximal UCD has low or no citrulline, while ASS1, ASL, and ARG1 deficiency cause elevation of citrulline. However, citrullinemia type 2 caused by citrin, a mitochondrial Ca^{++} binding aspartate/glutamate carrier (AGC, SLC25A13) deficiency also results in a citrulline level similar to that of ASL, called citrullinemia type 2. Thus, if just based on the elevated metabolite of citrulline, citrin deficiency should be included as a differential diagnosis. Citrin deficiency causes neonatal intrahepatic cholestasis, a very different clinical phenotype from that of UCD. Similarly, the hyperornithinemia-hyperammonemia-homocitrullinuria (HHH) syndrome caused by defects in ornithine translocase, a mitochondrial ornithine transporter, SLC25A15, is a differential diagnosis for hyperornithinemia and hyperammonemia. Thus, for an expanded panel, Solute Carrier Family 25 Mitochondrial Aspartate/Glutamate Carrier) Member 13 (SLC25A13) and Solute Carrier Family 25 (Mitochondrial Carrier; Ornithine Transporter), Member 15 (SLC25A15) are included in the panel (Table 1). Furthermore, deficiency in carbonic anhydrase VA, CA5A, a mitochondrial carbonic anhydrase responsible for providing bicarbonate substrate for CPS1, results in hyperammonemia and UCD has recently been reported to be an important gene for UCD [10]. Thus, a comprehensive UCD panel should contain all 9 genes. The same approach for gene inclusion can be applied to other panels (Table 1). Examples of panels and their development will be individually described in Sect. 3 below.

2.2 *Enrichment of Target Genes*

After target genes have been selected for a panel (Fig. 1), the next step is to enrich the target genes from the total DNA extract. Enrichment of target genes is usually accomplished by two major methods; PCR amplification using target sequence specific primers, and capture by hybridization to target probes. These methods have been described and compared previously [11] and are described in individual chapters in this book. Enrichment by capture has its advantages as; (a) avoiding allele drop out (i.e. -failure of PCR amplification due to variants within the primer binding sequences, including deletion of the region), (b) determination of phase if the variants in question are within the same NGS read (i.e. if two variants are in the same copy of the chromosome or in opposite copies), (c) simultaneous detection of copy number variations (CNVs) with sufficient read depth, (d) determination of breakpoint sequences if junction is in exons or regions captured by probes, (e) avoiding numerous PCRs if the gene contains a large number of exons, for example, *DMD* and *TTN*, (f) flexibility of target size from a single exon, single gene to the whole exome containing >200,000 exons. However, regions containing repetitive sequences or homologous sequences outside the target region, or pseudogenes, may require long range PCR or gene specific PCR to resolve these problems. In this chapter, we focus on capture based NGS panels.

2.3 *Massively Parallel Sequencing: Clinical Validation*

Once the target genes have been enriched the constructed DNA template library is loaded onto a next generation sequencing machine (Fig. 1). Several different massively parallel sequencing platforms are available for clinical application [11]. Different laboratories develop different panels, use different target gene enrichment methods, and apply different NGS platforms for clinical diagnosis. Therefore, in a CLIA certified laboratory, all methods require clinical validation [12, 13]. Performance parameters in individual laboratories should include test sensitivity, specificity, accuracy, and positive predictive values, etc. If there are exons or regions that are not sufficiently covered by capture NGS, it should be clarified which regions. Furthermore, if the laboratory claims 100% coverage, then, the approaches to obtain 100% coverage should be specified.

2.4 *Low Coverage and Homologous Regions*

After validation of the sequence analysis, it may be noted that certain regions exhibit insufficient coverage. This usually refers to coverage below 20×. The number of bases or exons with insufficient coverage depends on the overall average coverage. As shown in Fig. 2, by examining the coverage depth of the same 162 genes in the Mitome200 panel (capture library contains 500 nuclear genes) and WES (capture library contains 23,000 genes), the number of exons that are insufficiently covered is significantly reduced from about 200 in WES (average coverage is about 100×) to about 20 in Mitome200 panel (average coverage depth is about 1000×). Increasing the capture probes in these regions does not help to increase the average coverage depth, and increasing the coverage depth beyond 1000× does not seem to improve

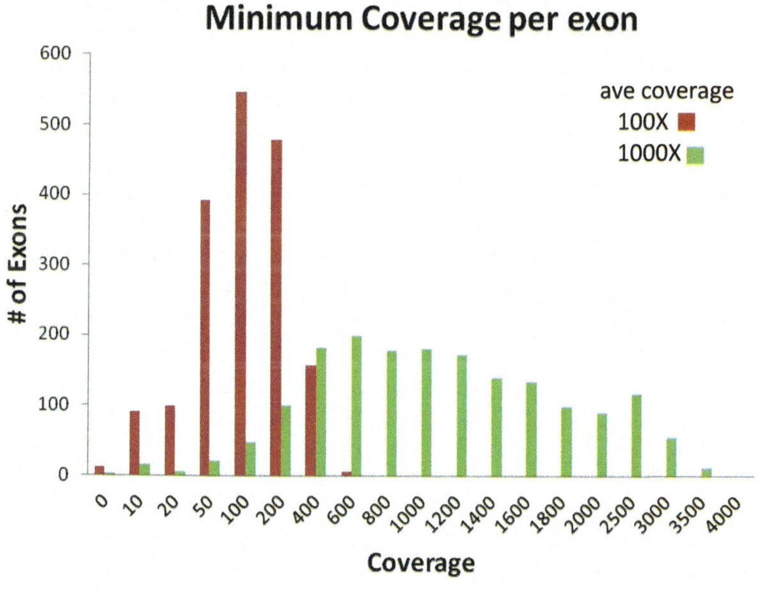

Fig. 2 Distribution of average coverage depth versus number of exons. The number of insufficiently covered exons increases as the average coverage depth decreases, from about 20 for 1000× coverage to ~200 at 100× coverage

the number of insufficiently covered exons. This is because these exons are either GC rich or contains repetitive sequences or certain intrinsic sequence structures that prevent efficient sequencing. Regions containing homologous sequences outside the regions of interest, such as pseudogenes or copy genes may also present problems in capture, sequencing, or alignment. To solve the problems of insufficient coverage, the majority of these regions can be analyzed by using sequence specific primers for PCR based gene enrichment followed by sequencing. If the number of low coverage exons is high, the gaps may be filled by multiplex PCR followed by NGS. However, highly repetitive regions or homologous pseudogenes are more difficult to resolve. Specific methods may have to be used [14, 15].

2.5 Variant Calls: Single Nucleotide Variants (SNVs) and CNVs

The NGS sequence results (Fig. 1) are analyzed by using analytical pipelines that may vary among different laboratories. Some may use commercially available software, some may use in-house developed software [13, 16]. Regardless of the software used, the goal is to reduce the number of variant calls to a manageable number for prioritization and annotation. Most of these algorithms are used for SNV calls, not for CNVs. However, with deep coverage, Feng et al. has reported the clinical utility of NGS for CNV calls [17]. Traditionally, if after sequencing, CNV is suspected, a targeted array CGH is usually performed to remove the suspicion of a CNV. The ability to detect both SNV and CNV in one single step is a big leap towards the "comprehensive" diagnostic approach. Public databases can be used to facilitate sequence analysis and variant annotation. However, due to variations among different laboratories, individual laboratories should build their own database to make the filtering process more efficient. Efficient quality read filtering should find a balance between the detection of true positives and actually filtering out the noise. It is also becoming apparent that a large, well maintained database of recurrent false positives and real annotated variants is a very valuable resource. Finally, interpretation and reporting of variants require American Board of Medical Genetics certified laboratory directors who explain the causative variants of the disease based on their expertise in the correlation of molecular findings with clinical phenotype. Details of guidelines of variant classification and interpretation have been documented [18, 19].

3 Examples of Panels

3.1 Overview of Panels

Examples of clinical panels are listed in Table 1 with their names, number of genes included, total number of exons and total target size in base pairs (bp). The last column lists the exons of the genes that routinely do not provide sufficient read coverage depth (<20×) or exons that require a special approach. As indicated, the

MSUD panel is a small panel focused on a specific disease which is caused by defects in one of 4 genes with a total of 44 coding sequences (CDS), a target size of 7300 bp and no patching required for low coverage exons. The hypoglycemia panel is a large panel targeted to a phenotype that involves multiple pathways (87 genes, 1069 CDS, a target size of 197,864 bp and 29 CDS that require patching due to low coverage, high homology, or problematic regions). The remaining 11 panels in Table 1 include from 5 to 36 genes with characteristics intermediate to those of the MSUD and hypoglycemia panels. Small panels do not necessarily require less patching. For example, the cobalamin pathway (with a target size of 43,567 bp) panel does not need PCR/sequencing for patching, while CoQ10 deficiency panel (5 genes, 8665 bp) and the peroxisome disorder panel (22 genes, 44,411 bp), each needs to fill in three low coverage regions (Table 1). In general, capture of a single gene followed by NGS is not an efficient way, unless the gene is large or contains large number of exons, such as *DMD*. In addition, if the target size is too small, the percentage of off-target sequences will be significant, thus, it will affect the overall sequence quality. Therefore, regardless of the size of the target regions, the size of a capture library is preferably maintained at around 100 kb to 800 kb. Multiple small panels can be combined by using one common capture library. However, sequence analysis and reporting may focus on the target genes only [2, 4–6, 20, 21].

Any aspects of metabolism could be targeted for an NGS panel. In addition to the clinically validated metabolic pathway/disease NGS panels included in Table 1, other examples ranging from very specific to very general, such as 3-Methyl Glutaconic Aciduria, Aminoacidemia, Congenital Adrenal Hyperplasia (CAH), Non-Ketoic Hypoglcinemia, Sulfur/Molybdenum Metabolism Disorders, Mucopolysaccharidosis, Lysosomal Storage Disease (LSD), organic acidopathies, and complex dual genome mitochondrial disorders. The availability of these NGS based panel tests and many more can be found in GENETests at https://www.genetests.org/.

3.2 *Brief Description of Some Clinically Validated Panels*

Table 2 lists the gene symbols and gene names in each panel, number of exons of each gene, their reference mRNA sequence number and mode of inheritance.

The simplest type of panel is the specific metabolic pathway or disease targeted panel. These types of panels are often used to confirm or determine the defective gene in the metabolic pathway where the problem lies as indicated by the presence/absence or accumulation of a diagnostic metabolite. Below are some examples of currently clinically available panels for molecular diagnosis of various metabolic pathways.

3.2.1 Maple Syrup Urine Disease (MSUD)

MSUD is an example of a specific disease targeted panel. Four genes are typically included in this disease panel: Branched Chain Keto Acid Dehydrogenase E1, Alpha Polypeptide (BCKDHA), Branched Chain Keto Acid Dehydrogenase E1, Beta

Polypeptide (BCKDHB), Dihydrolipoamide Branched Chain Transacylase E2 (DBT) & Dihydrolipoamide Dehydrogenase (DLD) [22]. Protein products of these four genes make up the branched-chain alpha-keto acid dehydrogenase complex, which is necessary to break down the branched amino acids; isoleucine, leucine, and valine. These are essential amino acids that are present in almost all foods. The inability to metabolize these amino acids due to BCKDH complex defects leads to toxic buildup of isoleucine, leucine, valine, and the presence of the diagnostic compound alloisoleucine. The characteristic maple syrup sweet smell is from sotolone (4,5-dimethyl-3-OH-2[5H]-furanone), a derivative of isoleucine. Major clinical features of MSUD include lethargy, coma, opisthotonous, convulsions, and development delay. Similar NGS panel approaches can be applied to other alpha-keto acid dehydrogenase complexes, such as glycine cleavage enzyme (GCE) and pyruvate dehydrogenase complex (PDHC). Defects in the mitochondrial glycine cleavage system have been found to cause nonketotic hyperglycinemia (NKH), characterized by severe encephalopathy. The GCE is a complex of four proteins: P protein (a pyridoxal phosphate-dependent glycine decarboxylase; GLDC), H protein (a lipoic acid-containing protein; GCSH), T protein (a tetrahydrofolate-requiring aminomethyl transferase AMT; or GCST), and L protein (a lipoamide dehydrogenase; DLD). Most patients with GCE have a defect in the GLDC gene. The PDHC also contains four similar proteins; the alpha and beta subunits of the decarboxylase: PDHA1 and PDHB; the lipoic acid acetyl transferase: DLAT; and the dihydrolipoyl dehydrogenase: DLD. Among the four proteins, only the DLD is shared by all alpha ketoacid dehydrogenase complexes. The other three proteins are unique for individual enzyme complexes.

3.2.2 Urea Cycle Disorders (UCD)

Urea Cycle Disorders (UCD) are an example of specific pathway targeted panel. The rationale and development of this panel has been described in Sect. 2.1 above. The majority of mutations in UCD genes are point mutations that can be easily detected by NGS. However, point mutations account for only about 61% of OTC deficiency. About15% of the *OTC* mutations are due to CNVs, particularly, large deletions. A deep coverage NGS based UCD panel can detect both SNVs and CNVs simultaneously, without having to run exon targeted aCGH for the detection of exonic deletions [17]. The six genes; *CPS1, NAGS, OTC, ASS1, ASL, and ARG1*; UCD panel can be expanded to include *SLC25A13*, *SLC25A15*, and *CA5A*, for differential diagnosis. The Urea Cycle removes toxic ammonia (formed from nitrogenous waste) from the blood by converting it to urea which is removed from the body via urine. It is critical to reduce high levels of ammonia in the blood and to make definitive molecular diagnosis as soon as possible for prompt and precise life-saving patient management [23].

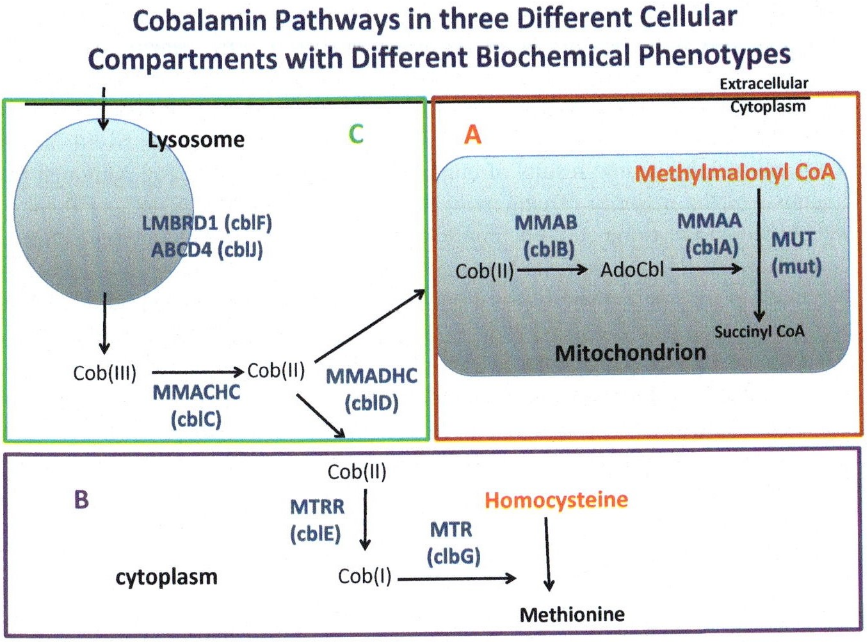

Fig. 3 The cobalamin pathways in three different cellular compartments with different biochemical phenotypes

3.2.3 The Cobalamin Pathway

This pathway involves the absorption and conversion of hydroxyl-cobalamin (Cbl, vitamin B12) to active forms of the cofactors, 5′-deoxyadenosyl cobalamin (AdoCbl) and methyl cobalamin (MeCbl), as well as the reactions that require these cofactors for catalysis (Fig. 3). Methylmalonyl-CoA mutase uses AdoCbl to convert methylmalonylCoA to succinylCoA, and the methionine synthase requires MeCbl to catalyze the methylation of homocysteine to methionine. Disorders of cobalamin metabolism include defects of absorption, transport and intracellular metabolism of cobalamin, and may impair the function of either or both enzymes. Diagnosis of these inborn errors has traditionally depended on metabolic screening tests such as urine organic acid and plasma amino acid analyses, as well as complementation assays. In general, defects affecting AdoCbl synthesis; such as mitochondrial cobalamin reductase and adenosyltransferase (*cblA, cblB)* or the mutase itself *(mut*); result in methylmalonic acidemia and acidura (MMA) (Fig. 3, box A). Defects affecting the cytoplasmic synthesis of MeCbl (*cblE, cblG*) result in homocystinuria (HC) (Fig. 3 box B). Defects affecting the release of hydroxylcobalamin from the lysosome (*cblF, cblJ*) or reduction of Co(III) to Co(II) in cytoplasm (*cblC, cblD*), result in combined MMA and HC (Fig. 3, box C). In addition, mutations in the gene for transcobalamin, *TCN2,* also lead to cobalamin transport defects and can present with MMA and HC. Thus, a basic panel for the cobalamin pathway and methylmalonic aciduria may contain *MMAA (cblA), MMAB (cblB), MMACHC (cblC), MMADHC (cblD), MTRR (cblE), LMBRD1 (cblF), MTR (cblG),* and *ABCD4 (cblJ)*. Before panel NGS was developed, the clinical diagnosis of these disorders was carried out by stepwise single gene analysis of these candidate genes based on the abnormal metabolite and results of complementary studies [21, 24]. Although the diagnosis of the majority of patients with an elevation of MMA, HC, or both, in their blood and/or urine could be confirmed by the analysis of these nine genes, a significant fraction of patients remained undiagnosed [24]. Thus, the panel is expanded to include potential causal genes that are not expressed in cultured skin fibroblasts, such as *GIF, AMN*, and *CUBN* (a large gene) (Table 3); and *SUCLA2, SUCLG1 and SUCLG2*, that may cause persistently mildly elevated MMA, as the differential genes (Table 3) [24]. This panel can be further extended to a total of 24 genes that include some pathway related genes and genes for differential diagnosis, such as *HCFC1* and *CD320* (a transcobalamin receptor). Indeed, patients undiagnosed by complementation assays were found to have mutations in these genes [24]. This illustrates how a panel can be continually expanded as new genes are being discovered.

3.2.4 Glycogen Storage Diseases (GSDs) and Other Similar Metabolic Pathways

Glycogen Storage Diseases (GSDs) are a group of genetically and clinically heterogeneous inborn errors of glycogen synthesis or catabolism, including glucose metabolism, glycolysis and gluconeogenesis. As listed in Tables 1 and 2, the genes are involved in glycogen synthesis, glycogen breakdown, glycolysis, and gluconeogenesis (for example *PCK1* and *PCK2*). It also includes glucose transporter (SLC2A2) and glucose-6-P transporter (SLC37A4). LAMP2 is a lysosomal associated membrane protein 2 that is responsible for autophagic myopathy and Danon disease. Both skeletal and cardiac muscles may be involved, resulting in a clinical phenotype resembling that of GSDII Pompe disease. Thus, it is included as a differential. Clinical utility of NGS based analysis of GSD panel has been reported [2]. Although GSDs can be divided into muscle forms and liver forms (Table 4), due to significant clinical overlap and considerable clinical variability, it is more effective as a single panel for all GSDs.

A similar NGS approach can be applied to other metabolic pathways as well, such as fatty acid beta oxidation (FAO), peroxisomal disorders (PD), and congenital disorders of glycosylation (CDG), listed in Table 2. CDG is more clinically and genetically heterogeneous compared to others, as it involves O- and N- glycosylation and a rapidly expanding number of genes. Clinically, it overlaps with neuromuscular and mitochondrial disorders both of which are by far the most genetically complex disease groups. The clinical utility of NGS in the diagnosis of CDG has been reported [1, 25].

The NGS based analysis of genes involving other disorders such as lysosomal storage diseases and mucopolysaccharidosis can be designed similarly.

Table 3 Extended cobalamin panel

	Gene symbol	Gene name	Phenotype
1	*TCN2*	Transcobalamin II	Transcobalamin II deficiency
2	*MMAA*	Methylmalonic aciduria, cblA type	Methylmalonic aciduria, vitamin B12 responsive
3	*MMAB*	Methylmalonic aciduria, cblB type	Methylmalonic aciduria, vitamin B12 responsive, due to defect in synthesis of AdoCbl
4	*MMACHC*	Methylmalonic aciduria and homocystinuria, cblC type	Methylmalonic aciduria and homocystinuria, cblC type
5	*MMADHC*	Methylmalonic aciduria and homocystinuria, cblD type	Methylmalonic aciduria and homocystinuria, cblD type
6	*MTRR*	Methionine synthase reductase	Homocystinuria-megaloblastic anemia, cblE type
7	*LMBRDI*	LMBR1 domain-containing protein 1	Methylmalonic aciduria and homocystinuria, cblF type
8	*MUT*	Methylmalonyl-CoA mutase	Methylmalonic aciduria, mut(0) type
9	*IVD*	Isovaleryl-CoA Dehydrogenase	Isovaleric acidemia
10	*MTHFR*	5,10-methylenetetrahydrofolate reductase	Homocystinuria due to MTHFR deficiency
11	*ABCD4*	ATP-binding cassette, subfamily D, member 4	Methylmalonic aciduria and homocystinuria, cblJ type
12	*ACSF3*	Acyl-CoA synthetase family, member 3	Combined malonic and methylmalonic aciduria
13	*HCFC1*	Host cell factor C1	Methylmalonic acidemia and homocysteinemia, cblX type, mental retardation
14	*MCEE*	Methylmalonyl-CoA epimerase	Methylmalonyl-CoA epimerase deficiency
15	*MTR*	Methionine synthase	Homocystinuria-megaloblastic anemia, cblG type
16	*GIF*	Gastric intrinsic factor	Intrinsic factor deficiency
17	*SLC46A1*	Solute carrier family 46 (folate transporter), member 1	Hereditary folate malabsorption
18	*CUBN*	Cubilin	Megaloblastic anemia
19	*AMN*	Amnionless	Megaloblastic anemia
20	*CBS*	Cystathionine beta-synthase	Homocystinuria, B6-responsive and nonresponsive types
21	*CD320*	Transcobalamin receptor	Methylmalonic aciduria due to transcobalamin receptor defect
22	*SUCLA2*	Succinate-CoA ligase, ADP-forming, beta subunit	Mitochondrial DNA depletion syndrome 5 (encephalomyopathic with or without methylmalonic aciduria)
23	*SUCLG1*	Succinate-CoA ligase, alpha subunit	Mitochondrial DNA depletion syndrome 9 (encephalomyopathic type with methylmalonic aciduria)
24	*SUCLG2*	Succinate-CoA ligase, GDP-forming, beta subunit	Mitochondrial DNA depletion syndrome

Gene symbol and full name are provided in the first two columns. Phenotype that results from mutations in the gene is listed in the third column

Table 4 Liver and muscle forms of GSD

GSD types	Genes	Liver panel	Muscle panel
GSD 0A	*GYS2*	√	
GSD 0B	*GYS1*		√
GSD IA	*G6PC*	√	
GSD IB	*SLC37A4*	√	
GSD II	*GAA*	√	√
GSD III	*AGL*	√	√
GSD IV	*GBE1*	√	
GSD V	*PYGM*		√
GSD VI	*PYGL*	√	
GSD VII	*PFKM*		√
GSD IX A	*PHKA2*	√	
GSD IX B	*PHKB*	√	√, mild
GSD IX C	*PHKG2*	√	
GSD IX D	*PHKA1*		√
GSD X	*PGAM2*		√
GSDXIV	*PGM1*		√

3.2.5 Maturity-Onset Diabetes of the Young (MODY)

Maturity-Onset Diabetes of the Young (MODY) is a group of monogenic forms of diabetes. MODY patients have decreased insulin levels resulting in less control of blood sugar and high blood sugar (hyperglycemia) that can lead to eye, kidney, nerve, and blood vessel damage. MODY is often misdiagnosed as Type I or Type 2 Diabetes. Although most cases of MODY can be ascribed to mutations in a handful of genes (Table 2), up to about 30 genes have been implicated. Targeted MODY panels have been found to be very useful both to differentiate from Type I or Type 2 diabetes and also to identify the rarer causes of MODY [26].

3.2.6 Hyperinsulinism and Hypoglycemia

While decrease in insulin level can cause hyperglycemia, increase in insulin level leads to reduced blood sugar level, a condition called hypoglycemina when blood glucose levels fall below the normal fasting glucose levels of 3.5–5.5 mmol/L. Hyperinsulinemic hypoglycemia (HH) is due to the disturbance of the tightly regulated insulin secretion such that insulin continues to be secreted despite hypoglycemia, HH is caused by defects in genes regulating insulin secretion from the beta cells (Table 2). [27] It is interesting to note that several genes appear in both hyperinsulinism and MODY panels. For example, when KCNJ11 (Potassium Inwardly-Rectifying Channel, Subfamily J, Member 11), a potassium ATP channel loses its function, it causes autosomal recessive hyperinsulinemic hypoglycemia (OMIM 601820), whereas dominant activating mutations in KCNJ11 cause neonatal diabetes mellitus or MODY (https://www.ncbi.nlm.nih.gov/omim).

In addition to HH, defects in metabolic pathways such as GSD, CDG, FAO, PD, and mitochondrial disorders may also result in hypoglycemia [28]. Thus, a hypoglycemia panel could be made by combining the appropriate targeted panels and hyperinsulinism (Table 2) to increase the detection rate should a more definitive metabolic disorder not be indicated by biochemical or clinical studies. There is a considerable degree of genetic heterogeneity in any panel based on a common clinical finding such as hypoglycemia.

4 Points for Consideration

4.1 Single Gene Versus Panel Analysis, Panel Overlap

If a patient presents with defined clinical features that can be explained by a single gene, it is probably most efficient to simply sequence the gene. OTC deficiency is a good example. An affected male infant presents with neonatal hyperammonemia and lethargy with almost zero plasma citrulline. Sanger sequencing of the *OTC* gene provides high diagnostic yield for classical cases with point mutations (SNVs). However, approximately 15–20% of the *OTC* defects are due to exonic or large deletions that cannot be detected by Sanger alone. Clinically validated deep NGS can simultaneous detect both SNVs and CNVs, Nevertheless, hyperammonemia and low plasma citrulline levels could also be due to CPS1, NAGS, or CA, that cause proximal UCD. Female patients may also have OTC deficiency due to skewed X-inactivation. Hyperammonemia and high citrulline could be due to defects in distal UCD genes. Due to phenotype overlaps and atypical cases, a panel approach may be more efficient.

There is also significant genetic and clinical overlap for metabolic panels such as metabolic myopathies that may be caused by defects in GSD, FAO, mitochondrial, and other individual nuclear genes. CDG has significant clinical overlap with mitochondrial and/or neuromuscular disorders. Thus, for comprehensive coverage and effective diagnosis, it is advantageous to have a large capture library including probes for closely related pathways and disorders.

4.2 Panel Size and Capture Library Size

Panel size may vary with the purpose of the panel. Small targeted panels containing only the genes of the highest proven clinical value can be very efficient. These panels are very effective for the critically ill patient for whom the most likely results need to be determined as quickly as possible. If, for example, 90% of the pathogenic variants that have been reported for a certain clinical phenotype/disease are in 7 genes then these 7 genes would make a good tier 1 panel. Examples are proximal UCD and severe combined immunodeficiency (SCID). If negative or when more time can be taken, testing more genes with less common pathogenic variants may make up the remaining 10% of the causative genes.

The size of a targeted panel depends on the number and size of the targeted genes. A gene may be as small as a couple hundred base pairs with a single exon or as large as the *DMD* gene that contains 79 coding exons and a mRNA of about 16 kb. The largest proteins known to date is titin at 4200 kD and a cDNA of more than 110 kb. Thus, the size of a targeted panel can range from a few kilo-bases to the entire exome that is 23,000 genes and about 50 Mb. In general the size of a capture library is equal or larger than the size of a specific panel. Currently, the largest probe hybridization based capture library for clinical molecular diagnosis is the whole exome. Since any panels are included in a particular capture library, panel size, in general, does not affect sequencing quality or coverage depth. Panel size depends on the design and can affect the diagnostic yield. It is critical to combine clinicians and geneticists with the genetic disease expertise who can advise on which genes to be included in a panel. Potential disease causing novel genes may be included if there is a reason to suspect its involvement to allow expansion of phenotype and genotype. Genes causing similar clinical phenotype may also be included in order to rule out a disease that is phenotypically difficult to distinguish.

One strategy is to create an all-inclusive comprehensive panel. However, inflated panel size may not be beneficial if the genes included are superfluous to clinical diagnosis leading to difficulties in interpretation of numerous variants of uncertain significance. Another strategy is to include genes that are believed to be eminently associated with the disease, but they could be novel with few or no patients reported. Expert clinicians can facilitate this type of new gene or pathogenic variant discovery. Clinicians and laboratory geneticists working together can provide expertise for intelligent panel design that is much needed for clinical diagnosis and prompt treatment.

The size of the capture probe library can affect the overall coverage depth, the quality of the reads, the consistency of the coverage depth of an individual exon, and the coefficient of variation of the average coverage depth. If the size is too small, compared to the total genomic DNA; for example only 23 genes out of the 23,000 genes (0.01%) is to be captured, a large fraction of the captured is expected to be non-specific off-target sequence. Although in the largest whole exome probe library, the off-target capture will be significantly reduced, the variation and error rate will be high due to the large pool of probes for non-specific hybridization. The optimal size of a capture probe library is about 100–800 kb.

4.3 Advantages of Panels over WES

Targeted panel NGS has several advantages over WES: (a) Deep coverage provides much better sequencing qualities. (b) Since the target size is usually small, it is possible to fully characterize the insufficiently covered coding regions, such that an alternative method can be specifically designed for these difficult to sequence regions to ensure 100% coverage. (c) Consistent average coverage depth of individual exons provides opportunities for sensitive CNV detection that complements with SNV detection making the panel NGS a comprehensive approach for mutation detection [17]. There is no need to perform array CGH for large deletions. (d) Short turn-around time (TAT) and lower cost compared to WES can be achieved. This is critical when an early and accurate genetic diagnosis is required for proper management of a severely affected patient. (e) Since it is a targeted panel, all genes are medically relevant so there is no hassle to interpret numerous incidental findings as in the case of WES. The only disadvantage is that it does not cover the whole exome, thus, there is no opportunity to discover new genes. However, in a clinical setting, the main purpose is to quickly make a definitive diagnose. WES generates a wealth of data. More than 80% of WES data currently cannot be interpreted. Targeted panel NGS concentrates resources on generating data that are significant and interpretable.

5 Conclusion

In summary, targeted NGS panels can play a significant role in the elucidation of genetic causes in both rare and heterogeneous disorders. Targeted panels focus on genes known or strongly suspected to be linked to a particular clinical phenotype or disease. This focusing allows for a quicker TAT, a much more manageable data set, and the feasibility for 100% coverage of all selected genes without the complication of explaining incidental findings. In this chapter we have discussed the overall workflow of panel based NGS as a platform for targeted metabolic panel tests, our experience with panel design, the selection of genes, the size of panel and capture library, panel overlap, as well as the advantages of panel NGS over WES. The analysis of targeted panels of metabolic pathways fits a unique niche between the single gene Sanger sequencing and the whole exome or genome sequencing, which is optimal for timely, efficient clinical diagnosis.

References

1. Jones, M.A., Rhodenizer, D., da Silva, C., et al.: Molecular diagnostic testing for congenital disorders of glycosylation (CDG): detection rate for single gene testing and next generation sequencing panel testing. Mol. Genet. Metab. **110**(1–2), 78–85 (2013)
2. Wang, J., Cui, H., Lee, N.C., et al.: Clinical application of massively parallel sequencing in the molecular diagnosis of glycogen storage diseases of genetically heterogeneous origin. Genet. Med. **15**(2), 106–114 (2013)
3. Sule, G., Campeau, P.M., Zhang, V.W., et al.: Next-generation sequencing for disorders of low and high bone mineral density. Osteoporos. Int. **24**(8), 2253–2259 (2013)
4. Wang, J., Zhang, V.W., Feng, Y., et al.: Dependable and efficient clinical utility of target capture-based deep sequencing in molecular diagnosis of retinitis pigmentosa. Invest. Ophthalmol. Vis. Sci. **55**(10), 6213–6223 (2014)
5. Tian, X., Liang, W.C., Feng, Y., et al.: Expanding genotype/phenotype of neuromuscular diseases by comprehensive target capture/NGS. Neurol. Genet. **1**(2), e14 (2015)

6. Yu, H., Zhang, V.W., Stray-Pedersen, A., et al.: Rapid molecular diagnostics of severe primary immunodeficiency determined by using targeted next-generation sequencing. J. Allergy Clin. Immunol. **138**(4), 1142–1151 (2016.) e1142
7. Applegarth, D.A., Toone, J.R., Lowry, R.B.: Incidence of inborn errors of metabolism in British Columbia, 1969-1996. Pediatrics. **105**(1), e10 (2000)
8. Vernon, H.J.: Inborn errors of metabolism: advances in diagnosis and therapy. JAMA Pediatr. **169**(8), 778–782 (2015)
9. Wang, J., Zhan, H., Li, F.Y., Pursley, A.N., Schmitt, E.S., Wong, L.J.: Targeted array CGH as a valuable molecular diagnostic approach: experience in the diagnosis of mitochondrial and metabolic disorders. Mol. Genet. Metab. **106**(2), 221–230 (2012a)
10. van Karnebeek, C.D., Sly, W.S., Ross, C.J., et al.: Mitochondrial carbonic anhydrase VA deficiency resulting from CA5A alterations presents with hyperammonemia in early childhood. Am. J. Hum. Genet. **94**(3), 453–461 (2014)
11. Cui, H.: Chapter 3. Methods of gene enrichment and massively parallel sequencing technologies. In: Wong, L.J.C. (ed.) Next Generation Sequencing: Translation to Clinical Diagnostics. Springer, New York. ISBN: 978-1-4614-7000-7 (2013)
12. Zhang, V.W., Wong, L.J.C., et al.: Chapter 15. Methods of gene enrichment and massively parallel sequencing technologies. In: Wong, L.J.C. (ed.) Next Generation Sequencing: Translation to Clinical Diagnostics. Springer, New York (2013)
13. Lubin, I.M., Kalman, L., Gargis, A.S.: Chapter 14. Guidelines and approaches to compliance with regulatory and clinical standards: quality control procedures and quality assurance. In: Wong, L.J.C. (ed.) Next Generation Sequencing: Translation to Clinical Diagnostics. Springer, New York (2013)
14. Li, J., Dai, H., Feng, Y., et al.: A comprehensive strategy for accurate mutation detection of the highly homologous PMS2. J. Mol. Diagn. **17**(5), 545–553 (2015)
15. Wang, J., Yu, H., Zhang, V.W., et al.: Capture-based high-coverage NGS: a powerful tool to uncover a wide spectrum of mutation types. Genet. Med. **18**(5), 513–521 (2016)
16. Yu, F., Coarfa, C.: Chapter 4. Sequence alignmnent, analysis, and bioinformatic pipelines. In: Wong, L.J.C. (ed.) Next Generation Sequencing: Translation to Clinical Diagnostics. Springer, New York (2013)
17. Feng, Y., Chen, D., Wang, G.L., Zhang, V.W., Wong, L.J.: Improved molecular diagnosis by the detection of exonic deletions with target gene capture and deep sequencing. Genet. Med. **17**(2), 99–107 (2015)
18. Wang, J., Landsverk, M.: Chapter 6. Algorithms and guidelines for interpretation of DNA variants. In: Wong, L.J.C. (ed.) Next Generation Sequencing: Translation to Clinical Diagnostics. Springer, New York (2013)
19. Wang, J., Schmitt, E.S., Landsverk, M.L., et al.: An integrated approach for classifying mitochondrial DNA variants: one clinical diagnostic laboratory's experience. Genet. Med. **14**(6), 620–626 (2012)
20. Pupavac, M., Tian, X., Chu, J., et al.: Added value of next generation gene panel analysis for patients with elevated methylmalonic acid and no clinical diagnosis following functional studies of vitamin B12 metabolism. Mol. Genet. Metab. **117**(3), 363–368 (2016)
21. Chu, J., Pupavac, M., Watkins, D., et al.: Next generation sequencing of patients with Mut methylmalonic aciduria: validation of somatic cell studies and identification of 16 novel mutations. Mol. Genet. Metab. **118**(4), 264–271 (2016)
22. Jaafar, N., Moleirinho, A., Kerkeni, E., et al.: Molecular characterization of maple syrup urine disease patients from Tunisia. Gene. **517**(1), 116–119 (2013)
23. Martin-Hernandez, E., Aldamiz-Echevarria, L., Castejon-Ponce, E., et al.: Urea cycle disorders in Spain: an observational, cross-sectional and multicentric study of 104 cases. Orphanet J. Rare Dis. **9**, 187 (2014)
24. Qin, L., Wang, J., Tian, X., et al.: Detection and quantification of mosaic mutations in disease genes by next-generation sequencing. J. Mol. Diagn. **18**(3), 446–453 (2016)

25. Jones, M., Hegde, M.: Chapter 8. Molecular diagnosis of congenital disorders of glycosylation (CDG). In: Wong, L.J.C. (ed.) Next Generation Sequencing: Translation to Clinical Diagnostics. Springer, New York (2013)
26. Ellard, S., Lango Allen, H., De Franco, E., et al.: Improved genetic testing for monogenic diabetes using targeted next-generation sequencing. Diabetologia. **56**(9), 1958–1963 (2013)
27. Nessa, A., Rahman, S.A., Hussain, K.: Hyperinsulinemic hypoglycemia – the molecular mechanisms. Front. Endocrinol. (Lausanne). **7**, 29 (2016)
28. Douillard, C., Mention, K., Dobbelaere, D., Wemeau, J.L., Saudubray, J.M., Vantyghem, M.C.: Hypoglycaemia related to inherited metabolic diseases in adults. Orphanet J. Rare Dis. **7**, 26 (2012)

The Next Generation Sequencing Based Molecular Diagnosis of Visual Diseases

Xia Wang and Richard Alan Lewis

Abstract Vision is one of the most important senses of human beings. It is estimated that 285 million people worldwide suffer from visual impairments. Although the etiology of visual impairments is both complex and multifactorial, genetic factors have been shown to play important roles. Both common and rare genetic variants have been associated with various types of inherited visual disorders. Studying the genetics of visual impairments can help to confirm or to refine clinical diagnosis, lead to better prognosis, guide the family planning, and allow targeted treatment(s). Recently developed next generation sequencing technology can generate enormous amount of sequencing data quickly at relatively low cost, and has great advantages compared to other sequencing methods. Due to the clinical and genetic heterogeneity of human visual disorders, next generation sequencing technology is a useful tool for the molecular diagnosis. In this review, we will discuss specifically the application of next generation sequencing in the molecular diagnosis of visual disorders with monogenic or oligogenic inheritance.

Keywords Next generation sequencing • Visual impairments • Molecular diagnosis • Monogenic and oligogenic inheritance

1 Introduction

Vision is one of the most important senses of human beings. More than 70% of what we learn in a lifetime comes to us through our sight [1]. It was estimated that, in 2010, 285 million people worldwide suffered from visual impairments, including 39 million with blindness and 246 million with low vision [2]. The major causes for visual impairments are uncorrected refractive errors and un-operated cataract, and the major causes for blindness are advanced cataracts and glaucoma (Table 1). Genetic factors play an important role in the etiology of visual disorders. For example, pathogenic variants in more than 230 genes are associated with retinal dystrophies

X. Wang, Ph.D. (✉) • R.A. Lewis, M.D., M.S.
Department of Molecular and Human Genetics, Baylor College of Medicine,
Houston, TX, USA
e-mail: xia.wang13@gmail.com

L.-J.C. Wong (ed.), *Next Generation Sequencing Based Clinical Molecular Diagnosis of Human Genetic Disorders*, DOI 10.1007/978-3-319-56418-0_4

Table 1 Prevalence and genetics of common visual impairments

Disorders	Frequency in general population	Genetics (loci/genes)
Uncorrected refractive errors	1/60 [2]	17 loci [105]
Cataract	1/70 [2]	>115 genes and loci [3, 106]
Glaucoma	1/1200 [2]	9 genes and 10 loci [91, 92]
Age-related macular degeneration	1/2400 [2]	19 loci [107]
Corneal Opacities	1/2400 [2]	>5 [108]
Diabetic Retinopathy	1/2400 [2]	34 loci [109]
Trachoma	1/2400 [2]	unknown
Retinitis Pigmentosa	1 /3000 to 1/4000 [110, 111]	64 genes and loci[a]
Retinoschisis	1/5000 to 1/25000 [112]	*RS1*[a]
Stargardt macular dystrophy	1/10,000 [113]	*ABCA4*[a]
Retinoblastoma	1/15000 to 20,000 [114]	*RB1*[a]

[a]https://sph.uth.edu/retnet/sum-dis.htm

thus far, and more than 110 genes are associated with congenital cataracts (Table 1, https://sph.uth.edu/Retnet/) [3]. Several common risk alleles/loci associated with "common complex eye disorders", such myopia, glaucoma, and the age-related macular degenerations, have been identified successfully (Table 1) [4–7]. Due to the clinical and genetic heterogeneity of visual disorders, next-generation sequencing (NGS) can be a useful tool to identify the underlying genetic causes of visual disorders. In this review, we will discuss specifically the application of NGS in the molecular diagnosis of visual disorders with either monogenic or oligogenic inheritance.

2 The Clinical Utility of Molecular Testing for Eye Diseases

Several clinical utilities are available for molecular testing of genetic eye diseases. First, molecular diagnosis helps to confirm or to refine the clinical diagnosis of the ocular disorder and expand the frequent systemic associations. Heritable eye diseases are clinically and genetically heterogeneous. Different genes can cause clinically indistinguishable appearances. Patients with the clinical diagnosis of an isolated, non-syndromic eye disorder may have other syndromic features not obviously present or easily identified during the initial ophthalmological evaluation [8, 9]. Therefore, molecular diagnosis helps to enhance an accurate prognosis and to improve patient management. For example, a patient diagnosed with non-syndromic infantile-onset retinal dystrophy was found to carry pathogenic variants in *IQCB1*, which is the causal gene for Senior-Loken syndrome. Because of the molecular finding, this patient's renal disease was anticipated before any kidney disease was known and treated/managed in a pre-planned manner [10]. For another example, identification of pathogenic variants in the *RB1* gene for patients with retinoblastoma reduces the otherwise costly and timely surveillance examinations [11]. Third, a firm molecular diagnosis can guide the family planning and reproductive decisions. For example, parents carrying mutations causing retinoblastoma, a malignant and potentially lethal tumor of the retina in infancy and childhood, may choose either preimplantation genetic diagnosis to ensure the delivery of a baby without such disease, or prenatal diagnosis to guide further management [12, 13]. Fourth, molecular diagnosis allows gene-specific treatment. For examples, several clinical trials have shown that gene replacement strategies are able to improve the vision of children with Leber congenital amaurosis (LCA) caused by mutations in *RPE65*, although the long-term effects need monitoring and perhaps improved techniques [14–16]. Finally, even identifying the unique cause alone of a visually impairing condition will relieve the anxiety and psychological burdens of many patients and their families who may believe that other environmental factors or misbehaviors caused the condition.

3 Methodology

This chapter focuses on NGS-based molecular diagnosis of eye diseases. However, other sequencing methods are still being used for the molecular diagnosis of eye diseases. Understanding the advantages and limitations of each method should help us to choose appropriate tests to meet the specific needs of different clinical situations. PCR followed by Sanger sequencing is considered the gold standard sequencing method and typically is used when the clinical presentation of the patient is specific and/or a small number of candidate genes associate with the differential diagnosis. For examples, eye diseases such as retinoblastoma, vitelliform macular dystrophies (VMD), familial exudative vitreoretinopathy (FEVR), and anophthalmia and microphthalmia (AM) have distinct clinical presentations and each has a small number of causal genes, and thus are suitable for Sanger sequencing [17–20]. However, Sanger sequencing is too costly when a large number of genes or exons may be associated with the disease. Also Sanger sequencing cannot be used to accurately detect CNV and low level mosaicism. In addition, unexpected SNPs on the primer can undermine the PCR based amplification of target regions (allele drop off). Microarray-based methods typically cover a large number of targeted mutations that are common for certain diseases. For examples, APEX-(arrayed primer extension) based tests covering hundreds of reported mutations have been used widely for the molecular diagnosis of several types of retinal dystrophies, such as LCA, Retinitis pigmentosa (RP), Bardet-Biedl syndrome (BBS), and Alström syndrome (ALMS) [21–23]. Although it is both cost and time effective, the diagnosis rate of array-based methods is relatively low. It detects only a fixed number of known variants and cannot identify novel or recently reported variants, which are frequently found in eye diseases that have high locus and allelic heterogeneity.

Recently, high resolution melting (HRM) has been used for high throughput genetic screening of RP [24]. It has low cost, rapid turnaround time, and is able to detect novel variants. However, it is suitable only for the analysis of medium-sized genes. Homozygosity mapping is a genetic method to identify disease-associated loci, especially in consanguineous families. It has been used widely to identify novel disease-causing genes and/or pathogenic variants in known genes underlying several diseases, including genetically heterogeneous eye diseases [25–27]. Because it involves genotyping of related family members and it helps only to map the disease-associated loci, homozygosity mapping is not commonly offered as a diagnostic test but is used to identify disease-associated genomic regions. Some clinical diagnostic laboratories, such Baylor Miraca Genetics Laboratories, do offer whole exome SNP array together with WES as a quality control method and to identify potential disease-associated loci.

NGS has been used widely for the molecular diagnosis of eye diseases. NGS methods can be classified into two categories: targeted NGS or whole genome sequencing (WGS). Targeted NGS methods enrich target genomic regions first before high throughput sequencing, while WGS require no prior target enrichment. Targeted NGS methods, including panel based NGS and whole exome sequencing (WES), are often used in molecular diagnostic studies of eye diseases. The first step of targeted NGS is to enrich the target genomic regions. The target enrichment strategies include multiplex PCR, molecular inversion probes (MIP), array hybridization capture, and in solution hybridization capture [28]. Different sequencing platforms, such as HiSeq, Ion Torrent, 454 GS, and single-molecular sequencing by Pacific Biosciences, were invented for high throughput sequencing [29–31]. For detailed comparison between different capture methods and different sequencing platforms, please refer to Part One review of this book and the references [32].

Compared with other methods, NGS has high throughput, low per-base cost, and the ability to detect novel variants and copy number variations (CNVs). It is particularly useful when the clinical presentation of patients is non-specific and many candidate genes/exons associate with the differential diagnosis. Specifically, panel-based NGS methods include a set of genes relevant to the specific diagnosis, while WES aims to interrogate the coding regions within the whole genome. Comparing with panel-based NGS, WES generally has higher total cost and lower average coverage depth. However, WES has the ability to detect mutations in recently identified disease genes or to discover novel genes associated with diseases. Several current limitations of NGS exist. First, the sequence quality of NGS at highly repetitive and/or GC rich regions is not optimal. Currently the NGS reads are not long enough to correctly map repetitive/homologous sequences. For example, *RPGR* gene, especially its open reading frame 15 (ORF15), is a mutational hot sport for X-linked RP [33]. However, the ORF15 has repetitive sequences that cannot be accurately sequenced and mapped by conventional targeted capture NGS [34]. To overcome this challenge, long range PCR followed by NGS and targeted alignment can be used to only amplify and map the target region. The GC-biases may be introduced during PCR amplification in the NGS library preparation process, and several improvements have been made to reduce such bias [35–37]. Second, NGS has relatively higher total sequencing costs. Third, NGS generates large volume of data and requires intensive computations for data analysis. Fourth, NGS has not yet been considered as the golden standard method to detect structural variations (SVs) and other types of genomic rearrangements. WGS has been reported to be more accurate than targeted NGS to call SVs and CNVs, partly because WGS has more uniform coverage then targeted NGS [38]. However, recent reports suggest that CNVs and Alu insertions can also be reliably detected by panel based NGS in clinical diagnostic laboratories [39]. Fifth, because NGS (especially WES or WGS) covers a large number of genes not apparently relevant to the patient's clinical presentations and is able to identify numerous genetic variants that may not be interpretable, potential ethical and psychological issues, including how to report incidental findings and how much emphasis to place on variants with uncertain clinical significance, may arise. This is less of a problem for targeted panel NGS which focus one genes directly relevant to the clinical presentations. Nevertheless, targeted NGS is becoming the most used method in the molecular diagnosis of eye diseases and is being rapidly improved.

4 Retinal Dystrophies

The retina is a complex tissue lining the inside of the back two-thirds of the eye that senses light and converts light into electrical signals. Inherited retinal dystrophies (RD) represent a group of heterogeneous eye disorders that are genetically determined and usually highly symmetric. The clinical presentations of RD vary from late onset mild visual impairment to congenital blindness. RD can be either non-syndromic, that is, occurring in the retina and adjacent tissues alone, or syndromic, thus are consistently associated with other systemic, extraocular features. To date, mutations in more than 200 genes are known to cause RD with either Mendelian or non-Mendelian (e.g., mitochondrial) inheritance (https://sph.uth.edu/Retnet/sum-dis.htm). The genes frequently mutated in each RD are summarized as Table 2. Due to wide clinical and genetic heterogeneities, more than 20 retinal diseases involved in the differential diagnosis of RD [40]. Collectively, it is estimated that RD affect 1 in every 2000 individuals, or more than 3 million people worldwild [41]. In the past few years, NGS has been applied widely for the molecular diagnosis of patient cohorts with various non-specific or specific RD, such as retinitis pigmentosa (RP) (Fig. 1), Stargardt macular dystrophy (STGD), Leber congenital amaurosis, familial exudative vitreoretinopathy, Usher syndrome (USH), and Bardet-Biedl Syndrome (BBS). The gene enrichment methods and results of recently published studies on NGS-based molecular diagnosis of RD are summarized as Table 3.

Table 2 Causal genes and estimated contributions to specific type of retinal dystrophies

Disorders	Fractions of total cases contributed by causal genes
AR RP including Usher and BBS (50–60% of RP)	*USH2A* (17%), *ABCA4* (5.6%), *PDE6B* (4–5%) [110]
AD RP (30–40% of RP)	*RHO* (25%), *RP1* and *PRPF31* (5%) [110]
X-linked RP (5–15% of RP)	*RPGR* (75%), *RP2* (7–10%) [110]
AR cone-rod dystrophy	*ABCA4* (16%), *EYS* (3%), *CERKL* (1%) [115]
AR cone dystrophy	*KCNV2* (13%), *ABCA4* (4%), *CNGB3* (3%) [115]
AD cone and cone-rod dystrophy	*GUCY2D* (8%), *CRX* (4%), *GUCA1A* (3%) [115]
X linked cone and cone-rod dystrophy	*RPGR* (53%), *OPN1LW/OPN1MW* (8%), *CACNA1F* (3%) [115]
Achromatopsia	*CNGB3* (59%), *CNGA3* (36%), *PDE6C* (3%) [115]
Leber congenital amaurosis	*CEP290* (15%), *GUCY2D* (12%), *CRB1* (10%) [116]
Bardet-Biedl syndrome	*BBS1* (23%), *BBS10* (20%), *BBS2* (8%) [117]
Usher syndrome type I	*MYO7A* (29–50%), *CDH23* (19–35%), *PCDH15* (11–19%) [118]
Usher syndrome type II	*USH2A* (55–90%), *GPR98* (3–6%) [118]

AR: autosomal recessive, *AD:* autosomal dominant

4.1 Diagnostic Rates

In clinical settings, the diagnostic rate for a clinically naïve diagnosis has become an important parameter for clinical geneticists to make decision on whom and where to send the test. The published molecular diagnostic rates depend on several factors (Table 3). First, the number of target genes covered by the panel NGS method affects the diagnostic rate. Gene panels covering updated lists of genes for specified types of RD tend to have higher diagnostic rates. In addition, it is well recognized that patients with certain clinical diagnoses may actually carry pathogenic variants in genes that not typically associated with that diagnosis [8, 9]. For example, patients diagnosed with LCA were frequently found to carry pathogenic variants in *ALMS1*, the causal gene for Alstrom syndrome. Therefore, a larger panel or even WES that covers more RD genes relevant to the differential diagnosis may have a better chance to detect variants in the "unexpected" genes and consequently have a higher diagnostic rate. Second, the methods and data quality, as well as analytical methods affect the diagnostic rate. Higher read depth, more accurate sequencing quality, and less human errors are the basis of high detection rate. Also, improved algorithms to detect both SNVs and CNVs using the same NGS data set can lead to higher diagnostic rate than methods that only focus on SNVs [42]. Third, the theoretical percentages of cases that can be explained by mutations in currently known disease genes vary among different diseases. For example, complete sequencing of the 9 known USH genes can explain as high as 72–90% of USH cases [43, 44]. Consistently, the published NGS diagnostic rates for USH cohorts are generally higher than other types of RD, ranging from 70–100% [45–47]. Whereas current known genes are estimated to account for 50% of FEVR cases, and the recent NGS

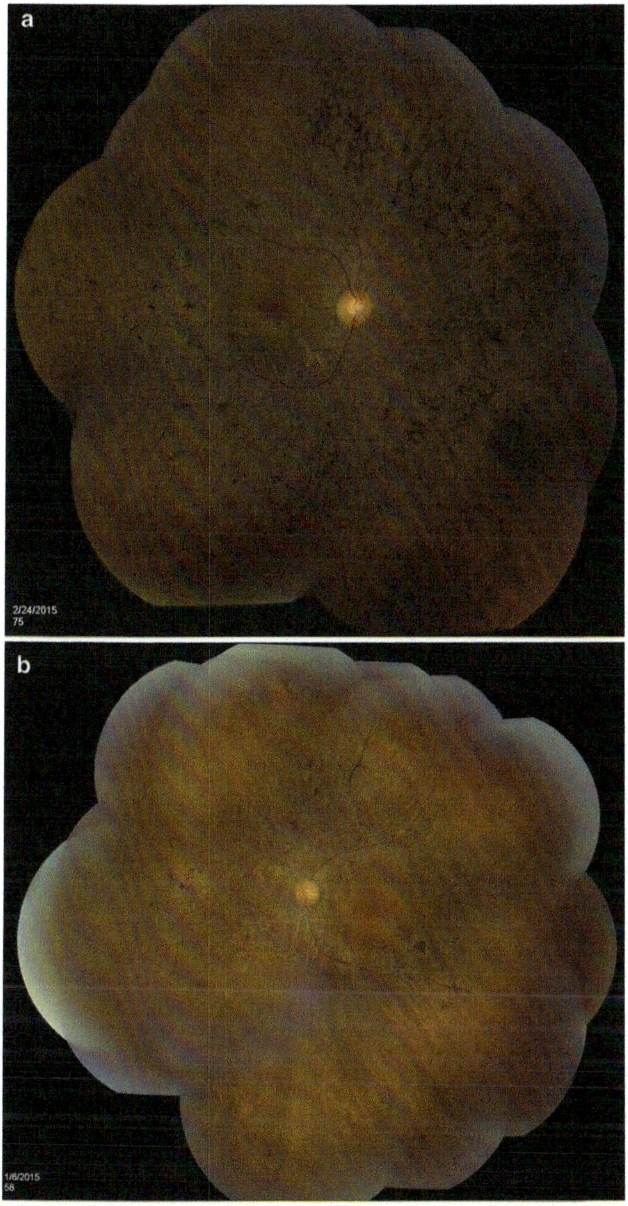

Fig. 1 (**a**) Photomontage of the left fundus of a 46 year old female with advanced retinitis pigmentosa. Note the pigment migration into the retina (creating the clumps and the linear formations around retinal vessels, so-called 'bone spicules'), the vascular attenuation, and the diffuse pallor of the optic nerve. No distinguishing characteristics separate which of the numerous genes is responsible for the progressive destruction of the retinal pigment epithelium. (**b**) Photomontage of the left fundus of a 60 year old female with advanced retinitis pigmentosa. Note the relative paucity of pigment deposits, but the atrophy of the retinal pigment epithelium, the vascular attenuation of both arteriolar and venular systems, and the diffuse pallor of the optic nerve

Table 3 Summary of published studies on NGS based molecular diagnosis of retinal dystrophies

	Clinical presentation	Cohort size	Target genes	Enrichment method	Diagnosis rate
Simpson et al. (2011) [119]	RP	5	45	Array hybridization	60% (3/5)
Neveling et al. (2012) [120]	RP	100	111	Array hybridization	36% (36/100)
Fu et al. (2013) [51]	arRP	31	163	Solution hybridization	39% (12/31)
Wang et al. (2014) [121]	RP	123	163	Solution hybridization	37% (45/123)
Wang et al. (2014) [39]	RP	65	66	Solution hybridization	82% (53/65)
Xu et al. (2014) [122, 123]	RP	157	189	Solution hybridization	53% (84/157)
Zhao et al. (2015) [50]	RP	82	186	Solution hybridization	60% (49/82)
Zernant et al. (2011) [124]	STGD	159	*ABCA4*	PCR	33% (53/159)
Strom et al. (2012) [125]	STGD	9	Whole Exome	Solution hybridization	67% (6/9)
Fujinami et al. (2013) [126]	STGD	79	*ABCA4*	PCR	47% (37/79)
Zhang et al. (2014) [127]	STGD	5	Whole Exome	Solution hybridization	100% (5/5)
Zaneveld et al. (2015) [128]	STGD	88	213	Solution hybridization	67% (59/88)
Coppieters et al. (2012) [65]	LCA	17	16	PCR	18% (3/17)
Wang et al. (2013) [8]	LCA	179	163	Solution hybridization	40% (72/179)
Licastro et al. (2012) [129]	Usher	12	Whole Exome/9[a]	Solution hybridization/ PCR[a]	42% (5/12)
Salvo et al. (2015) [49]	FEVR	92	163	Solution hybridization	49% (45/92)
Yoshimura et al. (2014) [46]	Usher type 1	17	9	PCR	94% (16/17)
Bujakowska et al. (2014) [45]	Usher type 1	47	Not mentioned	Solution hybridization	70% (33/47)
Qu et al. (2015) [47]	Usher	5	103	Array hybridization	100% (5/5)
Xing et al. (2014) [130]	BBS	5	144	Solution hybridization	100% (5/5)

(continued)

Table 3 (continued)

	Clinical presentation	Cohort size	Target genes	Enrichment method	Diagnosis rate
O'Sullivan et al. (2012) [131]	RD	50	105	[b]	42% (21/50)
Schorderet et al. (2012) [132]	RD	23	63	Solution hybridization	52% (12/23)
Shanks et al. (2012) [133]	RD	36	73	Array hybridization	25% (9/36)
Chen et al. (2013) [134]	RD	25	189	Array hybridization	56% (14/25)
Glockle et al. (2013) [41]	RD	170	105	Solution hybridization	91% (155/170)
Watson et al. (2014) [135]	RD	20	162	Solution hybridization	60% (12/20)
Huang et al. (2014) [136]	RD	179	252	Solution hybridization	55% (99/179)
Liu et al. (2015) [137]	RD	20	316	Array hybridization	55% (11/20)

[a]Two methods were used in one study
[b]Unclear

study on a FEVR cohort reported about 49% of diagnosis rate [48, 49]. Fourth, different ethnic groups have different mutation spectrums. For example, the diagnostic rate of RP patients is reported to be 40% in China but 60% in Northern Ireland [50, 51]. It has also been reported that Chinese and European ancestry LCA patients have different mutation spectrums [52]. It is possible that some populations may carry more variants in known disease loci than other populations, leading to higher diagnostic rate. Fifth, the diagnostic rate may be lower than expected if the patient cohort had been "prescreened" by other molecular diagnostic methods, such as Sanger sequencing on frequently mutated genes [8]. In the practice of medicine, clinical geneticists commonly order specific target gene tests first and then order NGS tests if the results of target gene tests are unrevealing. Therefore, several patient cohorts published in NGS studies have been "prescreened" to a certain extent. Lastly, the detail and specificity of the clinical characterization of the referred patients and the underlying genetic heterogeneity of the patient cohort can affect the success of a firm molecular diagnosis.

4.2 Inheritance Models

Most RD, whether ocular alone or part of a syndromic complex, are single gene disorders. However, several forms of oligogenic inheritance have been reported in RD and should be considered when interpreting the NGS data, especially for those genes in the same biological pathway. For example, in digenic inheritance (DI), two pathogenic variants in each of the two unlinked loci are required to cause diseases. The first human example of DI was reported in 1995 when, individuals from three families were found to have double heterozygous pathogenic variants in two unlinked photoreceptor-specific genes, *ROM1* and *PRPH2*, and developed RP [53]. It was later suggested that wild type *ROM1* and *PRPH2* proteins form functional heterotetramers, while null *ROM1* and mutant *PRPH2* lead to the reduced levels of *ROM1*/*PRPH2* heterotetramers, and the unstable *PRPH2* homotetramers likely cause disorganization of photoreceptor cell outer segments and consequently cause RP phentypes [54]. Schäffer 2013 provided a comprehensive summary of DI associated with various human disorders [55]. In triallelic inheritance (TI), three pathogenic variants in two different unlinked loci are required to cause disease. TI was first described in humans in BBS, when three pathogenic variants in *BBS2* and *BBS6* genes cause BBS [56]. Later, *BBS4* gene has also been reported to participate in TI of BBS [57]. However, the prevalence of TI cases in BBS is debated [58, 59]. In addition, modifiers may affect the severity of RD clinical presentations. For examples, a common allele p.A229T in *RPGRIP1L* gene is associated with photoreceptor cell loss in ciliopathies [60]. Polymorphisms of both *RPGRIP1L* and *IQCB1* gene can modify X-linked RP caused by RPGR mutations [61]. Variants in the *PDZD7* gene modify the retina dystrophy in Usher syndrome [62]. Lastly, It has been reported that patients may have pathogenic variants in more than one RD genes [41]. Indeed, several published reports of oligogenic inheritance lack further repeats or functional validations. Nevertheless, recognizing variant forms of oligogenic inheritance will improve the accuracy of interpretation in the molecular analyses of RD.

4.3 Pathogenic Variants

Most published NGS-based molecular diagnostic methods aim to capture both coding regions and the flanking intronic regions. Consequently, most studies focus on the analysis of coding changing variants, including nonsense, small insertions/deletions, canonical splice site, and missense changes. Professional societies, such as the American College of Medical Genetics and Genomics and the Association for Molecular Pathology, have published guidance to help the interpretation of sequence variants [63]. However, other non-coding variants have been reported to cause RD and should not be overlooked. First, deep intronic and synonymous changes are well known causes of RD. For example, the c.2991 + 1655A > G change in intron 26 of *CEP290* gene is one of the most frequent pathogenic variants found in LCA, accounting for about 21% of LCA cases. This intronic change creates a new splice-donor site and inserts a cryptic exon, which leads to a premature stop codon downstream of exon 26 [64]. To assure that this variant is included in the targeted NGS analysis, either specific primers or probes should be designed for the target capture, or directed Sanger sequencing should complement NGS [8, 65]. Similarly, several deep intronic and synonymous changes were identified in *ABCA4* gene and cause Stargardt disease. These changes lead to inclusion or deletion of sequence segments in the mRNA transcripts [66]. In addition, a deep intronic change c.7595-2144A > G was recently found in *USH2A* gene to cause Usher syndrome type 2 [67]. In all these examples, RNA sequencing was used to identify abnormal mRNA transcripts resulting from intronic or synonymous changes. To increase the diagnostic rate further, RNA sequencing may be used in parallel with DNA sequencing to identify novel intronic or synonymous changes that affect mRNA splicing, especially for cases of recessive disorders that have only one copy of a pathogenic variant without the second mutant allele and/or for cases that have candidate disease loci localized by genetic mapping. Second, exonic deletions/duplications and copy number variations (CNVs) have been reported to cause in RD. For example, deletions/duplications were reported to account for 26% of Usher syndrome cases with monoallelic mutations in *USH2A* gene [68]. Large rearrangements and duplications in *PCDH15* gene were identified in 13% of Usher syndrome Type 1 families who had been screened for mutations in the five known USH1 genes [69]. A common Alu insertions in *MAK* gene in Jewish RP patients have also been described before [39]. Although many clinical diagnostic laboratories offer deletion/duplication analysis by methods such as multiplex ligation-dependent probe amplification (MLPA) or array-based CGH analysis (aCGH), it will be ideal to detect deletions/duplications using the same NGS data set for single nucleotide variations (SNVs) analysis. Several bioinformatics approaches have been developed to detect CNVs from NGS data [9, 70]. However, the performance of CNVs analysis in clinical settings is still suboptimal because of high error rates caused by short reads, genome complexity, and large coverage variations of targeted NGS data. Recently, Feng, et al. reported the systematic detection of exon deletions with NGS data from clinical samples with RD and other inherited disorders [42]. CNV analysis with NGS data can be improved further by increasing both reads length of NGS and data quality, and advanced bioinformatics tools. Lastly, several genomic regions involved in RD may not be captured easily and analyzed by NGS. As one classic example, the ORF15 of the *RPGR* gene is a mutational hotspot in RP. However, the ORF15 sequence is highly repetitive and hard to be analyzed by conventional NGS [33, 34]. To overcome this limitation, long range PCR followed by NGS can be applied.

5 Retinoblastoma

Retinoblastoma is the single most common malignant ocular tumor in childhood, affecting about 1 in every 14,000 to 18,000 live births in the United States (National Cancer Institute). Retinoblastoma is caused predominantly by biallelic mutations in *RB1* gene. In the developed world, retinoblastoma presents most frequently as sudden onset of strabismus or as a "white pupil", either in the parent's observation or in an informal photograph. In heritable forms of retinoblastoma, affected individuals typically have tumors in each eye, usually multicentric, diagnosed at younger age, and germline *RB1* mutations. While in non-heritable form, children typically have unilateral tumors that include somatic mutations in *RB1*. Various types of mutations in *RB1*, including SNVs, CNVs, genomic rearrangements, and non-coding regulatory changes, have been reported in retinoblastoma tumors [71]. Promoter methylation of *RB1* also plays a role in the development of retinoblastoma [72]. In addition, genetic factors other than *RB1* gene, such as miRNAs, *MYCN* gene amplification, and polymorphisms in *MDM2* and *MDM4* gene, may contribute to the development of retinoblastoma [73–76].

The current molecular methods for the diagnosis of retinoblastoma include Sanger sequencing and allele-specific PCR for SNVs, MLPA and quantitative multiplex PCR (QMPCR) for insertions/deletions and CNVs, and a methylation assay to detect hypermethylated *RB1* promoters [11]. Due to the allelic heterogeneity of retinoblastoma, a stepwise molecular diagnostic approach has been proposed [77]. An mRNA analysis will determine the consequence of genomic variants affecting transcripts. Recently, NGS has been applied to the molecular diagnosis of retinoblastoma [78–80]. Compared with other methods, NGS can detect unbiasedly the low-level mosaic variants that are found frequently in sporadic retinoblastoma [78, 79]. NGS-based methods are also able to detect both SNVs and CNVs in retinoblastoma [80]. Because of the wide mutational spectrum of *RB1* and the current limitations of NGS, NGS combined with other complementary methods in a stepwise manner may be the optimal molecular strategy for retinoblastoma diagnosis.

6 Congenital Cataracts

Congenital cataracts (CCs) refers to opacification of parts of or the entire lens present at birth; when involving the entire cortex of the lens or some parts of the axial view of the incoming image, the cataract may lead to severe visual impairment. CCs are estimated to affect 1 to 6 babies in every 10,000 live births; at least half of CCs are associated with genetic factors [81–84]. CCs can present either as an isolated non-syndromic event or as one part of syndromic conditions. To date, more than 20 and 90 genes are associated with non-syndromic and syndromic CCs, respectively [3, 85]. The locus heterogeneity makes CCs as ideal genetic disorders for NGS-based molecular diagnoses.

Surprisingly, only a few mutational surveys on CC cohorts have been reported. Sanger sequencing and single strand conformation polymorphism (SSCP) have been used to screen for mutations in CC cohorts from India and from Denmark [86–88]. Recently, an NGS panel covering more than 110 CC genes was applied on 36 individuals with syndromic or non-syndromic bilateral CCs and was reported to detect the genetic causes of CCs in 75% of individuals, suggesting the high detection rate [3]. In this study, identification of unexpected genetic causes in patients with apparent CCs investigated at an early age had altered the clinical management and consequent genetic counseling. In another recent study, WES identified the genetic causes for half of the 18 families with CCs [89]. As a summary, the high genetic heterogeneity of CCs warrants NGS as the optimal molecular diagnostic method.

7 Glaucoma

Glaucoma is a visual impairment caused by retinal ganglion cell loss, optic nerve excavation and atrophy, and subsequent visual field defects [90]. Early-onset (adolescent) forms of glaucoma are typically inherited. Early-onset forms of glaucoma affect about 1 in every 2500 to 20,000 individuals [91]. Glaucoma is clinically heterogeneous and can be classified into several types, such as primary congenital glaucoma, juvenile/primary open-angle glaucoma, and primary angle-closure glaucoma. Currently, common risk variants in 10 loci and rare variants in 9 genes have been associated with glaucoma (Table 1) [91, 92]. Gene-gene/gene-environment interactions and epigenetics may also contribute to the pathogenesis of glaucoma [91]. Recently, Huang et al. applied WES and Sanger sequencing for the molecular diagnosis of 683 unrelated patients with primary glaucoma [93]. They used WES to analyze seven glaucoma genes and identified mutations in 20 out of 257 patients. In addition, they used Sanger sequencing on the *MYOC* gene and identified mutations in 5 of the remaining 426 patients. Compared to other inherited eye disorders, the diagnostic rate of WES based testing on glaucoma (8%) in this study is low. More genetic studies on glaucoma will garner better insight on the genetic contributions to glaucoma.

8 Challenges

NGS is able to generate enormous amounts of data, which in turn bring substantial analytical challenges. One emerging concern is that false positives may arise by chance as the amount of genomic data increases. Indeed, recent evaluations of previously reported "mutations" suggest that many of them are frequent in general populations and thus are not likely to cause diseases, including eye disorders [94–98]. To overcome this challenge, more detailed and stringent standards for the classification of variant pathogenicity are needed [63]. Further, curation of general or disease-specific variant databases will clarify the pathogenicity of variants currently felt to have uncertain or unknown significance [99–101].

Ethic, legal, and social issues also arise in the genomic era [102]. The practice of genetic testing should be carefully regulated, and genetic counseling should be provided by the attending physician or genetic counselors to maximize the benefit and reduce the risks to the consultant and the family [103]. Patients' understanding and needs should be met to provide optimal health care [104].

References

1. Marieb, E., Hoehn, K.: Human Anatomy & Physiology, 7th edn. Pearson Benjamin Cummings, San Francisco (2007)
2. Pascolini, D., Mariotti, S.P.: Global estimates of visual impairment: 2010. Br. J. Ophthalmol. **96**, 614–618 (2012)

3. Gillespie, R.L., et al.: Personalized diagnosis and management of congenital cataract by next-generation sequencing. Ophthalmology. **121**, 2124–2137.e1–2 (2014)
4. Miyake, M., et al.: Identification of myopia-associated WNT7B polymorphisms provides insights into the mechanism underlying the development of myopia. Nat. Commun. **6**, 6689 (2015)
5. Chen, Y., et al.: Common variants near ABCA1 and in PMM2 are associated with primary open-angle glaucoma. Nat. Genet. **46**, 1115–1119 (2014)
6. Cheng, C.-Y., et al.: Nine loci for ocular axial length identified through genome-wide association studies, including shared loci with refractive error. Am. J. Hum. Genet. **93**, 264–277 (2013)
7. Klein, R.J., et al.: Complement factor H polymorphism in age-related macular degeneration. Science. **308**, 385–389 (2005)
8. Wang, X., et al.: Comprehensive molecular diagnosis of 179 leber congenital amaurosis and juvenile retinitis pigmentosa patients by targeted next generation sequencing. J. Med. Genet. **50**, 674–688 (2013)
9. Werdich, X.Q., Place, E.M., Pierce, E.A.: Systemic diseases associated with retinal dystrophies. Semin. Ophthalmol. **29**, 319–328 (2014)
10. Ellingford, J.M., et al.: Pinpointing clinical diagnosis through whole exome sequencing to direct patient care: a case of senior-Loken syndrome. Lancet Lond. Engl. **385**, 1916 (2015)
11. Richter, S., et al.: Sensitive and efficient detection of RB1 gene mutations enhances care for families with retinoblastoma. Am. J. Hum. Genet. **72**, 253–269 (2003)
12. Dommering, C.J., et al.: Reproductive decision-making: a qualitative study among couples at increased risk of having a child with retinoblastoma. Clin. Genet. **78**, 334–341 (2010)
13. Xu, K., et al.: Preimplantation genetic diagnosis for retinoblastoma: the first reported liveborn. Am J. Ophthalmol. **137**, 18–23 (2004)
14. Bainbridge, J.W.B., et al.: Long-term effect of gene therapy on Leber's congenital amaurosis. N. Engl. J. Med. **372**, 1887–1897 (2015)
15. Jacobson, S.G., et al.: Improvement and decline in vision with gene therapy in childhood blindness. N. Engl. J. Med. **372**, 1920–1926 (2015)
16. MacLaren, R.E., et al.: Retinal gene therapy in patients with choroideremia: initial findings from a phase 1/2 clinical trial. Lancet Lond. Engl. **383**, 1129–1137 (2014)
17. Ottaviani, D., et al.: Spectrum of RB1 mutations in argentine patients: 20-years experience in the molecular diagnosis of retinoblastoma. Ophthalmic Genet. **34**, 189–198 (2013)
18. Meunier, I., et al.: Systematic screening of BEST1 and PRPH2 in juvenile and adult vitelliform macular dystrophies: a rationale for molecular analysis. Ophthalmology. **118**, 1130–1136 (2011)
19. Boonstra, F.N., et al.: Clinical and molecular evaluation of probands and family members with familial exudative vitreoretinopathy. Invest. Ophthalmol. Vis. Sci. **50**, 4379–4385 (2009)
20. Chassaing, N., et al.: Molecular findings and clinical data in a cohort of 150 patients with anophthalmia/microphthalmia. Clin. Genet. **86**, 326–334 (2014)
21. Henderson, R.H., et al.: An assessment of the apex microarray technology in genotyping patients with leber congenital amaurosis and early-onset severe retinal dystrophy. Invest. Ophthalmol. Vis. Sci. **48**, 5684–5689 (2007)
22. van Huet, R.A.C., et al.: The efficacy of microarray screening for autosomal recessive retinitis pigmentosa in routine clinical practice. Mol. Vis. **21**, 461–476 (2015)
23. Pereiro, I., et al.: Arrayed primer extension technology simplifies mutation detection in Bardet-Biedl and Alström syndrome. Eur. J. Hum. Genet. EJHG. **19**, 485–488 (2011)
24. Anasagasti, A., et al.: Genetic highthroughput screening in retinitis pigmentosa based on high resolution melting (HRM) analysis. Exp. Eye Res. (2013). doi:10.1016/j.exer.2013.10.011
25. Abu-Safieh, L., et al.: Autozygome-guided exome sequencing in retinal dystrophy patients reveals pathogenetic mutations and novel candidate disease genes. Genome Res. **23**, 236–247 (2013)
26. Beryozkin, A., et al.: Identification of mutations causing inherited retinal degenerations in the israeli and palestinian populations using homozygosity mapping. Invest. Ophthalmol. Vis. Sci. **55**, 1149–1160 (2014)

27. Littink, K.W., et al.: Homozygosity mapping in patients with cone-rod dystrophy: novel mutations and clinical characterizations. Invest. Ophthalmol. Vis. Sci. **51**, 5943–5951 (2010)
28. Mamanova, L., et al.: Target-enrichment strategies for next-generation sequencing. Nat. Methods. **7**, 111–118 (2010)
29. Metzker, M.L.: Sequencing technologies - the next generation. Nat. Rev. Genet. **11**, 31–46 (2010)
30. Mardis, E.R.: Next-generation sequencing platforms. Annu. Rev. Anal. Chem. Palo Alto Calif. **6**, 287–303 (2013)
31. Liu, L., et al.: Comparison of next-generation sequencing systems. J. Biomed. Biotechnol. **2012**, 251364 (2012)
32. Wong, L.-J.C.: Next Generation Sequencing. Springer, New York (2013)
33. Vervoort, R., et al.: Mutational hot spot within a new RPGR exon in X-linked retinitis pigmentosa. Nat. Genet. **25**, 462–466 (2000)
34. Huang, X.-F., Wu, J., Lv, J.-N., Zhang, X., Jin, Z.-B.: Identification of false-negative mutations missed by next-generation sequencing in retinitis pigmentosa patients: a complementary approach to clinical genetic diagnostic testing. Genet. Med. Off. J. Am. Coll. Med. Genet. **17**, 307–311 (2015)
35. Dabney, J., Meyer, M.: Length and GC-biases during sequencing library amplification: a comparison of various polymerase-buffer systems with ancient and modern DNA sequencing libraries. BioTechniques. **52**, 87–94 (2012)
36. Benjamini, Y., Speed, T.P.: Summarizing and correcting the GC content bias in high-throughput sequencing. Nucleic Acids Res. **40**, e72 (2012)
37. Aird, D., et al.: Analyzing and minimizing PCR amplification bias in Illumina sequencing libraries. Genome Biol. **12**, R18 (2011)
38. Gilissen, C., et al.: Genome sequencing identifies major causes of severe intellectual disability. Nature. **511**, 344–347 (2014)
39. Wang, J., et al.: Dependable and efficient clinical utility of target capture-based deep sequencing in molecular diagnosis of retinitis pigmentosa. Invest. Ophthalmol. Vis. Sci. **55**, 6213–6223 (2014)
40. Berger, W., Kloeckener-Gruissem, B., Neidhardt, J.: The molecular basis of human retinal and vitreoretinal diseases. Prog. Retin. Eye Res. **29**, 335–375 (2010)
41. Glöckle, N., et al.: Panel-based next generation sequencing as a reliable and efficient technique to detect mutations in unselected patients with retinal dystrophies. Eur. J. Hum. Genet. EJHG. **22**, 99–104 (2014)
42. Feng, Y., Chen, D., Wang, G.-L., Zhang, V.W., Wong, L.-J.C.: Improved molecular diagnosis by the detection of exonic deletions with target gene capture and deep sequencing. Genet. Med. Off. J. Am. Coll. Med. Genet. **17**, 99–107 (2015)
43. Bonnet, C., et al.: Complete exon sequencing of all known usher syndrome genes greatly improves molecular diagnosis. Orphanet J. Rare Dis. **6**, 21 (2011)
44. Roux, A.-F., et al.: Survey of the frequency of USH1 gene mutations in a cohort of usher patients shows the importance of cadherin 23 and protocadherin 15 genes and establishes a detection rate of above 90%. J. Med. Genet. **43**, 763–768 (2006)
45. Bujakowska, K.M., et al.: Targeted exon sequencing in usher syndrome type I. Invest. Ophthalmol. Vis. Sci. **55**, 8488–8496 (2014)
46. Yoshimura, H., et al.: Massively parallel DNA sequencing facilitates diagnosis of patients with usher syndrome type 1. PLoS One. **9**, e90688 (2014)
47. Qu, L.-H., Jin, X., Xu, H.-W., Li, S.-Y., Yin, Z.-Q.: Detecting novel genetic mutations in Chinese usher syndrome families using next-generation sequencing technology. Mol. Genet. Genomics MGG. **290**, 353–363 (2015)
48. Gilmour, D.F.: Familial exudative vitreoretinopathy and related retinopathies. Eye Lond. Engl. **29**, 1–14 (2015)
49. Salvo, J., et al.: Next-generation sequencing and novel variant determination in a cohort of 92 familial exudative vitreoretinopathy patients. Invest. Ophthalmol. Vis. Sci. **56**, 1937–1946 (2015)

50. Zhao, L., et al.: Next-generation sequencing-based molecular diagnosis of 82 retinitis pigmentosa probands from Northern Ireland. Hum. Genet. **134**, 217–230 (2015)
51. Fu, Q., et al.: Next-generation sequencing-based molecular diagnosis of a Chinese patient cohort with autosomal recessive retinitis pigmentosa. Invest. Ophthalmol. Vis. Sci. **54**, 4158–4166 (2013)
52. Wang, H., et al.: Comprehensive molecular diagnosis of a large chinese leber congenital amaurosis cohort. Invest. Ophthalmol. Vis. Sci. **56**, 3642–3655 (2015)
53. Kajiwara, K., Berson, E.L., Dryja, T.P.: Digenic retinitis pigmentosa due to mutations at the unlinked peripherin/RDS and ROM1 loci. Science. **264**, 1604–1608 (1994)
54. Goldberg, A.F., Molday, R.S.: Defective subunit assembly underlies a digenic form of retinitis pigmentosa linked to mutations in peripherin/rds and rom-1. Proc. Natl. Acad. Sci. U. S. A. **93**, 13726–13730 (1996)
55. Schäffer, A.A.: Digenic inheritance in medical genetics. J. Med. Genet. **50**, 641–652 (2013)
56. Katsanis, N., et al.: Triallelic inheritance in Bardet-Biedl syndrome, a Mendelian recessive disorder. Science. **293**, 2256–2259 (2001)
57. Katsanis, N., et al.: BBS4 is a minor contributor to Bardet-Biedl syndrome and may also participate in triallelic inheritance. Am. J. Hum. Genet. **71**, 22–29 (2002)
58. Abu-Safieh, L., et al.: In search of triallelism in Bardet-Biedl syndrome. Eur. J. Hum. Genet. EJHG. **20**, 420–427 (2012)
59. Mykytyn, K., et al.: Identification of the gene (BBS1) most commonly involved in Bardet-Biedl syndrome, a complex human obesity syndrome. Nat. Genet. **31**, 435–438 (2002)
60. Khanna, H., et al.: A common allele in RPGRIP1L is a modifier of retinal degeneration in ciliopathies. Nat. Genet. **41**, 739–745 (2009)
61. Fahim, A.T., et al.: Polymorphic variation of RPGRIP1L and IQCB1 as modifiers of X-linked retinitis pigmentosa caused by mutations in RPGR. Adv. Exp. Med. Biol. **723**, 313–320 (2012)
62. Ebermann, I., et al.: PDZD7 is a modifier of retinal disease and a contributor to digenic usher syndrome. J. Clin. Invest. **120**, 1812–1823 (2010)
63. Richards, S., et al.: Standards and guidelines for the interpretation of sequence variants: a joint consensus recommendation of the American College of Medical Genetics and Genomics and the Association for Molecular Pathology. Genet. Med. Off. J. Am. Coll. Med. Genet. **17**, 405–424 (2015)
64. den Hollander, A.I., et al.: Mutations in the CEP290 (NPHP6) gene are a frequent cause of leber congenital amaurosis. Am. J. Hum. Genet. **79**, 556–561 (2006)
65. Coppieters, F., et al.: Massively parallel sequencing for early molecular diagnosis in leber congenital amaurosis. Genet. Med. Off. J. Am. Coll. Med. Genet. **14**, 576–585 (2012)
66. Braun, T.A., et al.: Non-exomic and synonymous variants in ABCA4 are an important cause of Stargardt disease. Hum. Mol. Genet. **22**, 5136–5145 (2013)
67. Vaché, C., et al.: Usher syndrome type 2 caused by activation of an USH2A pseudoexon: implications for diagnosis and therapy. Hum. Mutat. **33**, 104–108 (2012)
68. Steele-Stallard, H.B., et al.: Screening for duplications, deletions and a common intronic mutation detects 35% of second mutations in patients with USH2A monoallelic mutations on Sanger sequencing. Orphanet J. Rare Dis. **8**, 122 (2013)
69. Aller, E., et al.: Identification of large rearrangements of the PCDH15 gene by combined MLPA and a CGH: large duplications are responsible for usher syndrome. Invest. Ophthalmol. Vis. Sci. **51**, 5480–5485 (2010)
70. Pirooznia, M., Goes, F.S., Zandi, P.P.: Whole-genome CNV analysis: advances in computational approaches. Front. Genet. **6**, 138 (2015)
71. Dommering, C.J., et al.: RB1 mutation spectrum in a comprehensive nationwide cohort of retinoblastoma patients. J. Med. Genet. **51**, 366–374 (2014)
72. Greger, V., Passarge, E., Höpping, W., Messmer, E., Horsthemke, B.: Epigenetic changes may contribute to the formation and spontaneous regression of retinoblastoma. Hum. Genet. **83**, 155–158 (1989)

73. Castéra, L., et al.: MDM2 as a modifier gene in retinoblastoma. J. Natl. Cancer Inst. **102**, 1805–1808 (2010)
74. de Oliveira Reis, A.H., et al.: Influence of MDM2 and MDM4 on development and survival in hereditary retinoblastoma. Pediatr. Blood Cancer. **59**, 39–43 (2012)
75. Zhao, J.-J., et al.: Identification of miRNAs associated with tumorigenesis of retinoblastoma by miRNA microarray analysis. Childs Nerv. Syst. ChNS Off. J. Int. Soc. Pediatr. Neurosurg. **25**, 13–20 (2009)
76. Rushlow, D.E., et al.: Characterisation of retinoblastomas without RB1 mutations: genomic, gene expression, and clinical studies. Lancet Oncol. **14**, 327–334 (2013)
77. Ayari Jeridi, H., et al.: Genetic testing in Tunisian families with heritable retinoblastoma using a low cost approach permits accurate risk prediction in relatives and reveals incomplete penetrance in adults. Exp. Eye Res. **124**, 48–55 (2014)
78. Chen, Z., et al.: Enhanced sensitivity for detection of low-level germline mosaic RB1 mutations in sporadic retinoblastoma cases using deep semiconductor sequencing. Hum. Mutat. **35**, 384–391 (2014)
79. Amitrano, S., et al.: Next generation sequencing in sporadic retinoblastoma patients reveals somatic mosaicism. Eur. J. Hum. Genet. EJHG. (2015). doi:10.1038/ejhg.2015.6
80. Devarajan, B., et al.: Targeted next generation sequencing of RB1 gene for the molecular diagnosis of retinoblastoma. BMC Cancer. **15**, 320 (2015)
81. Francis, P.J., Berry, V., Bhattacharya, S.S., Moore, A.T.: The genetics of childhood cataract. J. Med. Genet. **37**, 481–488 (2000)
82. Apple, D.J., Ram, J., Foster, A., Peng, Q.: Elimination of cataract blindness: a global perspective entering the new millenium. Surv. Ophthalmol. **45**(Suppl 1), S1–196 (2000)
83. Gilbert, C., Foster, A.: Childhood blindness in the context of VISION 2020--the right to sight. Bull. World Health Organ. **79**, 227–232 (2001)
84. Rahi, J.S., Dezateux, C., British Congenital Cataract Interest Group: Measuring and interpreting the incidence of congenital ocular anomalies: lessons from a national study of congenital cataract in the UK. Invest. Ophthalmol. Vis. Sci. **42**, 1444–1448 (2001)
85. Shiels, A., Bennett, T.M., Hejtmancik, J.F.: Cat-map: putting cataract on the map. Mol. Vis. **16**, 2007–2015 (2010)
86. Ponnam, S.P.G., et al.: Mutational screening of Indian families with hereditary congenital cataract. Mol. Vis. **19**, 1141–1148 (2013)
87. Hansen, L., et al.: Comprehensive mutational screening in a cohort of Danish families with hereditary congenital cataract. Invest. Ophthalmol. Vis. Sci. **50**, 3291–3303 (2009)
88. Santhiya, S.T., et al.: Mutation analysis of congenital cataracts in Indian families: identification of SNPS and a new causative allele in CRYBB2 gene. Invest. Ophthalmol. Vis. Sci. **45**, 3599–3607 (2004)
89. Sun, W., Xiao, X., Li, S., Guo, X., Zhang, Q.: Exome sequencing of 18 Chinese families with congenital cataracts: a new sight of the NHS gene. PLoS One. **9**, e100455 (2014)
90. Allingham, R.R., Liu, Y., Rhee, D.J.: The genetics of primary open-angle glaucoma: a review. Exp. Eye Res. **88**, 837–844 (2009)
91. Wang, R., Wiggs, J.L.: Common and rare genetic risk factors for glaucoma. Cold Spring Harb. Perspect. Med. **4**, a017244 (2014)
92. Li, Z., et al.: A common variant near TGFBR3 is associated with primary open angle glaucoma. Hum. Mol. Genet. **24**, 3880–3892 (2015)
93. Huang, X., et al.: Mutation analysis of seven known glaucoma-associated genes in Chinese patients with glaucoma. Invest. Ophthalmol. Vis. Sci. **55**, 3594–3602 (2014)
94. Cassa, C.A., Tong, M.Y., Jordan, D.M.: Large numbers of genetic variants considered to be pathogenic are common in asymptomatic individuals. Hum. Mutat. **34**, 1216–1220 (2013)
95. Andreasen, C., et al.: New population-based exome data are questioning the pathogenicity of previously cardiomyopathy-associated genetic variants. Eur. J. Hum. Genet. EJHG. **21**, 918–928 (2013)

96. Xue, Y., et al.: Deleterious- and disease-allele prevalence in healthy individuals: insights from current predictions, mutation databases, and population-scale resequencing. Am. J. Hum. Genet. **91**, 1022–1032 (2012)
97. Strom, S.P., Gorin, M.B.: Evaluation of autosomal dominant retinal dystrophy genes in an unaffected cohort suggests rare or private missense variants may often be benign. Mol. Vis. **19**, 980–985 (2013)
98. Pozo, M.G.-D., et al.: Re-evaluation casts doubt on the pathogenicity of homozygous USH2A p.C759F. Am. J. Med. Genet. A. **167**, 1597–1600 (2015)
99. Landrum, M.J., et al.: ClinVar: public archive of relationships among sequence variation and human phenotype. Nucleic Acids Res. **42**, D980–D985 (2014)
100. Ran, X., et al.: 'RetinoGenetics': a comprehensive mutation database for genes related to inherited retinal degeneration. Database J. Biol. Databases Curation. **2014**, bau047 (2014)
101. van Huet, R.A.C., et al.: The RD5000 database: facilitating clinical, genetic, and therapeutic studies on inherited retinal diseases. Invest. Ophthalmol. Vis. Sci. **55**, 7355–7360 (2014)
102. Clayton, E.W.: Ethical, legal, and social implications of genomic medicine. N. Engl. J. Med. **349**, 562–569 (2003)
103. Stone, E.M., et al.: Recommendations for genetic testing of inherited eye diseases: report of the American Academy of ophthalmology task force on genetic testing. Ophthalmology. **119**, 2408–2410 (2012)
104. McKibbin, M., et al.: Current understanding of genetics and genetic testing and information needs and preferences of adults with inherited retinal disease. Eur. J. Hum. Genet. EJHG. **22**, 1058–1062 (2014)
105. Hysi, P.G., Wojciechowski, R., Rahi, J.S., Hammond, C.J.: Genome-wide association studies of refractive error and myopia, lessons learned, and implications for the future. Invest. Ophthalmol. Vis. Sci. **55**, 3344–3351 (2014)
106. Ritchie, M.D., et al.: Electronic medical records and genomics (eMERGE) network exploration in cataract: several new potential susceptibility loci. Mol. Vis. **20**, 1281–1295 (2014)
107. Fritsche, L.G., et al.: Seven new loci associated with age-related macular degeneration. Nat. Genet. **45**(433–439), 439e1–439e2 (2013)
108. Ciralsky, J., Colby, K.: Congenital corneal opacities: a review with a focus on genetics. Semin. Ophthalmol. **22**, 241–246 (2007)
109. Hosseini, S.M., et al.: The association of previously reported polymorphisms for microvascular complications in a meta-analysis of diabetic retinopathy. Hum. Genet. **134**, 247–257 (2015)
110. Hartong, D.T., Berson, E.L., Dryja, T.P.: Retinitis pigmentosa. Lancet Lond. Engl. **368**, 1795–1809 (2006)
111. Haim, M.: Epidemiology of retinitis pigmentosa in Denmark. Acta Ophthalmol. Scand. Suppl. 1–34 (2002)
112. Functional implications of the spectrum of mutations found in 234 cases with X-linked juvenile retinoschisis. The Retinoschisis Consortium. Hum. Mol. Genet. **7**, 1185–1192 (1998)
113. Maugeri, A., et al.: The 2588G-->C mutation in the ABCR gene is a mild frequent founder mutation in the western European population and allows the classification of ABCR mutations in patients with Stargardt disease. Am. J. Hum. Genet. **64**, 1024–1035 (1999)
114. Kivelä, T.: The epidemiological challenge of the most frequent eye cancer: retinoblastoma, an issue of birth and death. Br. J. Ophthalmol. **93**, 1129–1131 (2009)
115. Roosing, S., et al.: Causes and consequences of inherited cone disorders. Prog. Retin. Eye Res. **42**, 1–26 (2014)
116. den Hollander, A.I., Roepman, R., Koenekoop, R.K., Cremers, F.P.M.: Leber congenital amaurosis: genes, proteins and disease mechanisms. Prog. Retin. Eye Res. **27**, 391–419 (2008)
117. Forsythe, E., Beales, P.L.: Bardet-Biedl syndrome. Eur. J. Hum. Genet. EJHG. **21**, 8–13 (2013)
118. Millán, J.M., et al.: An update on the genetics of usher syndrome. J. Ophthalmol. **2011**, 417217 (2011)

119. Simpson, D.A., Clark, G.R., Alexander, S., Silvestri, G., Willoughby, C.E.: Molecular diagnosis for heterogeneous genetic diseases with targeted high-throughput DNA sequencing applied to retinitis pigmentosa. J. Med. Genet. **48**, 145–151 (2011)
120. Neveling, K., et al.: Next-generation genetic testing for retinitis pigmentosa. Hum. Mutat. **33**, 963–972 (2012)
121. Wang, F., et al.: Next generation sequencing-based molecular diagnosis of retinitis pigmentosa: identification of a novel genotype-phenotype correlation and clinical refinements. Hum. Genet. **133**, 331–345 (2014)
122. Xu, Y., et al.: Mutations of 60 known causative genes in 157 families with retinitis pigmentosa based on exome sequencing. Hum. Genet. **133**, 1255–1271 (2014)
123. Xu, Y., et al.: Mutation analysis in 129 genes associated with other forms of retinal dystrophy in 157 families with retinitis pigmentosa based on exome sequencing. Mol. Vis. **21**, 477–486 (2015)
124. Zernant, J., et al.: Analysis of the ABCA4 gene by next-generation sequencing. Invest. Ophthalmol. Vis. Sci. **52**, 8479–8487 (2011)
125. Strom, S.P., et al.: Molecular diagnosis of putative Stargardt disease probands by exome sequencing. BMC Med. Genet. **13**, 67 (2012)
126. Fujinami, K., et al.: ABCA4 gene screening by next-generation sequencing in a British cohort. Invest. Ophthalmol. Vis. Sci. **54**, 6662–6674 (2013)
127. Zhang, X., et al.: Molecular diagnosis of putative Stargardt disease by capture next generation sequencing. PLoS One. **9**, e95528 (2014)
128. Zaneveld, J., et al.: Comprehensive analysis of patients with Stargardt macular dystrophy reveals new genotype-phenotype correlations and unexpected diagnostic revisions. Genet. Med. Off. J. Am. Coll. Med. Genet. **17**, 262–270 (2015)
129. Licastro, D., et al.: Molecular diagnosis of usher syndrome: application of two different next generation sequencing-based procedures. PLoS One. **7**, e43799 (2012)
130. Xing, D.-J., et al.: Comprehensive molecular diagnosis of Bardet-Biedl syndrome by high-throughput targeted exome sequencing. PLoS One. **9**, e90599 (2014)
131. O'Sullivan, J., et al.: A paradigm shift in the delivery of services for diagnosis of inherited retinal disease. J. Med. Genet. **49**, 322–326 (2012)
132. Schorderet, D.F., Iouranova, A., Favez, T., Tiab, L., Escher, P.: IROme, a new high-throughput molecular tool for the diagnosis of inherited retinal dystrophies. Biomed. Res. Int. **2013**, 198089 (2013)
133. Shanks, M.E., et al.: Next-generation sequencing (NGS) as a diagnostic tool for retinal degeneration reveals a much higher detection rate in early-onset disease. Eur. J. Hum. Genet. EJHG. **21**, 274–280 (2013)
134. Chen, X., et al.: Targeted sequencing of 179 genes associated with hereditary retinal dystrophies and 10 candidate genes identifies novel and known mutations in patients with various retinal diseases. Invest. Ophthalmol. Vis. Sci. **54**, 2186–2197 (2013)
135. Watson, C.M., et al.: Mutation screening of retinal dystrophy patients by targeted capture from tagged pooled DNAs and next generation sequencing. PLoS One. **9**, e104281 (2014)
136. Huang, X.-F., et al.: Genotype-phenotype correlation and mutation spectrum in a large cohort of patients with inherited retinal dystrophy revealed by next-generation sequencing. Genet. Med. Off. J. Am. Coll. Med. Genet. **17**, 271–278 (2015)
137. Liu, X., et al.: Molecular genetic testing in clinical diagnostic assessments that demonstrate correlations in patients with autosomal recessive inherited retinal dystrophy. JAMA Ophthalmol. **133**, 427–436 (2015)

Application of Next-Generation Sequencing to Hearing Loss

Lisa Dyer, Xinjian Wang, Kejian Zhang, John Greinwald Jr., and C. Alexander Valencia

Abstract Hearing loss is the most common birth defect and sensorineural disorder in humans. Hearing loss can be syndromic or nonsyndromic and can be further divided based on mode of inheritance. With over 100 genes responsible for the etiology of hearing loss, Sanger sequencing, being a traditional gene-by-gene approach, has become an unfavorable strategy in terms of cost and time. Next-generation sequencing technology offers the advantage of sequencing multiple genes in parallel with lower cost and higher time-efficiency and has quickly become a fundamental tool for targeted panel-based hearing loss diagnostics. In this chapter, we discuss the application of next-generation sequencing to hearing loss, address successes and challenges of the approach, examine the current role of whole exome sequencing and comment on the future of whole genome sequencing. We conclude that the application of next generation sequencing will tremendously broaden our knowledge of hearing loss; the outcome of which includes rapid and accurate diagnosis that would result in an earlier and more effective intervention.

The original version of this chapter was revised. An erratum to this chapter can be found at DOI 10.1007/978-3-319-56418-0_17

L. Dyer, Ph.D. • K. Zhang, M.D., MBA
Division of Human Genetics, Cincinnati Children's Hospital Medical Center, Cincinnati, OH, USA

Department of Pediatrics, University of Cincinnati College of Medicine, Cincinnati, OH, USA

X. Wang, Ph.D.
Division of Human Genetics, Cincinnati Children's Hospital Medical Center, Cincinnati, OH, USA

J. Greinwald Jr., M.D.
Ear and Hearing Center, Division of Pediatric Otolaryngology-Head and Neck Surgery, Cincinnati Children's Hospital Medical Center, Cincinnati, OH, USA

Department of Otolaryngology-Head and Neck Surgery, University of Cincinnati College of Medicine, Cincinnati, OH, USA

C. Alexander Valencia, Ph.D. (✉)
Department of Pediatrics, University of Cincinnati College of Medicine, Cincinnati, OH, USA

Division of Human Genetics, Cincinnati Children's Hospital Medical Center, 3333 Burnet Ave., MLC7016 R1028, Cincinnati, OH 45229, USA
e-mail: alexander.valencia@cchmc.org

L.-J.C. Wong (ed.), *Next Generation Sequencing Based Clinical Molecular Diagnosis of Human Genetic Disorders*, DOI 10.1007/978-3-319-56418-0_5

Keywords Next-generation sequencing • Hearing loss • Hearing loss diagnosis by NGS • Nonsyndromic hearing loss • Microdroplet-PCR • Panel-based hearing loss diagnosis

1 Introduction

Hearing loss is the most common birth defect and sensorineural disorder in humans with an approximately 30 million individuals in developed countries having hearing loss in both ears [14]. Hearing loss results from obstructions in the transmission of the sound anywhere between the outer ear and auditory cortex in the brain. In a normal condition, the sound signal that is collected by the outer ear is amplified by the middle ear for transmission to the cochlea, which then converts this energy into electrical signals that is ultimately transmitted to the brain through the auditory nerves [25]. Based on the defective anatomical structure involved, hearing loss can be classified as conductive, sensorineural or mixed. Conductive hearing loss is a defect in conducting sound waves through outer and middle ear due to abnormalities of outer ear, tympanic membrane (eardrum), or ossicles of the middle ear. Sensorineural hearing loss (SNHL) is due to a defect located anywhere from cochlea to the auditory cortex. Mixed hearing loss is a combination of both conductive and sensorineural abnormalities. Depending on the age at onset, hearing loss can be classified as prelingual, present before speech development, or postlingual, present after speech development. Severity of the hearing loss is measured by decibels (dB), can be graded from mild (26–40 dB) to profound (>90 dB), affecting from low to high frequencies [25].

In this chapter, we introduce syndromic and nonsyndromic hearing loss alongside the importance of hearing loss detection. Then, we focus our attention on covering the recent advances of hearing loss diagnosis in light of the application of next-generation sequencing (NGS) to this field.

2 Syndromes that Include Hearing Loss

One in 500 newborns is affected with bilateral permanent sensorineural hearing loss in developed countries, this number is increased to 2.7 per 1000 before the age of 5 years and 3.5 per 1000 during adolescence [18]. Approximately two-thirds of hearing prelingual or congenital loss is due to genetic factors and in the remaining one-third of cases, it is caused by environmental factors [14, 19]. The environmental factors that cause hearing loss include both prenatal and postnatal infections, use of ototoxic drugs and exposure to excessive noise. The majority of the inherited form of hearing loss is monogenic, and it can be syndromic or nonsyndromic. In the syndromic forms, hearing loss is accompanied by other physical manifestations, and it accounts for about 30% of the inherited hearing loss ([15]; Fig. 1A). Over 400

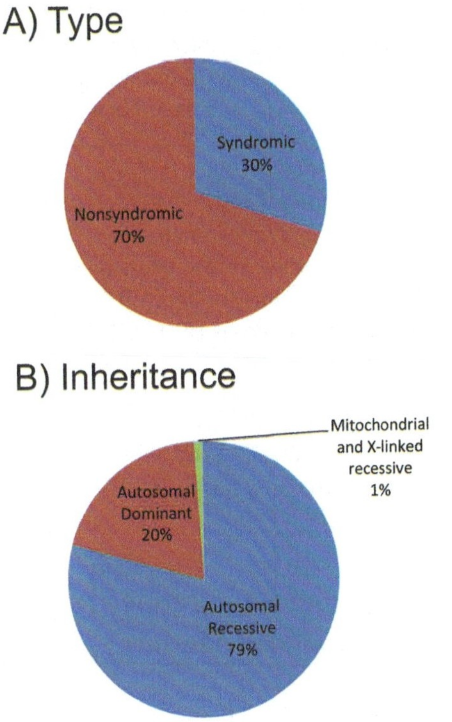

Fig. 1 (**A**) Hearing loss type. (**B**) Percentage of Hearing loss inhertance patterns

syndromes have been reported with hearing loss, and some of the common forms of syndromic hearing loss including Usher, Pendred, Jervell and Lange-Nielsen, Waardenburg, Branchio-oto-renal, and Stickler syndromes are among the many [14]. The nonsyndromic forms of hearing loss, with no other physical findings, accounts for about 70% inherited hearing loss (Fig. 1A). They are categorized into four different groups according to their mode of inheritance: (1) autosomal recessive, (2) autosomal dominant, (3) X-linked and (4) maternal inheritance due to mutations in mitochondrial genes (Fig. 1B). The autosomal recessive hearing loss is the most common type occurring in about 79% of patients, followed by autosomal dominant in about 20%. The X-linked and mitochondrial hearing loss are less common and accounting for only about 1% of the patients [6, 28, 29].

3 Nonsyndromic Hearing Loss

The nonsyndromic hearing loss (NSHL) is extremely heterogeneous and so far, over 150 loci responsible for nonsyndromic hearing loss that have been mapped. These loci are designated as DFN that is derived from abbreviation of DeaFNess followed by mode of transmission; DFNA refers to loci for autosomal dominant forms, DFNB refers to loci for autosomal recessive and DFNX to X-linked forms. The numbers following the designation are chronological order of locus identification (DFNB1 refers to first autosomal recessive locus). To date, 60 autosomal recessive (ARNSHL), 32 autosomal dominant (ADNSHL), 4 X-linked and 2 mitochondrial nonsyndromic genes have been identified (http://hereditaryhearingloss.org). Many of these genes cause more than one form of hearing loss. For example, *SLC26A4, CDH23, MYO7A, DFNB31, USH1C* and others cause both syndromic and nonsyndromic forms, *TMC1, GJB2, GJB6, MYO7A* and others may cause both autosomal dominant and autosomal recessive forms of hearing loss. Mutations in the *GJB2*, encoding connexin 26, that causes DFNB1 is the most common cause of hearing loss and accounts for about 50% of the cases with autosomal recessive hearing loss in many populations [9, 15]. The remaining cases are attributable to the mutations in other genes, and among others *SLC26A4, MYO7A, OTOF, CDH23* and *TMC1* are more prevalent [14]. Mutations in the rest of the genes are very rare, many of them have been found to cause hearing loss in one or two consanguineous families [15, 33]. Except *WFS1, KCNQ4, GJB2* and *COCH*, most of the genes causing autosomal dominant hearing loss are not a common cause of hearing loss [14].

4 Significance of Early Hearing Loss Recognition and Genetic Testing

Early detection and intervention for children with hearing loss offers opportunities for improving the language and speech development, thereby facilitating the acquisition of normal social, cognitive skills. Currently, 43 states and the District of Columbia and Puerto Rico have mandated newborn hearing screening programs (http://www.nidcd.nih.gov/health/statistics/pages/quick.aspx). There are two different types of newborn hearing tests utilized. The first is termed the automated otoacoustic emission (AOAE) screening test and is carried out by measuring the vibration of the hair cells in response to sounds emitting from a probe placed in the baby's ear. A microphone within the probe can detect the echo generated from a healthy cochlea. The second is termed the auditory brainstem response test (ABR) in which electrodes are placed on the baby's head to detect responses to sounds played to the ear. More than 95% of babies born in the United States are screened for hearing loss. Screening programs have proven to be effective as 77% of children confirmed to have permanent hearing loss have been enrolled in follow-up programs. However, the newborn screening method is not able to distinguish the primary cause of hearing loss. As 2/3 of hearing loss stems from a genetic cause, the elucidation of genetic basis of hearing loss is crucial for the clinical management of patients and their family. In addition, determination of genetic etiology in a large cohort of patients will provide better understanding of genotype-phenotype correlations, which could help to develop specific therapeutic interventions. For syndromic hearing loss, candidate genes for molecular diagnosis are possible to be selected based on associated symptoms, whereas this approach is not viable for nonsyndromic hearing loss as the phenotype caused by most of the genes is

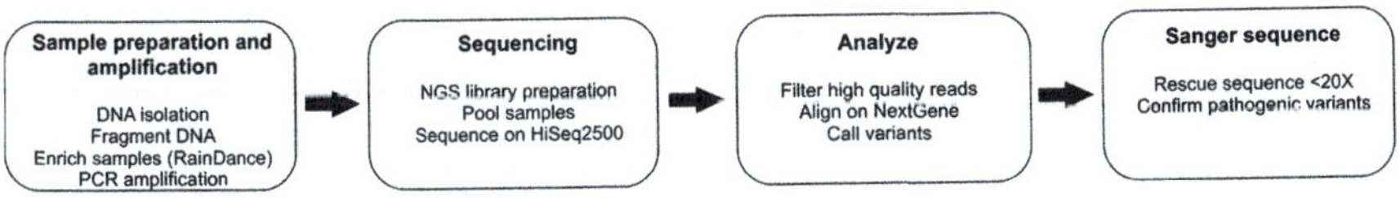

Fig. 2 Next-generation sequencing workflow of the clinical molecular genetics laboratory at the Cincinnati Children's Hospital Medical Center

indistinguishable. Therefore, sequential screening of all hearing loss genes has been widely applied to identify the genetic cause. Currently, genetic testing for hearing loss is conducted using different diagnostic algorithms at several institutions worldwide as listed by the genetic testing registry database (http://www.ncbi.nlm.nih.gov/gtr/). Mutation screening of coding and flanking intronic regions of the candidate genes using Sanger sequencing is the most common approach in the vast majority of these laboratories. However, with over 100 hearing loss genes, the extreme genetic heterogeneity of non-syndromic and syndromic hearing loss makes this strategy unfavorable in terms of cost and time. NGS technology offers the advantage of sequencing multiple genes in parallel with lower cost and higher time-efficiency and has quickly become a fundamental tool for panel-based hearing loss diagnostics. In the following sections, we describe the NGS panel workflow and compare the two main approaches, amplification-based and hybridization-based targeted enrichment of hearing loss genes.

5 NGS Workflow and Enrichment Types

NGS is composed of four main components, sample preparation and amplification (target enrichment), sequencing by an NGS platform, data analysis and Sanger confirmation. Our laboratory's workflow is represented in Fig. 2. The process of NGS begins by isolating and fragmenting genomic DNA from a patient sample, and unless whole genome sequencing (WGS) is performed, the genes of interest (e.g. hearing loss genes) need to be enriched for sequencing and amplified. Next, the fragmented DNA are prepared into a NGS library, the samples are pooled and sequenced using the sequencer HiSeq 2500 (Illumina, San Diego, CA). After sequencing, the high quality sequence reads are filtered and simultaneously mapped to human reference sequence. The DNA sequence variants that are detected by NGS analysis software and variants are further confirmed by Sanger sequencing. In addition, a laboratory may choose to rescue any low and uncovered regions to increase the clinical sensitivity of a panel by Sanger sequencing.

The two most widely used target enrichment strategies for clinical use are amplification-based or oligonucleotide hybridization-based methods [17, 26].

Amplification-based enrichment has been widely utilized as a one reaction, one amplicon approach for Sanger sequencing of single genes. In order to enrich a large number of amplicons for NGS applications, this method must be multiplexed. To overcome this limitation of standard PCR, RainDance Technologies (RDT) developed the Thunderstorm platform that uses microdroplets each containing a single specific primer pair, DNA, and the reagents necessary for subsequent thermocycling. Following amplification, the amplicons are pooled and are ready for library construction and sequencing. Because each PCR reaction is housed in a single microdroplet, multiple primer pairs do not interfere with each other or compete for reagents. In contrast, hybridization-based enrichment methods use hybridization probes that are complementary to the target regions either by an on-array capture approach or an in-solution capture approach. The on-array capture approach allows fragmented library DNA to hybridize to immobilized probes on an array. Any unbounded DNA is washed away and the targeted DNA is eluted. Compared to amplification-based capture approach, this method is less laborious but requires costly equipment, such as the hybridization and elution stations. In-solution capture has overcome these two limitations; the general hybridization principle is similar but is performed in solution with excess probes over template DNA and does not require expensive equipment. The probes are biotinylated and by using streptavidin-labeled magnetic beads allow for capture and purification of the target regions. Next, we describe the application of both of these enrichment strategies to NGS hearing loss panels.

6 Amplification-Based HL Panels

A European group designed a primer library including 646 specific primer pairs for exons and most of the UTR of 15 ARNSHL genes, using conventional PCR. All amplicons were pooled in an equal molar concentration and analyzed using Roche 454 NGS technology (Table 1; [10]). This platform generated the coverage of 95% targeted bases at 30× depth. Among five patients with congenital genetic deafness, causative mutations were identified in four patients. Among these, two novel mutations in *CDH23* and *OTOF* were found in three patients that were also characterized as interesting regions in a previous linkage study. Similarly, Licastro et al. used a long-PCR-based enrichment and NGS to develop a diagnostic panel for Usher syndrome genes [16]. Molecular diagnosis in Usher syndrome is hindered by significant genetic heterogeneity, the large size of some of the Usher genes, numerous polymorphic variants in genes such as *MYO7A* and *USH7A*, and digenic inheritance was also proposed in some Usher syndrome cases [5]. At least 11 loci and 9 causative genes have been reported associated with three subtypes of Usher syndrome. Current diagnostic strategies for Usher syndrome include Sanger sequencing of Usher genes, which is a demanding procedure in terms of both cost and time, or microarrays-based genotyping method that only detect previously reported mutations. This study showed this NGS platform had 94% coverage of target bases at 25×. Eleven pathogenic mutations in *MYO7A, CLRN1, GPR98, USH2A,* and *PCDH15* were identified in ten out of the twelve Usher patients, while genetic causation of two patients still stay negative, indicating a positive diagnostic rate of 84% in this study.

The main advantage of microdroplet-PCR-based technology, such as Raindance, is being able to combine high-throughput automation and highly sensitive, specific

Table 1 Performance of next-generation sequencing platforms for mutation detection of hearing loss genes

	Capture-based HL gene panel			PCR-based HL gene panel					Whole Exome sequencing		
Target	54 known HL genes in human	54 known HL genes in human	246 known HL genes in human and mouse	15 known HL genesin human	34 known HL genes in human	9 known USH genes in human	24 known HL genes in human	24 known HL genes in human	Human Exome	Human Exome	Human Exome
Target enrichment method	NimbleGen solid-based enrichment	Agilent SureSelect solution based enrichment	Agilent SureSelect solution based enrichment	PCR-based amplification enrichment	RDT microdroplet PCR enrichment (RainStorm)	Long-PCR enrichment	RDT microdroplet PCR enrichment (RainStorm)	RDT microdroplet PCR enrichment (ThunderStorm)	Agilent SureSelect 50 Mb enrichment	Illumina TrueSeq 62 Mb enrichment	Agilent SureSelect 50 Mb enrichment
Sequencing platform	Roche 454 GS FLX	Illumina GAII	Illumina GAII	Roche 454 GS FLX	Illumina HiSeq 2000	Roche 454 GS FLX and Illumina GAII	Illumina HiSeq 2000	Illumina HiSeq 2500	Illumina HiSeq 2000	Illumina HiSeq 2000	SOLiD 3 or 4 Systems
Mean target coverage	71×	903×	757×	~150×	1585×	NA	NA	2373×	NA	NA	NA
%Target coverage at calling threshold	96%	95%	95%	95%	95%	94%	94%	98%	57%	65%	~50%
Minimum calling threshold	3×	40×	10×	30×	30×	25×	40×	20×	20×	20×	25×
Sensitivity	93.98%	99.72%	NA	NA	99.00%	NA	>99.99%	>99.99%	NA	NA	NA
Specificity	97.92%	>99.00%	NA	NA	99.40%	NA	>99.99%	>99.99%	NA	NA	NA
References	[23]	[23]	[7]	[10]	[22]	[16]	[24]	Unpublished	[22]	[22]	[16]

HL hearing loss, *RDT* RainDance Technologies, *NA* not available.

and uniform amplification using targeted specific primers [27]. Recently, two groups used a similar strategy, applying RDT microdroplet-PCR enrichment (RainStorm) and sequencing on the Illumina HiSeq 2000 sequencer, to develop NGS hearing loss panels for 34 ARNSHL genes and 24 well-studied SNHL genes, respectively [22, 24]. The two NGS platforms targeted all exons and flanking intron regions of the hearing loss genes. Schrauwen et al. presented an overall mean coverage depth in the target area of 1585× and 95% of the bases were covered at 30×, while the panel developed by Sivakumaran et al. had a 94% of the targeted bases covered at 40×. Sanger sequencing was used to verify the known variants. These two NGS panels both achieved >99% sensitivity and specificity, indicating that the enrichment is a reliable platform for mutation detection of hearing loss genes. To detect the concordance between NGS panel and Sanger sequencing, Sivakumaran et al. used the NextGENe software (SoftGenetics, LLC) which detected a total of 394 variants in five genes, *GJB2, CDH23, MYO7A, EYA1* and *OTOF*, that had been sequenced by Sanger sequencing to confirm the accuracy. The results showed a > 99.99% concordance between NGS and Sanger sequencing by evaluating more than 30,000 bp in the 5 SNHL genes, except only one C > T substitution in *MYO7A* detected by NGS was not identified by Sanger sequencing due to a misalignment issue. Small indels were detected in the NGS data including a 22-bp deletion in intron 27 of *MYO7A*. Since the acceptable false-positive and false-negative rates are more stringent for clinical diagnostic use, the authors favorably suggested to set a minimal 40× of coverage as threshold at the target bases for RDT microdroplet PCR enrichment (RainStorm) [24]. This hearing loss panel, otoseq, is commercially available at Cincinnati Children's Hospital Medical Center's (CCHMC) molecular laboratory and includes the sequencing and analysis of 23 genes (Table 2).

In the Schrauwen et al. panel design, all genes were selected from an ARNSHL gene list and twenty four patients with prelingual, moderate to profound hereditary nonsyndromic hearing loss in autosomal recessive inheritance, and without *GJB2* mutations were carefully selected. Nine out of 24 patients (37.5%) were confidently diagnosed. Six patients were found to have homozygous mutations and three patients had compound heterozygous mutations. The results also suggested a possible digenic finding in *OTOF* and *SLC26A4* genes in one patient. However, these two genes perform completely different functions in the inner ear and proteins are expressed at different ear locations which weakened the evidence of digenic inheritance in this patient. The convincing follow-up functional and family study are important steps to confirm this hypothesis [22].

7 Hybridization-Based HL Panels

Shearer et al. developed a comprehensive diagnostic platform named as OtoSCOPE, that targets the exons of 54 known NSHL genes including Usher syndrome genes (Table 2; [23]). In this study, two hybridization capture-based enrichment approaches, NimbleGen solid-phase enrichment and Agilent SureSelect (SS) solution-based capture enrichment were paired with 454 GS FLX (Roche, Branford, CT)

Table 2 Genes and associated disorders included in the OtoSeq (hearing loss) panel offered by the Cincinnati Children's Hospital Medical Center's clinical molecular genetics laboratory

Gene	Full gene name	Inheritance	OMIM phenotype
CDH23	Cadherin 23	AR, DR	Usher Syndrome, type 1D/F
CLRN1	Clarin 1	AR	Usher Syndrome, type 3A
EYA1	Eyes absent 1	AD	Branchiootorenal syndrome
FOXI1	Forkhead box I1	AR	Hearing loss with enlarged vestibular aqueduct
GJB2	Connexin 26	AD, AR, DD	Hearing loss
GJB6	Gap junction protein, beta-6	AD, AR, DD	Hearing loss
GPR98	G-protein coupled receptor 98	DD, AR	Usher syndrome, type 2C
KCNJ10	Potassium channel, subfamily J, member 10	AR	Hearing loss with enlarged vestibular aqueduct
MYO6	Myosin VI	AD, AR	Hearing loss
MYO7A	Myosin VIIA	AD, AR	Hearing loss; Usher Syndrome, type 1B
OTOF	Otoferlin	AR	Auditory Neuropathy; Hearing loss, autosomal recessive 9
PCDH15	Protocadherin 15	AR, DR	Hearing loss; Usher Syndrome
POU3F4	Pou domain, class 3, transcription factor 4	XLR	Hearing loss
SIX1	Sine oculus homeobox 1	AD	Branchiootorenal syndrome
SIX5	Sine oculus homeobox 5	AD	Branchiootorenal syndrome
SLC26A4	Solute Carrier Family 26, Member 4	AR	Pendred Syndrome; Hearing loss with enlarged vestibular aqueduct
TMC1	Transmembrane Channel-Like Protein 1	AD, AR	Hearing loss
TMIE	Transmembrane Inner Ear-Expressed Gene	AR	Hearing loss
TMPRSS3	Transmembrane Protease, Serine 3	AR	Hearing loss
USH1C	Usher Syndrome, Type 1C	AR	Hearing loss; Usher Syndrome, Type 1C
USH1G	Usher Syndrome, Type 1G	AR	Usher syndrome, type 1G
USH2A	Usher Syndrome, Type 1A	AR	Usher syndrome, type2A; Retinitis pigmentosa
WHRN	Whilrlin	AR	Hearing loss; Usher syndrome, type 2D

AR Autosomal Recessive, *AD* Autosomal Dominant, *DD* Digenic Dominant, *DR* Digenic Recessive

pyrosequencing and Illumina GAII cyclic reversible termination sequencing, respectively. By comparing these two platforms, SS-Illumina was shown to be superior in terms of scalability, cost and sensitivity, generating a 13-fold higher average depth of coverage on targeted bases (903× vs 71×), with 95.3% of targeted hearing loss genes covered at 40× threshold (Table 2). Highly heterozygous SNPs in the target regions were confirmed by Sanger sequencing to determine the sensitivity and specificity, which were both greater than 99% for the SS-Illumina platform. Beside the use of three patients as positive controls, NSHL mutations were found in *STRC, MYO6, KCNQ4, MYN14* and *CDH23* genes in 5 out of 6 idiopathic sensorineural hearing loss patients, including three novel mutations. However, the variants found in one patient were ruled out as causative mutations by segregation analysis. Similarly, a capture-based enrichment approach was used in another study that was designed to detect 246 genes responsible for either human or mouse deafness. With a 95% of coverage of the targeted bases at 10×, pathogenic mutations were identified and Sanger confirmed in 6 of the 11 probands and their families in *CDH23, MYO15A, TECTA, TMC1,* and *WFS1* [7].

Even though successes have been shown, hybridization capture-based enrichment is known to have restrictions in capture of GC-rich or repetitive elements as well as gene family members that share sequence homology. The presence of repetitive or high GC content sequences can be missed due to poor annealing or secondary structure. This can lead to incomplete selection, selection bias and uneven capture efficiency. This may result in reduced sensitivity and specificity that are highly required in diagnostic testing [10, 22]. On the other hand, PCR amplification-based enrichment has been employed in several hearing loss studies to address sequence homology, high GC content by chemical dissociation of the double helix and uniform amplification.

8 NGS Panel Clinical Validation, Implementation and Challenges

To establish NGS as a diagnostic tool, each laboratory must optimize, validate the performance, and establish acceptable quality control (QC) parameters [21]. Prior to implementing a new NGS-based targeted assay, its analytical accuracy, precision, sensitivity and specificity must be verified as a validation of performance parameters. Accuracy is defined as the closeness of agreement between the measured value and the value that is accepted either as a conventional true value or an accepted reference value. Precision is the closeness of agreement among repeat measurements of samples and must meet laboratory-developed test (LDT) expectation. Sensitivity and specificity are statistical measures of the true positive rate and the true negative rate, respectively. For genetic tests, analytic sensitivity is the likelihood that an assay will detect a sequence variation when present within the analyzed genomic region. Specificity refers to the probability that an NGS assay will not detect sequence variations when none are present within the analyzed genomic region. Specificity will vary by technology and must be evaluated by the laboratory director to ensure acceptable quality prior to initiating a new clinical test. In addition to these four parameters, an accurate estimate of adequate depth of coverage, set forth by establishing an appropriate coverage threshold by the individual laboratory for their LDT is necessary to make accurate variant calls.

To validate the performance parameters for the 24 gene hearing loss NGS panel OtoSeq (Table 2), nine previously sequenced patient's DNA samples were sequenced using the directseq protocol from RainDance Technologies according to the steps outlined in Fig. 2. At the time of the validation there were no standard reference controls that could be used for accuracy. However, the College of American Pathologists (CAP) now has a reference DNA material but they provide names of about 200 genes whose variants need to be reported back to CAP. Unfortunately, this material is useful for exome validations and less for panels because the genes included may not be in the laboratory's panel. For the validation, samples that had many and different variant types were selected. For the previous hearing loss RainDance panel (RainStorm instrument), the depth of coverage percentage was 94% at 40×. The current and newest hearing loss RainDance directseq (Thunderstorm instrument) has the depth of coverage required to meet the laboratory QC set to >98% at 20× with an average coverage of 2373×. The average percent coverage for each of the nine samples at 3×, 5×, 10×, 20×, and 40× were 98.76%, 98.74%, 98.72%, 98.68% and 98.65%, respectively. This coverage was decreased from 40× to 20× because the new sequencing strategy, RainDance directseq, does not require concatenation processing and the data is much cleaner (less false positive), thus making variant calls more accurate at even 20× coverage. The accuracy was calculated by comparing previous RainDance panel data and Sanger confirmed variants against the RainDance directseq data. The average accuracy of the nine samples for RainDance and RainDance directseq was 98.8%. Furthermore, a comparison of Sanger confirmed variants with RainDance directseq yielded an accuracy of 100%. The discordance of 1.2% between the RainDance and RainDance directseq reflects the existence of low coverage regions, however, these regions are typically rescued via Sanger sequencing to increase the accuracy of NGS to 100%. For precision, the laboratory compared the inter-run concordance of several samples using RainDance directseq and it was determined to be an average of 100%. Using NGS data, the average analytical sensitivity [TP/(TP + FN) × 100%; TP = true positive, FN = false negative] and specificity [TN/(TN + FP) × 100%; TN = true negative, FP = false positive] were calculated, for the nine samples, to be 100% and 100%, respectively. To ensure maximum sensitivity, all low coverage regions (<20×) are rescued by Sanger sequencing and variants of reports are always Sanger sequencing confirmed.

False positive variants are a problem in NGS platforms, and two strategies for excluding these are employed after the validation; (1) exclude all variants that were not Sanger confirmed and (2) visually inspect NGS data to see if obvious false positive variants can been found. For example, polyT, C, G, A stretches, deletion/duplication regions and GC rich areas may influence the calls, thus making it false positive. Deletion/duplication calls often need to be Sanger confirmed to ensure their presence and correct location. All previous false positive variants and known SNPs are cataloged in an internal OtoSeq NGS database to help in the interpretation of future clinical cases. Specifically, recurrent variants with high frequency can be excluded. Thus, the establishment of a panel database that catalogs false positives and SNPs as well as true variants with pathogenicity calls is key for clinical implementation of a multigene NGS panel test.

Following the validation the laboratory director reviews that data and decides the acceptable QC parameters. One of the most important QC measures for NGS tests is the establishment of a standard coverage for each run, since the quality of the variant call increases with coverage. For OtoSeq, acceptable QC values include a minimum coverage of each variant call in a targeted panel is >20× and a minimum mean coverage of >98% of bases at 20×. To detect germline or somatic mosaicism, a threshold for higher coverage is necessary and the laboratory must determine their NGS assay's limit of detection through a sensitivity evaluation. However, at this time no clinical laboratories are currently offering such test for patients with hearing loss.

NGS platforms offer the versatility to detect a variety of variants and structural abnormalities such as single-nucleotide polymorphisms (SNPs), and small deletions and duplications, albeit, there is size limitation for deletion/duplication is detection. For example, larger abnormalities may be missed such as partial or whole gene deletions or duplications. To address this issue a deletion/duplication assay by comparative genomic hybridization has been developed and offered clinically at some institutions. Each clinical laboratory must understand the detection limits of NGS and find clever ways to address them by the use of other complementary technologies.

NGS hearing loss panels have proven to be a cost-effective method for solving the genetic heterogeneity observed in hearing loss. In addition, a large number of samples can be processed with this approach and results in the ability to detect new causative variants and rare benign variation enhancing our understanding of human genetic variation. One quarter of the known genes now recognized to cause nonsyndromic hearing loss have been discovered in the past 5 years via NGS technology [1]. Because of the rapid expansion of newly discovered hearing loss genes it may be necessary for a clinical laboratory to add or modify the genes to their established NGS panel. The addition of new genes requires redesign and revalidation of the laboratory's custom enrichment library, a process that is time-consuming and costly. Each laboratory must consider the value, namely, clinical utility, of incorporating additional genes into an NGS panel. In addition, due to the dynamic nature of this technology, suppliers of NGS are constantly improving their chemistry and data output. Laboratories must constantly validate any changes made to the existing panel. In the near future, laboratories may consider the application of whole exome sequencing (WES) or whole genome sequencing (WGS) as a way to circumvent the aforementioned issues.

9 Triumphs and Challenges of WES

Most described pathogenic variants for Mendelian disorders are located in exons or splice sites, which disturb the amino acid sequence of the affected protein. Current estimates suggest 85% of disease variants are harbored in the protein coding regions, although these regions only make up 1–2% of the human genome. Because of the high concentration of pathogenic variants located in exons, whole exome sequencing (WES) is replacing traditional methods of gene sequencing and has emerged as an efficient and more cost-effective diagnostic and research tool for many inherited diseases [3]. WES has had considerable success as a hearing loss gene discovery tool in consanguineous families with ARNSHL and small families that are too small for linkage analysis [13]. WES has now identified more than 20 new non-syndromic hearing loss genes. With over 30,000 single nucleotide variants identified per human exome effective filtering methods are necessary to remove common benign variations. Homozygosity mapping combined with WES has proven to be a beneficial filtering strategy to identify novel ARNSHL genes. Rehman et al. and Diaz-Horta et al. used homozygous mapping and WES to identify 12 homozygous pathogenic variants in 20 known consanguineous families from Turkey and Iran with a history of ARSNHL who screened negative for *GJB2* variants [12, 20]. The Agilent SureSelect 50 Mb exome capture kit was used and samples were sequenced on an Illumina HiSeq 2000. Variants in 39 known ARSNHL genes were filtered according to autosomal recessive and compound heterozygous inheritance models and less than 1% minor allele frequency. The average depth of coverage was 53× for the ARSNHL genes and 84% at 20× coverage. Owing to the extreme locus heterogeneity in ARSNHL, unique pathogenic variants were detected in all 12 positive exome families. Similarly, Behlouli et al. identified a homozygous nonsense variant in a novel hearing loss gene, *EPS8*, encoding an actin-binding protein of cochlear hair cell stereocilia in two siblings with isolated profound hearing loss born to consanguineous Algerian parents [4]. To date, approximately half of the newly discovered ARNSHL genes have been identified by homozygosity mapping and WES [1, 30]. Interestingly, Zheng et al. was the first to identify a novel ADNSHL gene, *CEACAM16*, in a relatively small family with progressive bilateral sensorineural postlingual moderate hearing loss that initially presents in adolescence by using whole genome linkage combined with WES [35]. Since this initial report, multiple other ADNSHL genes have been identified by linkage and WES including *P2RX2*, *OSBPL2*, *TBC1D24*, and *TNC* [2, 31, 32, 34]. These studies have highlighted that WES for the discovery of hearing loss genes has been successful. Compared to panel-based NGS, WES has the benefit of post in-silico analysis when there are new insights of novel genes, and there is no cost, time or revalidation necessary to add a new gene.

Despite these benefits and gene discovery accomplishments made by WES in the field of hearing loss, WES has well described shortcomings. WES suffers from the same capture-based drawbacks as panel-based NGS. As all the protein-coding regions of the genome are not yet well-defined, current exome capture kits only target exons that are annotated [3]. Although the chance is small, causative variants may lie in a region not currently targeted. Additionally, exons rich in GCs will not be captured efficiently leading to drop-out of these exons. In addition, adequate coverage of known hearing loss genes by WES was not optimal for a clinical diagnostic assay. For example, comparison between the performance of RDT deafness NGS libraries to the Agilent SureSelect 50 Mb Exome, Illumina TrueSeq 62 Mb exome and Agilent SureSelect exome kits only had 50 to 65% of the targeted hearing loss gene exons covered at 20× compared to 98% at 20× for the RainDance directseq hearing loss library (Table 1).These results suggest that several exome captures do not provide sufficient target enrichment for the well-recognized hearing loss genes. However, recent improvement in capture technology have increased the coverage of hearing loss genes (24 genes) to >96% at 20× using the NimbleGen SeqCap EZ V3 kit. Thus, WES or the WES-phenotype slice, in which analysis is restricted to genes or candidate genes known to cause hearing loss may be indicated in a patient with a negative hearing loss NGS panel or has additional extra-sensory features.

10 Are We Ready for WGS? Considerations from a Clinical Laboratory Viewpoint

Compared to NGS gene panels and WES, some of the advantages of WGS include increased time efficiency per sample as there is no target enrichment step, increased detection of structural variation such as copy number variants (CNVs), and the identification of deep intronic variants. By bypassing the enrichment step in WGS, this allows simpler sample processing, a reduction in false positive calls and better coverage of the genome. The identification of structural variation such as copy number variants and the mapping of chromosome rearrangement breakpoints are now possible by integrating advanced computational algorithms in the analysis pipelines.

In WGS, the average coverage for each sequenced base is lower as compared to WES and NGS panels. For example, average coverage for WGS, WES and NGS panels are approximately 30×, 100×, and over 1000×, respectively. The differences in coverage can translate into different confidence levels for each variant call [8]. Incorporating WGS into clinical practice has legitimate challenges including, but not limited to, the cost and quantity of data storage, data interpretation, and computational power for processing large amounts of data. Currently, storage of the data is expensive and one reasonable option may be to store the patient's DNA and rerun the sample if re-analysis is requested. Stored data may be less of a concern than clinical interpretation, as there is no "gold standard" for whole genome interpretation. In addition to advancements in sequencing technology, there must be headway in centralized variant databases to aid in variant interpretation to increase efficiency and reduce the time and cost for manual curation. Presently, the National Center for Biotechnology Information (NCBI) has made great strides towards a centralized database for variant interpretation. This database, ClinVar, allows clinical laboratories to submit the clinical significance and their interpretation of variants that they have identified in their laboratory. As this database continues to collect an increasing number of variants daily, this will be an excellent resource for variant curation. Moreover, another hurdle for clinical interpretations WGS variants is that human curation will still be necessary despite stringent bioinformatic filtering. The application of WGS needs to be demonstrated in the gene discovery and clinical diagnosis of hearing loss.

Dewey et al. highlights on many of hurdles WGS must overcome before being routinely utilized in clinical practice [11]. In this study, 12 adult whole genomes were sequenced via two different platforms, Illumina and Complete Genomics, and genome coverage and sequencing platform concordance were measured along with how long a person spent curating candidate disease variants and the cost of clinical interpretation. Minimum threshold coverage (≥10 reads) was not obtained for a median of 10–19% (Illumina range 5–34%; Complete Genomics range 18–21%) of genes associated with Mendelian inheritance. However, genotype concordance between platforms for single nucleotide variants (SNVs) was high with approximately 99% overall similarity. For small insertion/deletion variants that were associated with disease there was less consistency between calls with a range of 10–75% concordance. The phenotype software program used identified 90–127 rare SNVs, insertion/deletions and large structural variants based on each participant's phenotype. Manual curation was still required. The estimated cost of WGS plus variant curation, computing infrastructure and data storage was calculated to be $14,815 [11].

In order to reach $1000 per genome, it is evident that some advancements need to be made in sequencing platform and variant curation. One such advancement is Illumina's HiSeq X Ten sequencer that is capable of sequencing 16 full human genomes at 30× coverage per three-day run and generating 1.8 TB of data. The machine's cost is currently estimated to be 10 million dollars. This cost is most likely too expensive for the majority of clinical laboratories currently offering NGS clinical services to invest in. If this machine becomes the gold standard for NGS, will clinical laboratories now outsource their sequencing? If so, how will the laboratory receive 1.8 TB of data per run and how will this data be stored? These are obvious questions in which we will find out the answers in the very near future.

11 Summary

Hearing loss is one of the most common birth defect and sensorineural disorder in humans and NGS platforms provide a great tool that can lower cost by increasing capacity without compromising diagnostic standards. It becomes possible to screen a large number of known deafness genes, by the implementation of NGS technologies, at a price that would only allow a few genes to be analyzed by Sanger sequencing. This inclusion of more genes to panels has also made it possible to detect novel mutations in genes that were rarely tested, thus, expanding the list of mutations associated hearing loss. Technically, comparisons of NGS platforms developed for hearing loss illustrated that several of them are suitable for clinical implementation. It is up to individual laboratories to perform thorough validations of such platforms to assess performance parameters, and other diagnostic features including speed, cost efficiency and scalability (samples per batch), to see if they meet the acceptable QC thresholds. Application of NGS in clinical laboratories have and will greatly improve the diagnostic rate of hearing loss, and improve earlier intervention in patients with hearing loss. With continued improvements in sequencing platforms, computational algorithms, and variant interpretation, WES is offered as a complimentary test to NGS panels and each meets unique clinical scenario needs. Subsequently, earlier implementation of educational services and medical treatment will positively change the life quality of the patients.

References

1. Atik, T., Bademci, G., Diaz-Horta, O., et al.: Whole-exome sequencing and its impact in hereditary hearing loss. Genet. Res. **97**, e4 (2015). doi:10.1017/S001667231500004X
2. Azaiez, H., Booth, K.T., Bu, F., et al.: TBC1D24 mutation causes autosomal-dominant nonsyndromic hearing loss. Hum. Mutat. **35**, 819–823 (2014). doi:10.1002/humu.22557
3. Bamshad, M.J., Ng, S.B., Bigham, A.W., et al.: Exome sequencing as a tool for Mendelian disease gene discovery. Nat. Rev. Genet. **12**, 745–755 (2011). doi:10.1038/nrg3031
4. Behlouli, A., Bonnet, C., Abdi, S., et al.: EPS8, encoding an actin-binding protein of cochlear hair cell stereocilia, is a new causal gene for autosomal recessive profound deafness. Orphanet J. Rare Dis. **9**, 55 (2014). doi:10.1186/1750-1172-9-55
5. Bonnet, C., Grati, M., Marlin, S., et al.: Complete exon sequencing of all known Usher syndrome genes greatly improves molecular diagnosis. Orphanet J. Rare Dis. **6**, 21 (2011). doi:10.1186/1750-1172-6-21
6. Brownstein, Z., Avraham, K.B.: Deafness genes in Israel: implications for diagnostics in the clinic. Pediatr. Res. **66**, 128–134 (2009). doi:10.1203/PDR.0b013e3181aabd7f
7. Brownstein, Z., Friedman, L.M., Shahin, H., et al.: Targeted genomic capture and massively parallel sequencing to identify genes for hereditary hearing loss in Middle Eastern families. Genome Biol. **12**, R89 (2011). doi:10.1186/gb-2011-12-9-r89
8. Clark, M.J., Chen, R., Lam, H.Y.K., et al.: Performance comparison of exome DNA sequencing technologies. Nat. Biotechnol. **29**, 908–914 (2011). doi:10.1038/nbt.1975
9. Cohen, M., Phillips 3rd, J.A.: Genetic approach to evaluation of hearing loss. Otolaryngol. Clin. N. Am. **45**, 25–39 (2012). doi:10.1016/j.otc.2011.08.015
10. De Keulenaer, S., Hellemans, J., Lefever, S., et al.: Molecular diagnostics for congenital hearing loss including 15 deafness genes using a next generation sequencing platform. BMC Med. Genet. **5**, 17 (2012). doi:10.1186/1755-8794-5-17
11. Dewey, F.E., Grove, M.E., Pan, C., et al.: Clinical interpretation and implications of whole-genome sequencing. JAMA. **311**, 1035–1045 (2014). doi:10.1001/jama.2014.1717
12. Diaz-Horta, O., Duman, D., Foster, J., et al.: Whole-exome sequencing efficiently detects rare mutations in autosomal recessive nonsyndromic hearing loss. PLoS One. **7**, e50628 (2012). doi:10.1371/journal.pone.0050628
13. Gao, X., Dai, P.: Impact of next-generation sequencing on molecular diagnosis of inherited non-syndromic hearing loss. J. Otolaryngol. **9**, 122–125 (2014). doi:10.1016/j.joto.2014.11.003
14. Hilgert, N., Smith, R.J.H., Van Camp, G.: Forty-six genes causing nonsyndromic hearing impairment: which ones should be analyzed in DNA diagnostics? Mutat. Res. **681**, 189–196 (2009). doi:10.1016/j.mrrev.2008.08.002
15. Kochhar, A., Hildebrand, M.S., Smith, R.J.H.: Clinical aspects of hereditary hearing loss. Genet. Med. **9**, 393–408 (2007). doi:10.1097/GIM.0b013e3180980bd0
16. Licastro, D., Mutarelli, M., Peluso, I., et al.: Molecular diagnosis of Usher syndrome: application of two different next generation sequencing-based procedures. PLoS One. **7**, e43799 (2012). doi:10.1371/journal.pone.0043799
17. Mamanova, L., Coffey, A.J., Scott, C.E., et al.: Target-enrichment strategies for next-generation sequencing. Nat. Methods. **7**, 111–118 (2010). doi:10.1038/nmeth.1419

18. Morton, C.C., Nance, W.E.: Newborn hearing screening--a silent revolution. N. Engl. J. Med. **354**, 2151–2164 (2006). doi:10.1056/NEJMra050700
19. Raviv, D., Dror, A.A., Avraham, K.B.: Hearing loss: a common disorder caused by many rare alleles. Ann. N. Y. Acad. Sci. **1214**, 168–179 (2010). doi:10.1111/j.1749-6632.2010.05868.x
20. Rehman, A.U., Morell, R.J., Belyantseva, I.A., et al.: Targeted capture and next-generation sequencing identifies C9orf75, encoding taperin, as the mutated gene in nonsyndromic deafness DFNB79. Am. J. Hum. Genet. **86**, 378–388 (2010). doi:10.1016/j.ajhg.2010.01.030
21. Rehm, H.L., Bale, S.J., Bayrak-Toydemir, P., et al.: Working Group of the American College of Medical Genetics and Genomics Laboratory Quality Assurance Commitee: ACMG clinical laboratory standards for next-generation sequencing. Genet. Med. **15**, 733–747 (2013). doi:10.1038/gim.2013.92
22. Schrauwen, I., Sommen, M., Corneveaux, J.J., et al.: A sensitive and specific diagnostic test for hearing loss using a microdroplet PCR-based approach and next generation sequencing. Am. J. Med. Genet. A. **161A**, 145–152 (2013). doi:10.1002/ajmg.a.35737
23. Shearer, A.E., DeLuca, A.P., Hildebrand, M.S., et al.: Comprehensive genetic testing for hereditary hearing loss using massively parallel sequencing. Proc. Natl. Acad. Sci. U. S. A. **107**, 21104–21109 (2010). doi:10.1073/pnas.1012989107
24. Sivakumaran, T.A., Husami, A., Kissell, D., et al.: Performance evaluation of the next-generation sequencing approach for molecular diagnosis of hereditary hearing loss. Otolaryngol Head Neck Surg. **148**(6), 1007–1016 (2013). doi:10.1177/0194599813482294
25. Smith, R.J., Shearer, A.E., Hildebrand, M.S., Van Camp, G.: Deafness and hereditary hearing loss overview. In: Pagon, R.A., Bird, T.D., Dolan, C.R., et al. (eds.) GeneReviews™. University of Washington, Seattle (1993)
26. Teer, J.K., Mullikin, J.C.: Exome sequencing: the sweet spot before whole genomes. Hum. Mol. Genet. **19**, R145–R151 (2010). doi:10.1093/hmg/ddq333
27. Tewhey, R., Warner, J.B., Nakano, M., et al.: Microdroplet-based PCR enrichment for large-scale targeted sequencing. Nat. Biotechnol. **27**, 1025–1031 (2009). doi:10.1038/nbt.1583
28. Van Camp, G., Willems, P.J., Smith, R.J.: Nonsyndromic hearing impairment: unparalleled heterogeneity. Am. J. Hum. Genet. **60**, 758–764 (1997)
29. Vandebona, H., Mitchell, P., Manwaring, N., et al.: Prevalence of Mitochondrial 1555A→G Mutation in Adults of European Descent. N. Engl. J. Med. **360**, 642–644 (2009). doi:10.1056/NEJMc0806397
30. Vona, B., Müller, T., Nanda, I., et al.: Targeted next-generation sequencing of deafness genes in hearing-impaired individuals uncovers informative mutations. Genitourin. Med. **16**, 945–953 (2014). doi:10.1038/gim.2014.65
31. Xing, G., Yao, J., Wu, B., et al.: Identification of OSBPL2 as a novel candidate gene for progressive nonsyndromic hearing loss by whole-exome sequencing. Genitourin. Med. **17**, 210–218 (2015). doi:10.1038/gim.2014.90
32. Yan, D., Zhu, Y., Walsh, T., et al.: Mutation of the ATP-gated P2X(2) receptor leads to progressive hearing loss and increased susceptibility to noise. Proc. Natl. Acad. Sci. U. S. A. **110**, 2228–2233 (2013). doi:10.1073/pnas.1222285110
33. Zbar, R.I., Ramesh, A., Srisailapathy, C.R., et al.: Passage to India: the search for genes causing autosomal recessive nonsyndromic hearing loss. Otolaryngol. Head Neck Surg. **118**, 333–337 (1998)
34. Zhao, Y., Zhao, F., Zong, L., et al.: Exome sequencing and linkage analysis identified tenascin-C (TNC) as a novel causative gene in nonsyndromic hearing loss. PLoS One. **8**, e69549 (2013). doi:10.1371/journal.pone.0069549
35. Zheng, J., Miller, K.K., Yang, T., et al.: Carcinoembryonic antigen-related cell adhesion molecule 16 interacts with alpha-tectorin and is mutated in autosomal dominant hearing loss (DFNA4). Proc. Natl. Acad. Sci. U. S. A. **108**, 4218–4223 (2011). doi:10.1073/pnas.1005842108

Next-Generation Sequencing Based Clinical Molecular Diagnosis of Primary Immunodeficiency Diseases

Lijun Wang, C. Alexander Valencia, Rebecca A. Marsh, and Kejian Zhang

Abstract Primary immunodeficiency diseases (PIDs) are a group of disorders caused by defects in one or multiple components of the immune system. PID patients usually present with recurrent or severe infections and can be difficult to manage with conventional treatments. The types of infections in patients with PIDs are related to which arm of the immune system is affected, and often provide the first clues to the nature of the immunologic defect. Without appropriate therapy, many patients die in infancy or early childhood. Because of the clinical heterogeneity and broadly overlapping phenotypes among the PIDs, it is often challenging to reach a definitive clinical diagnosis. Compelling evidence has demonstrated that most PIDs are genetic disorders, and there are more than 240 genes that have been identified in association with different PIDs. Accurate gene sequencing in PIDs not only can bring a definite molecular diagnosis at an early stage, but can also improve the clinical prognosis of patients by facilitating initiation of appropriate therapies based on the underlying diagnosis. However, Sanger-based single gene sequencing is time-consuming and costly if multiple genes need to be analyzed sequentially due to genetic and phenotypic heterogeneity. Thus, it is not practical for the prompt definitive diagnosis of PIDs. Next-generation sequencing (NGS) is a recently developed, massively parallel sequencing technology, which can sequence all targeted

L. Wang
Department of Human Genetics, Emory University School of Medicine,
EGL Genetics/Eurofins Clinical Diagnostics, 2460 Mountain Industrial Boulevard,
Tucker, GA 30084, USA

C. Alexander Valencia • K. Zhang, MD, MBA (✉)
Division of Human Genetics, Cincinnati Children's Hospital Medical Center,
Cincinnati, OH, USA

Department of Pediatrics, University of Cincinnati, College of Medicine, 3333,
Burnet Ave MLC 7016, Cincinnati, OH 45229, USA
e-mail: kejian.zhang@cchmc.org

R.A. Marsh
Cancer and Blood Disease Institute, Cincinnati Children's Hospital Medical Center,
Cincinnati, OH, USA

Department of Pediatrics, University of Cincinnati, College of Medicine,
3333 Burnet Ave MLC 7016, Cincinnati, OH 45229, USA

L.-J.C. Wong (ed.), *Next Generation Sequencing Based Clinical Molecular Diagnosis of Human Genetic Disorders*, DOI 10.1007/978-3-319-56418-0_6

regions (multiple genes, whole exome, or whole genome) of the human genome simultaneously. In fact, the NGS technology has made it possible to sequence all known disease causing genes in one experiment. As a result, NGS has become a primary approach for both clinical molecular diagnosis and discovery of novel genes in Mendelian human disorders (Gilissen et al. Genome Biol 12:228, 2011; Shendure and Ji Nat Biotechnol 26:1135–1145, 2008). In this chapter, we describe the most recent applications of NGS technology to PIDs with a focus on clinical molecular diagnosis.

Keywords Clinical genetic diagnosis • Primary immunodeficiency diseases • Next-generation sequencing • Whole exome sequencing • Whole genome sequencing

Abbreviations

ACMG	American College of Medical Genetics
ALPS	Autoimmune lymphoproliferative syndrome
CID	Combined immunodeficiency
CTP	Cytidine triphosphate
FHL	Familial hemophagocytic lymphohistiocytosis
HGMD	Human Gene Mutation Database
HLH	Hemophagocytic lymphohistiocytosis
IUIS	International Union of Immunological Societies
NGS	Next-generation sequencing
PIDs	Primary immunodeficiency diseases
SCID	Severe combined immunodeficiency
SNP	Single-nucleotide polymorphism
WES	Whole exome sequencing
WGS	Whole genome sequencing
XLP	X-linked lymphoproliferative disease

1 Introduction

Inherited primary immunodeficiency diseases (PIDs) are a group of disorders which are caused by defects of the immune system. PIDs usually present with some common clinical manifestations such as recurrent or severe infections, including viral, bacterial, fungal and protozoal infections that are difficult to manage with conventional treatments. Patients may also suffer a variety of autoimmune or autoinflammatory complications. Compelling evidences have demonstrated that most of PIDs are caused by genetic defects, and therefore many patients develop severe diseases during the first years of life [1]. Although the incidence of PIDs varies in different countries with a range of 1 in 700 to 1 in 19,000, more and more evidence has suggested that PIDs are not rare disorders, and are more common than generally thought [2–4]. In the United States, approximately 1 in 1200 individuals are diagnosed with PIDs [4]. According to the classification of the International Union of Immunological Societies (IUIS) Expert Committee for PIDs, PIDs can be divided into nine different groups, including: (1) combined immunodeficiencies without syndromic features; (2) combined immunodeficiencies with syndromic features; (3) predominantly antibody deficiencies; (4) diseases of immune dysregulation; (5) congenital defects of phagocyte number, function, or both; (6) defects in innate immunity; (7) autoinflammatory disorder; (8) complement deficiencies, and (9) phenocopies of PID [5]. Because PIDs are a significant cause of premature death in children, early diagnosis and appropriate management are vital to save patients and to reduce any devastating permanent damage. Although the typical clinical features and basic laboratory evaluation for immunodeficiency are valuable, identification of specific gene mutations is considered as the most reliable method for establishing a definitive diagnosis. Up to date, approximately, 320 genes that are associated with PIDs have been reported in literatures, and 249 of them were recognized and classified by IUIS in 2014. (Table 1) [5]. Sanger-based single gene sequencing is time-consuming and expensive; thus, physicians often face a big challenge in terms of choosing a reasonable number of the most likely candidate genes from more than 240 PID-associated genes for sequencing. Given the fact that many of these disorders are clinically indistinguishable from each other; targeted functional studies are usually not clinically available. Sequencing all of the disease-related genes would be ideal for the molecular diagnosis of PIDs. In addition to clinical heterogeneity, there is also high degree of genetic heterogeneity that can cause Sanger sequencing of a manageable number of known targeted genes insufficient and inefficient in identifying novel mutations. Over the past 5 years, the clinical application of NGS technology has developed to address these limitations [6]. NGS is a massively parallel sequencing technology that can sequence all targeted regions (multiple genes, whole exome or whole genome) of the human genome in one set-up. Currently, there are three common NGS-based approaches: targeted NGS panels; whole exome sequencing (WES); and whole genome sequencing (WGS). The development of NGS technology has made it possible to sequence all known disease-causative genes simultaneously in clinical laboratories today. Furthermore, NGS has become a successful technology for the discovery of novel genes for Mendelian disorders. Indeed, the NGS-based target gene panels and WES have been rapidly adopted by clinical laboratories. WES has not only resulted in a tremendous progress in disease diagnosis but also has led to discoveries of many novel disease genes [7, 8]. In comparison to the targeted NGS panels, WES and WGS are more comprehensive, but much more expensive and time-consuming. Although the rapid development of NGS technology can ultimately overcome shortcomings of WES and WGS and make them cheaper and more accessible, interpretations of vast majority of gene variants in genes or regions of unknown clinical significance, as well as incidental findings remain challenging. As a result, gene discovery remains primarily for research purposes [9–11]. In this chapter, we describe the most recent applications of NGS in PIDs with a focus on clinical molecular diagnosis.

Table 1 249 PID associated genes were classified into nine categories by the International Union of Immunological Societies (IUIS) Expert Committee for Primary Immunodeficiency (April 2014)

Category	Number of genes	Genes
Combined immunodeficiency	48	*ADA, AK2, ARTEMIS,* ***CARD11, CD27****, CD3D, CD3E, CD3G, CD3Z,* ***CD40, CD40LG, CD8A,*** *CIITA, CORO1A, DNA ligase IV,* ***DOCK8,*** *IKBKB, IL-21R, IL-2RG, IL7RA,* ITK, *JAK3, LCK,* ***LRBA****, MAGT1, MALT1, OX40, PI3K-δ,* ***PIK3CD****, PNP,* ***PRKDC****, PTPRC, RAG1, RAG2, RFX5, RFXANK, RFXAP, RHOH,* ***RMRP, SH2D1A****, STK4, TAP1, TAP2, TAPBP, TRAC, UNC119, ZAP70*
Combined immunodeficiencies with associated or syndromic features	38	*ATM, BLM, CHD7, DKC1, DNMT3B (ICF1),* ***DOCK8,*** *FOXN1, IKAROS,* ***MCM4,*** *MRE11, MTHFD1, NBS1, NOLA2 (NHP2), NOLA3 (NOP10 PCFT), ORAI1, PMS2, POLE1,* ***RMRP****, RNF168, RTEL1, SEMA3E, SLC46A1, SMARCAL1, SP110, SPINK5, STAT3,* ***STAT5B,*** *STIM1, TBX1, TCN2, TERC, TERT, TINF2, TTC7A, TYK2,* ***WAS****, WIPF1, ZBTB24 (ICF2)*
Predominantly antibody deficiencies	26	*AICDA, BLNK, BTK, CD19, CD20,* ***CD21, CD40, CD40LG****, CD79a, CD79β, CD81,* ***CXCR4****, ICOS, IGHM, IGKC, IGLL1,* ***LRBA****, NFKB2,****PIK3CD****, PIK3R1,* ***PRKCD****, TCF3, TNFRSF13B (TACI), TNFRSF13C (BAFF-R), TWEAK, UNG*
Diseases of immune dysregulation	32	*ACP5, ADAR1, AIRE,* ***CARD11,*** *CASP10, CASP8,* ***CD27****, FADD, FOXP3,* ***IL-10, IL-10RA, IL-10RB****, IL-2RA, ITCH,* ***ITK****, LYST, PRF1,* ***PRKCD****, RAB27A, RNASEH2A, RNASEH2B, RNASEH2C, SAMHD1,* ***SH2D1A****,* ***STAT5B****, STX11, STXBP2,* ***TNFRSF6,*** *TNFSF6, TREX1, UNC13D, XIAP(BIRC4)*
Congenital defects of phagocyte number, function, or both	34	*ACTB, C16ORF57, CEBPE, COH1, CSF2RA, CTSC, CYBA, CYBB, ELANE, FPR1, FUCT1, G6PC3, G6PT1, GATA2, GFI1, HAX1, IFNGR1, IFNGR2, IL-12B, IL-12RB1, IRF8, ISG15,* ***ITGB2****, KINDLIN3, LAMTOR2(ROBLD3),* ***NCF1****, NCF2, NCF4, RAC2, SBDS,* ***STAT1,*** *TAZ, VPS45,* ***WAS***
Defects in innate immunity	22	*ACT1, APOL-I, CARD9,* ***CXCR4****, EVER1, EVER2, HOIL1, IKBA, IL-17F, IL-17RA, IRAK-4,* ***MCM4,*** *MYD88, NEMO (IKBKG), RPSA,* ***STAT1****, STAT2, TBK1, TLR3, TRAF3, TRIF, UNC93B1*
Autoinflammatory disorders	18	*CARD14, CIAS1, HOIL1,* ***IL-10, IL-10RA, IL-10RB,*** *IL-1RN, IL-36RN, LPIN2, MEFV, MVK, NLRP12, NOD2, PLCG2, PSMB8, PSTPIP1, SH3BP2, TNFRSF1;*
Complement deficiencies	31	*C1QA, C1QB, C1QC, C1R, C1S, C2, C3, C4A, C4B, C5, C6, C7,* ***C8A,*** *C8B, C9,* ***CD21,*** *CD46, CD59, CFB, CFD, CFH, CFHR1–5, CFI, CFP, COLEC11, FCN3,* ***ITGB2****, MASP1, MASP2, SERPING1, THBD*
Phenocopies of PID	4	*AIRE, KRAS, NRAS,* ***TNFRSF6***

Of note: genes that are associated with more than one category of PIDs are in bold

2 Next-Generation Sequencing (NGS) Approaches in Primary Immunodeficiency Diseases (PIDs)

NGS, the high throughput, massively parallel sequencing technology, allows sequencing multiple genes simultaneously. Therefore, NGS-based gene sequencing is particularly suitable for the molecular diagnosis of PIDs. Multiple NGS approaches have been established over the past few years in PIDs for both clinical diagnosis and research purpose. Three most common approaches that have been applied to PIDs are targeted NGS gene panel, WES and WGS [12–14]. The targeted NGS panel is designed to detect genes known to be associated with a particular clinical disease phenotype simultaneously, and enables clinicians to focus on a specific group of genes of interest. Thus, the targeted NGS panel allows deeper sequencing of genes relevant to diseases [15, 16]. WES is designed for diseases with non-specific clinical features and/or diagnosis to sequence the complete coding and flanking noncoding regions of human exomes, where approximately 85% of disease-causing mutations are located [17]. WES is becoming practical for clinically hard to diagnose Mendelian disorders due to reduced cost [18]. Notably, WES has demonstrated enormous potential in the discovery of novel disease-causative genes [19]. The WGS is aimed to sequence the complete DNA sequence of the whole genome, including the information in deep introns and other untranslated regions that are not covered by WES. In addition to the challenging interpretation of enormous amount of variants, there is still a distance for WGS to be time and cost effective. For these reasons, the WGS has not yet been applied to clinical use widely, although the recent progress of WGS application is promising [20, 21].

3 Broadly Targeted Next-Generation Sequencing (NGS) Approach for the Diagnosis of Primary Immunodeficiency Diseases (PIDs)

A targeted NGS panel analyzes only the genes known to be related to a particular disease phenotype, thus avoids analysis of unrelated or possibly related genes [22]. Since the NGS panel analyses focus on target genes of interest, it is possible to achieve deeper sequence coverage, higher sensitivity in mutation detection with higher accuracy [23, 24]. This approach has become the first-line testing in PIDs, and has been utilized successfully at identifying mutations in the known-disease genes in PIDs. At least three different target enrichment methods have been adopted by clinical laboratories: RainDance emulsion PCR (RainDance Technologies, Lexington, MA, USA), Hybridization-based (SureSelect, Agilent Technologies Inc., Santa Clara, CA, USA; SeqCap EZ system, Roche NimbleGen; Nextera and TruSeq capture systems, Illumina) and Haloplex PCR target enrichment (Agilent Technologies Inc., Santa Clara, CA, US) captures. Nijman et al. [12] developed a targeted NGS panel to facilitate a genetic diagnosis in any of 170 known PID-related genes. The NGS panel was performed on an AB SOLiD 5500XL sequencer (Applied Biosystems, Bleiswijk, The Netherlands). Two different types of enrichment approaches were adopted, yielding a high coverage at 20× with 93.77% in Array-based capture (Agilent SurePrint G3 1 M Custom CGH Microarray, Agilent Technologies Inc., Santa Clara, CA, US) and 91.78% in SureSelect capture (Agilent SureSelectXT Target Enrichment System, Agilent Technologies Inc., Santa Clara, CA, US) respectively. Forty PID patients with known mutations were analyzed, and 1500 variants per person were detected after the primary analysis. To prioritize variants for pathogenic properties, this group developed an internal classification pipeline by using Cartagenia BENCHlab NGS module (Cartagenia, Leuven, Belgium). Briefly, variants were first checked using an internal database, and then the Human Gene Mutation Database (HGMD) database. Variants were considered benign if the minor allele frequency was greater than 5% in the following databases: dbSNP, Exome Variant Server, and 1000Genomes. Synonymous variants and variants located more than 20 bp into flanking intronic sequences were discarded. Nonsense, frameshift, and canonical splice site variants were considered pathogenic. In addition, the Alamut mutation interpretation software (Interactive Biosoftware, Rouen, France) was applied for interpretation and classification of the variants. This pipeline analysis resulted in approximately 15–25 variants per patient for further in-depth expert evaluation. This study indicated that both capture designs had a high sensitivity (>99.5%) and specificity (>99.9%) for the detection of point mutations, but only 85% of success rate for the detection of small deletion/insertion variants. To evaluate the efficiency of this NGS panel for the reclassification of PID patients, 26 patients were selected for this test who had failed to receive a genetic diagnosis previously. These patients were composed of three groups: combined immunodeficiency (CID, n = 20), autoimmune lymphoproliferative syndrome-like disease (ALPS, n = 4), and hemophagocytic lymphohistiocytosis-like disease (HLH, n = 2). In three of the patients with CID and in one patient with HLH, a reclassification genetic diagnosis was established; three of these four patients (3 of CID, 1 of HLH) presented with atypical phenotypes based on the disease diagnosis criterion [12]. This study has demonstrated that the targeted NGS approach (using either the glass array capture or solution–based capture method), is accurate and efficient for the detection of mutations in PIDs-related genes and that a targeted NGS-based panel can be used as a first-line genetic test for PID patients. Of note, in this panel 9 genes had inadequate sequence coverage, including: *CARD9*, *C4A*, *C4B*, *CFD*, *ELANE (ELA2)*, *FCGR1A (CD64)*, *FCGR2B (CD32)*, *IKBKG (NEMO)*, and *NCF1*. The low coverage is likely due to high CG content, high homology of pseudogenes, or both. Therefore, this panel is not sensitive and applicable when mutations are suspected in any of these 9 genes [12].

Stoddard et al. [25] developed their targeted NGS panel combined Haloplex custom target enrichment and the Ion Torrent PGM technologies, which allow the rapid screening of large panels of genes [26]. This panel included 173 genes that were known or highly suspected to be associated with particular PIDs [5]. For capture design, 2455 target regions including the coding exons plus 10 flanking bases of 173 genes were submitted for DNA capture probe design using the Agilent SureDesign web-based application software. The final probe design was expected to yield 42,909 amplicons covering 99.53% of the submitted target regions. A custom designed HaloPlex Target Enrichment kit (Agilent Technologies) was used for the capture of the target regions, which included the four following steps: (1) digestion of genomic DNA with restriction enzymes; (2) hybridization of fragments with the complementary probes; (3) capture of target DNA using streptavidin beads and ligation of circularized fragments; and (4) PCR amplification of captured target libraries. The library templates were clonally amplified using the Ion One Touch 2™, followed by an enrichment process for the recovered template-positive ion sphere particles. A standard Ion PGM 200 Sequencing V2 protocol using Ion 318 V2 chips (Life Technologies) was performed for the NGS. In total, they utilized 11 healthy controls, thirteen PID patients previously evaluated, and 120 patients with undiagnosed PIDs. This NGS panel revealed variants with 98.1% sensitivity and >99.9% specificity. Moreover, a molecular diagnosis was made for 18 of 120 patients (15%) who previously lacked a genetic diagnosis, including 9 patients who presented with atypical clinical manifestations and had previously undergone extensive genetic and functional testing. Interestingly, although the Haloplex kit was able to provide >90% coverage for most target regions; there were low coverage regions for a few genes, including *HLA-DRB5*, *TNFRSF13C*, *UNC93B1*, *CD79A* and *NCF4*. The poor coverage in some regions of these genes was probably due to high GC-content, repeat regions, and highly homologous sequences. Like all the other sequencing strategies, this NGS panel was not able to detect large deletions, insertions and chromosomal abnormalities. Additional techniques are required to evaluate copy number variations. The HaloPlex Target Enrichment System enables fast, simple and efficient analysis of genomic regions of interest in a large number of samples. By combining single-tube target amplification and removing the need for library preparation, the total sample preparation time and cost is reduced by eliminating the need for dedicated instrumentation or automation. However, the restriction enzyme digests can result in unexpected coverage gaps especially when fragments are longer than the read length. While there were technical limitations of the above, Stoddard et al. [25] has demonstrated that their targeted NGS panel was a cost-effective, first-line genetic test for PIDs. This targeted NGS panel would be more appropriate to be applied first for the PID patients, who present with atypical or widely variable/nonspecific clinical phenotypes.

Similarly, Moens et al. [27] developed another targeted NGS panel using a selector-based target enrichment (HaloPlex system, Halo Genomics). The selector assays were designed to cover all exons and UTRs +/− 25 bp of 179 genes [27] of all known disease-causing genes in PIDs. The NGS panel sequencing was performed on Illumina's Genome Analyzer IIx (Illumina, San Diego, CA, USA). In this study, 33 patients were examined, 18 of which had at least one known causal mutation

Table 2 Summary of targeted-NGS panels in PIDs

Platform	Enrichment	Coverage	Panel gene number	Diagnosis yield	Reference
AB SOLiD 5500XL sequencer	Array-based Hybridization (In solution) based	94% 92%	170	15% (4/26)	Nijman et al. [12]
Ion Torrent	HaloPlex	89%	173	15% (18/120)	Stoddard et al. [25]
Illumina's Genome Analyzer IIx	Selector -based	88%	179	40% (6/15)	Moens et al. [27]
Ion Torrent Proton	AmpliSeq	96%	162	25% (35/139)	Al-Mousa et al. [28]

prior to the experiment. This HaloPlex based enrichment followed by Illumina sequencing provided a minimal coverage of 20 reads in an average of 88% target regions. The average read depth in the targeted region was 1304 ± 662 reads per base. By sequencing 18 individuals with known mutations, this NGS approach detected 20 out of 24 mutations (83%) and solved the diagnosis for 78% (14 out of 18 individuals) of patients by one single assay. Of the 4 missed mutations, 3 (2 missense, 1 splice variant) of them had low read depth and 1 (small deletion) was not covered by design. There were other regions of no/low coverage in this study, which might be due to overall low read depth with ~21% of the target genes contained one or more exons with <20X average read depth across all samples. Interestingly, the *CFD* gene showed a low overall coverage both in this NGS panel and a different NGS panel described previously [12]. Despite the shortcomings of targeted NGS panel, the majority of PID cases could be resolved by using this sequencing approach.

More recently, Al-Mousa *et al.* developed an unbiased targeted NGS approach for PIDs by using the Ion Torrent Proton NGS sequencing platform [28]. This comprehensive NGS panel included 162 genes that were associated with PIDs. To evaluate the panel's clinical utility, sensitivity and specificity, total of 261 suspected PID patients were tested. Of the 261 patients, 122 had known disease mutations and were used to assess the specificity and sensitivity. The actual coverage for the targeted regions (encoding regions and 10-bp flanking regions of associated introns) was 96.5%, and only 9 of the 162 genes had a coverage less than 90%. The sensitivity for the single nucleotide variant has reached 96%, the missed mutations were due to low read depth. The overall specificity for this panel was 88.2%. Interestingly, this NGS panel detected unknown mutations and resulted in genetic diagnosis in 35 of 139 unsolved cases.

Although there is some difference among the coverages when different platforms, enrichment methods and gene numbers in the different NGS panels were used, they all demonstrated the efficacy of NGS panels for the diagnosis of PIDs (Table 2).

4 Specific Targeted-NGS Sub-panels in Primary Immunodeficiency Diseases

The targeted-gene panels are aimed to establish definitive diagnosis in subgroup of PID patients who have similar clinical and cellular manifestations (Table 1). Compared to the previous large all-inclusive NGS panels (~170 genes), these specific targeted-NGS gene sub-panels (10–40 genes) would generate a much lower number of gene variants. In addition to being less time-consuming, the specific targeted NGS method can result in a higher molecular diagnostic yield. One of the reasons is that we can provide 100% coverage for the genes of sub-panels, in which the limited low coverage regions of the genes can be rescued by Sanger sequencing. The advantage of using small panels is particularly significant for some PIDs with defined clinical phenotype and fitting particular diagnostic criteria, for example, familial hemophagocytic lymphohistiocytosis (FHL). FHL is a rare, primary immunodeficiency disease characterized by an uncontrolled hyperinflammatory response [29]. Five FHL subtypes (FHL1, FHL2, FHL3, FHL4, and FHL5) have been described. Four genes in which mutations are causative have been identified: *PRF1* (FHL2), *UNC13D* (FHL3), *STX11* (FHL4), and *STXBP2* (FHL5) (Table 1, under the category of diseases of immune dysregulation). These genetic abnormalities affect granule-dependent lymphocyte cytotoxicity by impairing trafficking, docking, priming for exocytosis, or membrane fusion of cytolytic granules. The function of this pathway may also be severely impaired by the loss of functional perforin, the key pore-forming protein for the delivery of proapoptotic granzymes. Diverse mutations in this pathway all give rise to similar clinical phenotypes (albeit of variable severity). Although FHL has an autosomal recessive inheritance pattern, our recent study also has indicated that a digenic mode of inheritance may also exist as a result of a synergistic function effect within genes involved in cytotoxic lymphocyte degranulation (Fig. 1) [30, 31]. Many FHL patients develop the disease within first few months or years of life and, occasionally, in utero, although later childhood or adult onset is more common than previously suspected [32, 33]. Without treatments, most FHL patients die within 2 months of disease onset due to the progression of hemophagocytic lymphohistiocytosis (HLH). Although a possible diagnosis of FHL can be made based on 8 clinical criteria (fever, splenomegaly, bicytopenia, hypertriglyceridemia and/or hypofibrinogenemia, hemophagocytosis, low/absent NK-cell-activity, hyperferritinemia, and high-soluble interleukin-2-receptor levels) [34], the definitive diagnosis of a genetic form of HLH (FHL) is often challenging because of the lack of specificity of those diagnostic criteria and their poor correlation with the different types of defects in particular gene(s). In addition, there are some overlapping clinical features between FHL and a few other inherited immune disorders associated with highly lethal hemophagocytic syndromes, including X-linked lymphoproliferative disease (SH2D1A and XIAP), Pudlak syndrome *(AP3B1* and *BLOC1S6)*, Chediak-Higashi syndrome(LYST), Griscelli syndrome type 2 (RAB27A), X-linked immunodeficiency with magnesium defect, Epstein-Barr virus infection, and neoplasia (MAGT1), CD27 deficiency and Interleukin-2-inducible T-cell Kinase (ITK)

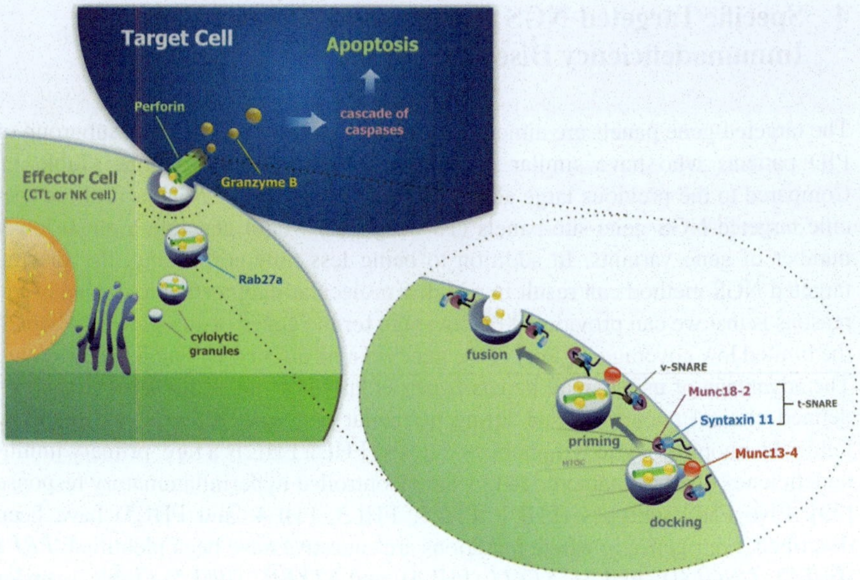

Fig. 1 The granule exocytosis pathway of cytotoxic lymphocytes in familial hemophagocytic lymphohistiocytosis (FHL) Perforin is the key delivery molecule for proapoptotic granzymes in the perforin-dependent cytotoxic lymphocytes and is associated with FHL2. Defects in other FHL-associated genes *(MUNC13-4, STX11, STXBP2 and RAB27A)* may also affect granule-dependent lymphocyte cytotoxicity by impairing trafficking, docking, priming for exocytosis, or membrane fusion of cytolytic granules. Synergistic effects of these different molecules in this cytotoxic pathway have also been observed

deficiency [35–45] . For this reason, a comprehensive genetic diagnostic panel is needed. The Molecular Genetics Laboratory at the Cincinnati Children's Hospital Medical Center (CCHMC) has developed a specific targeted NGS panel for FHL. This specifically targeted FHL panel has 14 genes; *AP3B1*, *BLOC1S6* (*PLDN*), *ITK*, *LYST*, *MAGT1*, *PRF1*, *RAB27A*, *SH2D1A*, *SLC7A7*, *STX11*, *STXBP2*, *TNFRSF7* (*CD27*), *UNC13D* and *XIAP* (*BIRC4*). All coding exons and 20 base nucleotides into the flanking intronic regions, as well as 5′ and 3′ untranslated regions (20 base nucleotides from 1st or last exon) were enriched using microdroplet PCR technology (RainDance Technologies Inc., USA) as previously published method [46], followed by NGS sequencing on the Illumina HiSeq 2500 instrument (Illumina Inc., USA). The resulting sequence reads were aligned against the reference DNA sequence followed by variant calls using NextGENe software (SoftGenetics, LLC, USA) [47]. PCR/Sanger sequencing was then used to fill the gaps of insufficiently covered regions and to confirm pathogenic and novel variants. The analytic sensitivity of this methodology is >99%. Although small deletions and insertions of <10 bases can be routinely detected in this panel, larger deletions or duplications would not be able to picked up by this technology. For this reason, we

Table 3 Summary of Data Quality Metrics of the FHL, ALPS and SCID targeted PID panels

Sample	No. reads	Read 1_ QF (%)	Read 2_QF (%)	Total	Aligned	% Target base >20x
CM43	7,477,516	6,425,427 (85.93)	6,224,399 (83.24)	12,649,826	11,509,198 (90.98)	98.55
CP1310	8,738,296	7,449,855 (85.26)	7,199,718 (82.39)	14,649,573	12,829,627 (87.58)	98.61
DA2343	12,288,088	10,422,619 (84.82)	10.103,394 (82.22)	20,526,013	18,113,818 (88.25)	98.66
DR2443	14,764,976	12,512,440 (84.74)	12,080,620 (81.82)	24,593,060	21,994,803 (89.43)	98.67
DV603	14,590,439	12,236,329 (83.87)	11,907,173 (81.61)	24,143,502	19,720,659 (81.68)	98.67
EL1979	10,406,290	8,809,832 (84.66)	8,527,994 (81.95)	17,337,826	15,794,657 (91.10)	98.61
FA1	15,911,339	13,130,769 (82.52)	12,821,601 (80.58)	25,952,370	18,043,233 (69.52)	98.69
HG533	9,905,099	8,386,478 (84.67)	8,141,958 (82.20)	16,528,436	14,455,987 (87.46)	98.60
JG2054	12,640,488	10,716,990 (84.78)	10,410,183 (82.36)	21,127,173	19,122,219 (90.51)	98.61
NT198	11,883,576	10,160,179 (85.50)	9,809,702 (82.55)	19,969,881	18,004,904 (90.16)	98.68
QB1596	7,788,648	6,572,817 (84.39)	6,351,195 (81.54)	12,924,012	11,884,876 (91.96)	98.51
TB1641	7,210,022	6,161,792 (85.46)	5,904,489 (81.89)	12,066,281	11,182,195 (92.67)	98.29

have developed targeted deletion and duplication analysis of each gene on this panel. The average read depth of the target regions for the different panels is more than 98% covered at >20X (Table 3). By reviewing the NGS results of panel test on the first 370 clinical cases from patients suspected of HLH,, single or bi-allelic HLH pathogenic variants were identified in 31 patients, 175 patients had variants of uncertain clinical significance, and 13 patients carried variants in more than two genes. Although the detection of exonic deletions and insertions using NGS data has been reported (Feng YM et al., GIM 2015 17:99 PMID 25032985, Wang J, et al. GIM PMID: 26,402,642), we have not validated our NGS for such purposes. Thus, we have developed targeted deletion and duplication analysis for each gene on this panel as a complementary testing. Patients with a normal NGS result or a single (heterozygous) mutation are usually recommended for the deletion and duplication test. Gross deletions and duplications have been identified in 5 patients in several FHL-related genes (unpublished data). Given the lower cost, faster turn-around time and higher yields of the detection of causative mutations than traditional gene-by-gene PCR/Sanger analyses, the HLH targeted NGS panel has been recommended as the first line test for patients presenting with FHL-like syndromes.

Similarly, another targeted subpanel for the diagnosis of autoimmune lymphoproliferative syndrome (ALPS) is also clinically available. ALPS is a disorder of T cell dysregulation caused by defective Fas-mediated apoptosis [48]. ALPS patients usually present with lymphadenopathy, hepatosplenomegaly, autoimmunity and increased rates of malignancy. The diagnosis of ALPS is based on a constellation of clinical findings, laboratory abnormalities, and identification of mutations in genes relevant for the tumor necrosis factor receptor superfamily member 6 (Fas) pathway of apoptosis [49]. However, it is always challenging to reach a definite diagnosis because of the ALPS heterogeneity, the disease variability and different expressivity. An ALPS NGS panel could serve as an important aid for the molecular diagnosis of ALPS. This targeted NGS ALPS panel includes the following 9 genes: *FAS*, *FASLG*, *CASP10*, *CASP8*, *FADD*, *KRAS*, *NRAS*, *MAGT1* and *ITK*. Among these genes, identified mutation(s) in *FAS*, *FASLG*, or *CASP10* genes can confirm the ALPS diagnosis [50]. Mutations in *FADD* are associated with the patients who have many of the biochemical markers of ALPS, but lack the characteristically clinical features of lymphadenopathy and splenomegaly [51]; biallelic mutations in *CASP8* result in a rare immunodeficiency characterized by lymphadenopathy and splenomegaly, marginal elevation of double negative T cells (DNTCs), defective FAS-mediated apoptosis, in addition to frequent bacterial and viral infections. Mutations in *NRAS* and *KRAS* may lead to an ALPS-like condition known as RAS-associated lymphoproliferative disease [50, 52]. Mutations in *ITK* and *MAGT1* are not associated with ALPS but are included in this panel as part of the differential diagnosis of lymphoproliferative disorders. The ALPS panel has shown a reasonable clinical sensitivity. Of the 80 patients tested recently (personal communication), 6 patients (7.5%) had pathogenic (4) or likely pathogenic variants (2), establishing a definite, molecular diagnosis. In addition, 5 patients (6.3%) had variants with unknown clinical significance (unpublished data).

Severe combined immunodeficiency (SCID) is a group of distinct congenital disorders that involves combined cellular and humoral immunodeficiency resulting from the lack of function or significant dysfunction of T lymphocytes and B lymphocytes. SCID is the most severe form of PIDs [53]. The patients with SCID usually develop disease between 3 to 6 months, typically present with recurrent or persistent infections (severe bacterial, viral or fungal infections) and failure to thrive [54, 55]. Although different forms of SCID are currently classified according to the presence or absence of T, B, and NK cells, the discovery of novel causative genes has added new complex clinical phenotypes [5, 56]. X-linked SCID (X-SCID) is the most common form of SCID affecting male infants. It is the result of defects in IL2RG gene, which encodes the common gamma chain, gamma c, of the leukocyte receptors for interleukin-2 and multiple other cytokines [57]. Puck et al. [57] has identified deleterious *IL2RG* mutations in 87 of 103 families (84.5%) with males affected with non-ADA–deficient SCID, suggesting a high frequency of *IL2RG* mutations in X-linked SCID. The remaining SCID disorders are caused by autosomal recessive mutations. The estimate prevalence is 1 in 50,000 births with a higher prevalence in males [56]. SCID is considered a pediatric emergency, and is often fatal by 6–12 months of age without treatments. For this reason, at least 34 states have already implemented or agreed to move forward with newborn screening for SCID. The screening is performed by assaying for T-cell receptor excision circles (TRECs). This test has led to an early identification of SCID patients, and made it possible for providing appropriate managements prior to serious damages in patients. Although the TRECs screening has not been adopted nationally, the outcome of the SCID screening has been very encouraging [58]. However, a follow-up of sequencing the SCID-related genes is required in order to establish a definite SCID diagnosis, which has been regarded as a gold standard. Currently, there is an available SCID NGS panel including 20 genes (*ADA; CD3D; CD3E; CD45(PTPRC); DCLRE1C; FOXN1; IL2RG; IL7R; JAK3; LIG4; NHEJ1; ORAI1; PNP; RAG1; RAG2; RMRP; STAT5B; STIM1; TBX1; ZAP70*). These genes are associated with either SCID and or SCID-type symptoms, such as Omenn syndrome, Cartilage-Hair hypoplasia, and Velocardiofacial syndrome [59–65]. Omenn syndrome is characterized by an absence of circulating B cells and an infiltration of the skin and the intestine by activated oligoclonal T lymphocytes. Along with immunodeficiency, Omenn syndrome presents with severe erythroderma, desquamation, alopecia, lymphadenopathy, eosinophilia and elevated IgE. Cartilage-Hair hypoplasia can be caused by mutations in *RMRP* gene, which is characterized by metaphyseal chondrodysplasia presenting with short stature and short limbs. In addition, many patients presented with SCID-type immunodeficiency [66, 67]. Because immunodeficiency secondary to thymic hypoplasia is common in Velocardiofacial syndrome and TBX1 is the most important gene for this syndrome [68], this gene has been included in the SICD panel too. This SCID panel revealed 60–90% of the reported mutations, and the sensitivity of DNA sequencing is over 99% for the detection of nucleotide base changes, small deletions and insertions in the genes of interest. In 50 patients performed on SCID panel recently at CCHMC, we identified 9 pathogenic variants, 4 likely pathogenic variants and 26 variants of unknown clinical significance. All the above variants are confirmed by Sanger sequencing. Overall, 10% of patients (5/50) reached a definite molecular diagnosis for SCID by either carrying two pathogenic variants or one X-linked pathogenic variant in males (unpublished data). Due to large exonic deletions have been reported in *ADA, DCLRE1C, IL2RG, JAK3, NHEJ1, PTPRC, RAG1, RAG2, RMRP, STAT5B* and *TBX1*, deletion/duplication testing should be indicated as follow-up test in patients with a single mutation in any of the above genes. To address this issue, Yu et al. have developed a target gene capture/NGS assay with deep coverage which facilitates simultaneous detection of single nucleotide variants and exonic copy number variants in one comprehensive assessment [69].

As the pathophysiology of the PIDs is better characterized and new genes are emerging with unprecedented speed [5, 70, 71], the targeted panels are required to be expanded and updated on a frequent basis in order to meet the diagnostic needs and improve clinical sensitivity. In fact, more than 30 new gene defects have been added by IUIS to the updated version regarding the classification of PIDs in 2014 since the previous classification in 2011 [5]. It is likely that novel PID-related genes will continue to be identified in the future with the rapid advances in NGS technology as well as the widespread use of WES and WGS.

5 Whole Exome Sequencing (WES) in Primary Immunodeficiency Diseases (PIDs)

The current gene panels can only offer rapid genetic diagnosis for PIDs caused by mutations in known genes. However, when facing a new phenotype, atypical phenotypes or the phenotypes that are difficult to be classified into any categories of PID, WES would be the most useful tool currently for the clinical molecular diagnosis as well as the discovery of novel genes associated with diseases. Using WES, Dickinson et al. [13] examined 4 unrelated patients with an immunodeficiency syndrome that involved loss of dendritic cells, monocytes, and B and natural killer cells (DCML deficiency) [72]. They identified novel disease-causing mutations in *GATA2* gene in all 4 patients. GATA2 is a transcription factor, which is composed of 2 highly conserved zinc finger domains that mediate protein-DNA and protein-protein interactions. GATA2 is required for stem cell homeostasis [73]. Furthermore, the functional studies indicated that haploinsufficiency and dominant-negative loss of GATA-2 function were potential mechanisms of pathogenesis in the DCML deficiency. This study again proved WES as a powerful tool for identifying disease-causing mutations in a small number of unrelated and sporadic cases of PIDs [13]. Hermansky-Pudlak syndrome (HPS) is a rare autosomal recessive disorder characterized by platelet dysfunction, oculocutaneous albinism, and life-threatening pulmonary fibrosis. By WES, Badolato et al. [35] identified a homozygous nonsense mutation (c.232C > T (p.Q78X)) in *PLDN* in a female with HPS-like primary immunodeficiency syndrome. In vitro, this *PLDN* mutation caused defective NK-cell degranulation and cytolysis, suggesting that the c.232C > T (p.Q78X) change in *PLDN* is pathogenic.

WES is particularly useful for patients who have no identifiable mutations after available NGS panels are exhausted. Patel et al. [74] reported an infant with low TRECs and non-SCID T lymphopenia. An early diagnosis of SCID with appropriate treatment and management, including the avoidance of exposure to viral infections or live virus vaccines, offering immune system restoring treatments or early hematopoietic stem cell transplantation (HSCT), would change the patient's prognosis significantly. A targeted NGS sequencing of SCID associated genes (*ADA*, *AK2*, *CD3D*, *CD3zeta*, *DCLRE1C*, *ILRG*, *IL7R*, *JAK3*, *LIG4*, *NHEJ1*, *PNP*, *PTPRC*, *RAC2*, *RAG1*, *RAG2*, *RMRP*, and *ZAP70*; GeneDx, Gaithersburg, MD) did not reveal any pathogenic mutations. By contrast, WES analysis identified two nonsense mutations; c. 842 T > G (p. L281X) and c.1030C > T (p. Q344X) in a compound heterozygous state in the *NBN* gene. These mutations were predicted to result in a loss of function, and were consistent with the absence of protein by immunoblotting and radiosensitivity testing on the patient lymphocytes. *NBN* encodes nibrin, which is a component of a molecular complex involved in the early recognition and subsequent repair of DNA damage [75]. Thus, the WES test led to a definite diagnosis of Nijmegen breakage syndrome (NBS). The clinical phenotype of NBS is variable, although most patients with NBS present with immunodeficiency [76]. The appropriate application of WES resulted in the definite diagnosis of NBS, avoiding a complicated differential diagnosis with a large group of immunodeficiency diseases. Despite that many new PID associated genes have been identified, more novel disease-causative genes are expected to be discovered in future by NGS technology [5].

One limitation of the targeted NGS panels is that the gene list has to be updated frequently, followed by clinical validation which is laborious and expensive. WES overcomes this limitation. For example, although over 14 different SCID genes have reported, no specific gene defects have yet detected in many patients with hereditary abnormalities [77]. In a nonconsanguineous patient with early onset profound combined immunodeficiency and immune dysregulation, Punwani et al. [78] did not find any mutations by a comprehensive NGS panel of known SCID genes. While WES analysis revealed compound heterozygous mutations; c.1019-2A > G and c.1060delC (p.Y353fs*18), in a new gene, *MALT1*. Functional studies indicated that both T cells and B cells were damaged and NF-κB signaling pathway was impaired. This study suggested that the immunodeficiency in this patient was due to MALT1 defect. Based on these molecular findings, the patient was effectively treated by HSCT. This example highlighted the importance of a definite diagnosis that not only established a definite diagnosis possible but also brought a successful outcome with appropriate intervention.

Due to its cost-effectiveness, WES has been broadly used for discovering new genetic etiologies of immunodeficiency. Zhang et al. [79] reported a new syndrome of severe atopy, recurrent infections, autoimmunity, vasculitis, renal failure, and lymphoma, associated with motor and neurocognitive impairments. Using WES combined with Sanger sequencing, they identified two mutations; c.1585G > C (p. E529Q) and c.1438_1442del (p. L480Sfs*10) in *PGM3* in a compound heterozygous state from one family, and a homozygous mutation, c.975 T > G (p.D325E) in another family. The further functional studies indicated reduced enzymatic activity and abnormal glycosylation which was resulted from the mutations. The mutations were segregated with the disease. Interestingly, all these patients showed hyposialylation of O-linked serum glycans, consistent with impaired PGM3 function. *PGM3* gene encodes phosphoglucomutase 3 (PGM3), which is a member of the hexose phosphate mutase family and catalyzes the reversible conversion of GlcNAc-6-phosphate (GlcNAc-6-P) to GlcNAc-1-P, required for protein glycosylation [80]. For the first time, these results defined a new *PGM3* –mediated disorder characterized by severe atopy, immune deficiency, autoimmunity, intellectual disability and hypomyelination.

Willmann et al. [81] studied a large consanguineous pedigree with two patients presenting with combined immunodeficiency including recurrent, severe bacterial and viral infections and *Cryptosporidium* infection. Combined WES with single-nucleotide polymorphism (SNP) array-based homozygosity mapping, they identified a single homozygous variant, c. 1694C > G (p. P565R) in *MAP3K14* on chromosome 17q21. This gene encodes NF-κB-inducing kinase (NIK), which is a serine/threonine protein-kinase. NIK binds to TRAF2 and stimulates NF-kappaB activity. Interestingly, the patients with mutated *NIK* exhibit B-cell lymphopenia, have decreased frequencies of class-switched memory B cells and hypogammaglobulinemia due to impaired B-cell survival, and impaired ICOSL expression.

In this study, the unexpectedly broad range of phenotypic aberrations (affecting B-, T- and NK-lineages) highlighted essential roles for NIK and adequate control of non-canonical NF-κB signaling for the generation and maintenance of the human immune system, thus, demonstrating the functional NIK deficiency as a novel, pervasive combined primary immunodeficiency syndrome. By WES, Martin et al. [82] identified a homozygous mutation (c.1692-1G > C) in *CTPS1* in 8 patients from 5 unrelated families with a novel and life-threatening immunodeficiency. All patients presented with early onset of severe infections mostly caused by herpes viruses, including EBV and varicella zooster virus (VZV) and also suffered from recurrent encapsulated bacterial infections, a spectrum of infections of a typical combined deficiency of adaptive immunity. *CTPS1* encodes CTP synthase 1, which is responsible for the catalytic conversion of uridine triphosphate to cytidine triphosphate (CTP). CTP is a building block required for the biosynthesis of DNA, RNA and phospholipids [83]. The CTP synthase activity may play an important role for DNA synthesis in lymphocytes [84]. This *CTPS1* mutation is predicted to affect a splice donor site at the junction of intron 17–18 and exon 18, leading to the expression of an abnormal transcript lacking exon 18. Functional studies demonstrated that CTPS1 deficiency led to an impaired capacity of activated T and B cells to proliferate in response to antigen receptor-mediated activation. As a result of WES test, a new type of PIDs was confirmed.

Despite the non-specific and overlapping clinical and laboratory features of PIDs, new gene defects are continuing to be identified by WES, leading to a rapid classification of new types of PIDs. One good example is the identification of the dedicator of cytokinesis 2 gene (*DOCK2*) [85]. *DOCK2* gene encodes a hematopoietic cell-specific, *Caenorhabditis elegans* Ced-5, mammalian DOCK180 and *Drosophila melanogaster* myoblast city (CDM) family protein that is indispensable for lymphocyte chemotaxis. DOCK2 is specifically expressed in hematopoietic cells, predominantly in the peripheral blood leukocytes, and may be involved in remodeling of the actin cytoskeleton required for lymphocyte migration, through the activation of RAC [86]. Dobbs et al. [85] performed WES and immunologic studies on five unrelated children, who presented with a distinctive type of combined immunodeficiency that is characterized by early-onset, invasive bacterial and viral infections; T-cell lymphopenia; impaired T-cell, B-cell, and NK-cell function; and defective interferon immunity in both hematopoietic and non-hematopoietic cells. They detected biallelic mutations in *DOCK2* in all 5 patients, and all mutations were predicted to be deleterious. The functional studies of DOCK2 deficiency in humans revealed an impaired RAC1 activation and defects in actin polymerization, T-cell proliferation, chemokine-induced lymphocyte migration, and NK-cell degranulation. Thus, they demonstrated DOCK2 deficiency as a new Mendelian disorder with pleiotropic defects of hematopoietic and non-hematopoietic immunity. Furthermore, normalization of immunologic abnormalities and resolution of infections were obtained in 3 patients after HSCT. This rescue of the clinical phenotype was possibly due to the generation of a source of cells producing interferon-α/β (e.g., plasmacytoid dendritic cells) and therefore complementing the defect in non-hematopoietic tissues. By contrast, the 2 patients without HSCT treatment died early in childhood. In conclusion, the definite molecular diagnosis helped physician develop a personalized treatment plan for the patient and resulted in better outcome in these patients. Although we only described a few examples of using WES for clinical diagnosis and discovery of new genetic defects in this chapter, the impressive outcomes have promised a broad application of WES in PIDs.

6 Selective Whole Exome Sequencing

Although many NGS panels are available, none of them has included all known disease causing genes for certain single or groups of diseases. With the advent of NGS technology; novel genes are consistently being identified; however, these genes are not able to be immediately added on to corresponding panels. There are several major reasons: (1) the genetic field is dynamic, (2) the complexity of NGS panel design and (3) a time-consuming clinical validation process. Thus, selective exome sequencing (SES) is an excellent alternative that can capture all of the currently known and future disease causing genes efficiently. The NGS technology and sequencing process for SES is similar to WES. The strategy for such SES approach is to sequence the whole exome but only analyze a group of genes of interest. The SES is best suited for patients with clearly defined, genetic heterogeneous conditions whereby a comprehensive gene panel is not available, or the patient has a single gene disorder for which clinical testing is not currently available. Moreover, SES sequencing permits the analysis of genes related to patient's phenotypes, thus, offers the flexibility of incorporating new clinical genes at any time. In comparison with the regular WES, the SES sequencing offers test results with shorter turnaround time because the analysis is restricted to specific number of genes, a "focus panel". Unlike WES trio analysis (a common WES strategy), SES testing is only performed on the proband and does not use samples from family members for the analysis, which reduces the cost of the test. Another difference with WES is that SES reports will not include any incidental findings because it is a targeted panel and non-panel genes will not be analyzed including those recommended by the American College of Medical Genetics (ACMG) guidelines. SES is not a replacement for established panel tests because we expect the sensitivity to be somewhat reduced by having some regions with lower or no coverage. However, the coverage is still acceptable. For example, the coverage of the comprehensive SES PID panel composed of 336 genes had an average coverage of 112X. It also had a 98, 98, 97, 96, 94 and 90.20% of coverage at 3X, 5X, 10X, 20X, 30X and 40X levels respectively (unpublished data). Similar coverage numbers were observed with other smaller sub-PID panels. For the aforementioned reasons, the SES sequencing approach is predicted to be one of the approaches that transform molecular diagnostics.

7 Whole Genome Sequencing (WGS) in Primary Immunodeficiency Diseases (PIDs)

WGS is designed to target the whole genome, which includes both the protein coding regions and the non-coding regions. Theoretically, WGS would allow characterization of all variants of the whole genome including the large deletions, duplications that WES fails to detect [10]. Consequently, the WGS will give rise to large numbers of gene variants. The clinical adoption of WGS has been challenging both due to the high cost and the difficult and time-consuming nature of interpretation of the variants particularly resulted from the non-coding regions [10].

Notably, WGS has been proven to be an effective tool for the molecular diagnosis of several genetic disorders, with no candidate gene variants detected by other NGS testing before [20]. The study was performed by Taylor et al. on 500 patients with diverse genetic disorders without disease-causing variants or candidate genes resulted from other NGS tests. On average, 82.7% of the genome including 88.2% of the exome was covered by at least 20X. To evaluate the clinical efficacy of WGS, only156 patients or families with Mendelian and immunological disorders were summarized. Overall, they identified disease-causing variants in 21% of cases, with the proportion increasing to 34% (23/68) for Mendelian disorders and 57% (8/14) in family trios. In addition, they detected 32 potentially clinically actionable variants in 18 genes unrelated to the referral disorder. Interestingly, there were two candidate pathogenic variants outside the coding fraction of the genome. One of them is in 5′ UTR of the *EPO* gene from two independent families with erythrocytosis and co-segregated with the disease; another one is a complex deletion of 1.4 kb of the X chromosome and insertion of 50 kb from chromosome 2p from a patient with X-linked hypoparathyroidism. This variant lay 81.5 kb downstream of *SOX3*, and is segregated with the phenotypes. The discovery of these two pathogenic candidates demonstrated the value of WGS for screening the noncoding genome. This study has implicated that WGS for clinical diagnosis is becoming realistic in spite of many challenges ahead.

WGS has shown a strong potential of identifying mutations in noncoding regions of genome in PIDs. Mousallem et al. [14] performed WGS on a girl with a clinical diagnosis of SCID, who presented at infancy with a history of failure to thrive and recurrent infections. Flow cytometry did not show any T or B cells, but revealed an elevated percentage of NK cells. T-cell proliferation studies showed no responses to mitogens. In this study, the WGS detected and defined the exact breakpoints of a homozygous 82-kb deletion spanning exons 1 to 4 in *DCLRE1C*. WGS analysis of the other two SCID patients revealed the same deletion mutation in *DCLRE1C* in addition to the splice mutation (c.362 + 2 T > A). The combined presence of this frame shift deletion and the c.362 + 2 T > A (IVS5 + 2 T > A) splice site mutation was predicted to be the cause of SCID in this patient. This study demonstrated a promising potential of using WGS to reveal size and precise breakpoint of large complex mutations that cannot be achieved by WES and other NGS testing.

8 Summary

Even though the application of NGS technology to clinical molecular diagnosis is still at an early stage, impressive results have been obtained in PIDs, leading to an accurate, rapid diagnosis. NGS panels can detect variants in most known disease genes at once, and has thus made comprehensive PID diagnosis easier and faster. Focused exome sequencing is a middle-of-the-road test with the advantages over NGS panels of being able to include larger more comprehensive panels and the ability to add newly discovered disease associated genes, and the promise of having a faster turnaround time than the full WES test. WES, originally a powerful tool for dissecting genetics, is now much more affordable and has been widely adopted in PID diagnostics. In addition to the high diagnostic yield, WES has identified a large number of novel genes that cause diseases. WES is greatly accelerating our elucidation in all of the genetic diseases including the PIDs. While the WGS for clinical use is still in its infancy, WGS has raised an exciting possibility that all of the gene variants can be detected at once. As knowledge is rapidly acquired with respect to the clinical significance of the millions of variants carried by each individual, the WGS may be employed as a routine genetic testing strategy for clinical diagnosis 1 day. Despite the limitations resulted from targeted NGS panels, WES or even WGS, NGS-based gene sequencing tests have clearly demonstrated their unique potentials for the most complicated diagnoses in PIDs. However, we have to be aware that each method has its own specific limitations. It is critical for clinicians and molecular geneticists to choose the most appropriate NGS-based tests in order to reach the best outcome. For atypical PID patients with Mendelian inheritance pattern or the patients with a negative result by targeted NGS panels, WES should be recommended. WGS has not been adopted formally by clinical laboratories, it may reveal many variants in the genes that are remained unknown. Thus, the interpretation to these gene variants should be taken cautiously, especially when these variants are in a noncoding region or regulatory region.

Despite the emerging challenges, NGS has proven revolutionary and has significantly impacted all the fields of genetic and genomics. The NGS-based technologies have not only empowered clinical molecular diagnostics, but also provided the best tool for dissecting the genetic bases of unknown diseases. In the next few years, there is no doubt that the application of NGS (target NGS panels and WES) will continue to be the leading force in clinical work and clinical and/ or basic genetic research. WGS has attracted a great deal of attention because of its potential for detecting gene defects of the whole genome. Therefore, in the near future, the WGS should also be considered as an ultimate approach to identify the unknown genetic disorders in clinic.

References

1. Ochs, H.D., Hitzig, W.H.: History of primary immunodeficiency diseases. Curr. Opin. Allergy Clin. Immunol. **12**, 577–587 (2012)
2. Joshi, A.Y., Iyer, V.N., Hagan, J.B., St Sauver, J.L., Boyce, T.G.: Incidence and temporal trends of primary immunodeficiency: a population-based cohort study. Mayo Clin. Proc. **84**, 16–22 (2009)
3. Bousfiha, A.A., Jeddane, L., Ailal, F., Benhsaien, I., Mahlaoui, N., Casanova, J.L., et al.: Primary immunodeficiency diseases worldwide: more common than generally thought. J. Clin. Immunol. **33**, 1–7 (2013)
4. Boyle, J.M., Buckley, R.H.: Population prevalence of diagnosed primary immunodeficiency diseases in the United States. J. Clin. Immunol. **27**, 497–502 (2007)
5. Al-Herz, W., Bousfiha, A., Casanova, J.L., Chatila, T., Conley, M.E., Cunningham-Rundles, C., et al.: Primary immunodeficiency diseases: an update on the classification from the international union of immunological societies expert committee for primary immunodeficiency. Front. Immunol. **5**, 162 (2014)
6. Chinen, J., Notarangelo, L.D., Shearer, W.T.: Advances in basic and clinical immunology in 2013. J. Allergy Clin. Immunol. **133**, 967–976 (2014)
7. Gilissen, C., Hoischen, A., Brunner, H.G., Veltman, J.A.: Unlocking mendelian disease using exome sequencing. Genome Biol. 12, 228 (2011)
8. Shendure, J., Ji, H.: Next-generation DNA sequencing. Nat. Biotechnol. 26, 1135–1145 (2008)
9. Beaulieu, C.L., Majewski, J., Schwartzentruber, J., Samuels, M.E., Fernandez, B.A., Bernier, F.P., et al.: FORGE Canada consortium: outcomes of a 2-year national rare-disease gene-discovery project. Am. J. Hum. Genet. **94**, 809–817 (2014)
10. Biesecker, L.G., Green, R.C.: Diagnostic clinical genome and exome sequencing. N. Engl. J. Med. **370**, 2418–2425 (2014)
11. Saunders, C.J., Miller, N.A., Soden, S.E., Dinwiddie, D.L., Noll, A., Alnadi, N.A., et al.: Rapid whole-genome sequencing for genetic disease diagnosis in neonatal intensive care units. Sci. Transl. Med. **4**, 154ra35 (2012)
12. Nijman, I.J., van Montfrans, J.M., Hoogstraat, M., Boes, M.L., van de Corput, L., Renner, E.D., et al.: Targeted next-generation sequencing: a novel diagnostic tool for primary immunodeficiencies. J. Allergy Clin. Immunol. **133**, 529–534 (2014)
13. Dickinson, R.E., Griffin, H., Bigley, V., Reynard, L.N., Hussain, R., Haniffa, M., et al.: Exome sequencing identifies GATA-2 mutation as the cause of dendritic cell, monocyte, B and NK lymphoid deficiency. Blood. **118**, 2656–2658 (2011)
14. Mousallem, T., Urban, T.J., McSweeney, K.M., Kleinstein, S.E., Zhu, M., Adeli, M., et al.: Clinical application of whole-genome sequencing in patients with primary immunodeficiency. J. Allergy Clin. Immunol. **136**(2), 476–9.e6 (2015)
15. Jones, M.A., Bhide, S., Chin, E., Ng, B.G., Rhodenizer, D., Zhang, V.W., et al.: Targeted polymerase chain reaction-based enrichment and next generation sequencing for diagnostic testing of congenital disorders of glycosylation. Genet. Med. Off. J. Am. Coll. Med. Genet. **13**, 921–932 (2011)
16. Ku, C.S., Naidoo, N., Pawitan, Y.: Revisiting mendelian disorders through exome sequencing. Hum. Genet. **129**, 351–370 (2011)
17. Botstein, D., Risch, N.: Discovering genotypes underlying human phenotypes: past successes for mendelian disease, future approaches for complex disease. Nat. Genet. **33**(Suppl), 228–237 (2003)
18. Yang, Y., Muzny, D.M., Reid, J.G., Bainbridge, M.N., Willis, A., Ward, P.A., et al.: Clinical whole-exome sequencing for the diagnosis of mendelian disorders. N. Engl. J. Med. **369**, 1502–1511 (2013)
19. Rabbani, B., Tekin, M., Mahdieh, N.: The promise of whole-exome sequencing in medical genetics. J. Hum. Genet. **59**, 5–15 (2014)

20. Taylor, J.C., Martin, H.C., Lise, S., Broxholme, J., Cazier, J.B., Rimmer, A., et al.: Factors influencing success of clinical genome sequencing across a broad spectrum of disorders. Nat. Genet. **47**(7), 717–726 (2015)
21. Veenstra, D.L., Piper, M., Haddow, J.E., Pauker, S.G., Klein, R., Richards, C.S., et al.: Improving the efficiency and relevance of evidence-based recommendations in the era of whole-genome sequencing: an EGAPP methods update. Genet. Med. Off. J. Am. Coll. Med. Genet. **15**, 14–24 (2013)
22. Robinson, P.N., Krawitz, P., Mundlos, S.: Strategies for exome and genome sequence data analysis in disease-gene discovery projects. Clin. Genet. **80**, 127–132 (2011)
23. Mardis, E.R.: Next-generation sequencing platforms. Annu Rev Anal Chem (Palo Alto, Calif). **6**, 287–303 (2013)
24. Raje, N., Soden, S., Swanson, D., Ciaccio, C.E., Kingsmore, S.F., Dinwiddie, D.L.: Utility of next generation sequencing in clinical primary immunodeficiencies. Curr. Allergy Asthma Rep. **14**, 468 (2014)
25. Stoddard, J.L., Niemela, J.E., Fleisher, T.A., Rosenzweig, S.D.: Targeted NGS: a cost-effective approach to molecular diagnosis of PIDs. Front. Immunol. **5**, 531 (2014)
26. Rothberg, J.M., Hinz, W., Rearick, T.M., Schultz, J., Mileski, W., Davey, M., et al.: An integrated semiconductor device enabling non-optical genome sequencing. Nature. **475**, 348–352 (2011)
27. Moens, L.N., Falk-Sorqvist, E., Asplund, A.C., Bernatowska, E., Smith, C.I., Nilsson, M.: Diagnostics of primary immunodeficiency diseases: a sequencing capture approach. PLoS One. **9**, e114901 (2014)
28. Al-Mousa, H., Abouelhoda, M., Monies, D.M., Al-Tassan, N., Al-Ghonaium, A., Al-Saud, B., et al.: Unbiased targeted next-generation sequencing molecular approach for primary immunodeficiency diseases. J. Allergy Clin. Immunol. **137**, 1780–1787 (2016)
29. Rosado, F.G., Kim, A.S.: Hemophagocytic lymphohistiocytosis: an update on diagnosis and pathogenesis. Am. J. Clin. Pathol. **139**, 713–727 (2013)
30. Zhang, K., Chandrakasan, S., Chapman, H., Valencia, C.A., Husami, A., Kissell, D., et al.: Synergistic defects of different molecules in the cytotoxic pathway lead to clinical familial hemophagocytic lymphohistiocytosis. Blood. **124**, 1331–1334 (2014)
31. Jordan, M.B., Allen, C.E., Weitzman, S., Filipovich, A.H., McClain, K.L.: How I treat hemophagocytic lymphohistiocytosis. Blood. **118**, 4041–4052 (2011)
32. Zhang, K., Filipovich, A.H., Johnson, J., Marsh, R.A., Villanueva, J.: Hemophagocytic lymphohistiocytosis, familial. In: Pagon, R.A., Adam, M.P., Ardinger, H.H., Wallace, S.E., Amemiya, A., Bean, L.J.H., et al. (eds.) GeneReviews(R). University of Washington, Seattle. All rights reserved, Seattle (WA) (1993)
33. Allen, M., De Fusco, C., Legrand, F., Clementi, R., Conter, V., Danesino, C., et al.: Familial hemophagocytic lymphohistiocytosis: how late can the onset be? Haematologica. **86**, 499–503 (2001)
34. Henter, J.I., Horne, A., Arico, M., Egeler, R.M., Filipovich, A.H., Imashuku, S., et al.: HLH-2004: diagnostic and therapeutic guidelines for hemophagocytic lymphohistiocytosis. Pediatr. Blood Cancer. **48**, 124–131 (2007)
35. Badolato, R., Prandini, A., Caracciolo, S., Colombo, F., Tabellini, G., Giacomelli, M., et al.: Exome sequencing reveals a pallidin mutation in a Hermansky-Pudlak-like primary immunodeficiency syndrome. Blood. **119**, 3185–3187 (2012)
36. Dotta, L., Parolini, S., Prandini, A., Tabellini, G., Antolini, M., Kingsmore, S.F., et al.: Clinical, laboratory and molecular signs of immunodeficiency in patients with partial oculo-cutaneous albinism. Orphanet J. Rare Dis. **8**, 168 (2013)
37. Li, F.Y., Lenardo, M.J., Chaigne-Delalande, B.: Loss of MAGT1 abrogates the Mg2+ flux required for T cell signaling and leads to a novel human primary immunodeficiency. Magnes. Res. Off. organ Int. Soc. Dev. Res. Magnes. **24**, S109–S114 (2011)

38. van Montfrans, J.M., Hoepelman, A.I., Otto, S., van Gijn, M., van de Corput, L., de Weger, R.A., et al.: CD27 deficiency is associated with combined immunodeficiency and persistent symptomatic EBV viremia. J. Allergy Clin. Immunol. **129**, 787–93.e6 (2012)
39. Yang, X., Miyawaki, T., Kanegane, H.: SAP and XIAP deficiency in hemophagocytic lymphohistiocytosis. Pediatr. Int. Off. J. Japan Pediatr. Soc. **54**, 447–454 (2012)
40. Sperandeo, M.P., Andria, G., Sebastio, G.: Lysinuric protein intolerance: update and extended mutation analysis of the SLC7A7 gene. Hum. Mutat. **29**, 14–21 (2008)
41. Shamsian, B.S., Rezaei, N., Alavi, S., Hedayat, M., Amin Asnafi, A., Pourpak, Z., et al.: Primary hemophagocytic lymphohistiocytosis in Iran: report from a single referral center. Pediatr. Hematol. Oncol. **29**, 215–219 (2012)
42. Jessen, B., Bode, S.F., Ammann, S., Chakravorty, S., Davies, G., Diestelhorst, J., et al.: The risk of hemophagocytic lymphohistiocytosis in Hermansky-Pudlak syndrome type 2. Blood. **121**, 2943–2951 (2013)
43. Cullinane, A.R., Curry, J.A., Carmona-Rivera, C., Summers, C.G., Ciccone, C., Cardillo, N.D., et al.: A BLOC-1 mutation screen reveals that PLDN is mutated in Hermansky-Pudlak syndrome type 9. Am. J. Hum. Genet. **88**, 778–787 (2011)
44. Linka, R.M., Risse, S.L., Bienemann, K., Werner, M., Linka, Y., Krux, F., et al.: Loss-of-function mutations within the IL-2 inducible kinase ITK in patients with EBV-associated lymphoproliferative diseases. Leukemia. **26**, 963–971 (2012)
45. Li, F.Y., Chaigne-Delalande, B., Su, H., Uzel, G., Matthews, H., Lenardo, M.J.: XMEN disease: a new primary immunodeficiency affecting Mg2+ regulation of immunity against Epstein-Barr virus. Blood. **123**, 2148–2152 (2014)
46. Sivakumaran, T.A., Husami, A., Kissell, D., Zhang, W., Keddache, M., Black, A.P., et al.: Performance evaluation of the next-generation sequencing approach for molecular diagnosis of hereditary hearing loss. Otolaryngol. Head Neck Surg. Off. J. Am. Acad. Otolaryngol. Head and Neck Surg. **148**, 1007–1016 (2013)
47. Tewhey, R., Warner, J.B., Nakano, M., Libby, B., Medkova, M., David, P.H., et al.: Microdroplet-based PCR enrichment for large-scale targeted sequencing. Nat. Biotechnol. **27**, 1025–1031 (2009)
48. Teachey, D.T., Seif, A.E., Grupp, S.A.: Advances in the management and understanding of autoimmune lymphoproliferative syndrome (ALPS). Br. J. Haematol. **148**, 205–216 (2010)
49. Bleesing, J.J.H., Johnson, J., Zhang, K.: Autoimmune lymphoproliferative syndrome. In: Pagon, R.A., Adam, M.P., Ardinger, H.H., Wallace, S.E., Amemiya, A., Bean, L.J.H., et al. (eds.) GeneReviews(R). University of Washington, Seattle. All rights reserved, Seattle (WA) (1993)
50. Oliveira, J.B., Bleesing, J.J., Dianzani, U., Fleisher, T.A., Jaffe, E.S., Lenardo, M.J., et al.: Revised diagnostic criteria and classification for the autoimmune lymphoproliferative syndrome (ALPS): report from the 2009 NIH international workshop. Blood. **116**, e35–e40 (2010)
51. Rieux-Laucat, F., Le Deist, F., Fischer, A.: Autoimmune lymphoproliferative syndromes: genetic defects of apoptosis pathways. Cell Death Differ. **10**, 124–133 (2003)
52. Niemela, J.E., Lu, L., Fleisher, T.A., Davis, J., Caminha, I., Natter, M., et al.: Somatic KRAS mutations associated with a human nonmalignant syndrome of autoimmunity and abnormal leukocyte homeostasis. Blood. **117**, 2883–2886 (2011)
53. Buckley, R.H.: Molecular defects in human severe combined immunodeficiency and approaches to immune reconstitution. Annu. Rev. Immunol. **22**, 625–655 (2004)
54. Griffith, L.M., Cowan, M.J., Notarangelo, L.D., Puck, J.M., Buckley, R.H., Candotti, F., et al.: Improving cellular therapy for primary immune deficiency diseases: recognition, diagnosis, and management. J. Allergy Clin. Immunol. **124**, 1152–60.e12 (2009)
55. Rivers, L., Gaspar, H.B.: Severe combined immunodeficiency: recent developments and guidance on clinical management. Arch. Dis. Child. **100**, 667–672 (2015)
56. Aloj, G., Giardino, G., Valentino, L., Maio, F., Gallo, V., Esposito, T., et al.: Severe combined immunodeficiencies: new and old scenarios. Int. Rev. Immunol. **31**, 43–65 (2012)

57. Puck, J.M., Pepper, A.E., Henthorn, P.S., Candotti, F., Isakov, J., Whitwam, T., et al.: Mutation analysis of IL2RG in human X-linked severe combined immunodeficiency. Blood. **89**, 1968–1977 (1997)
58. Verbsky, J., Thakar, M., Routes, J.: The Wisconsin approach to newborn screening for severe combined immunodeficiency. J. Allergy Clin. Immunol. **129**, 622–627 (2012)
59. Villa, A., Santagata, S., Bozzi, F., Giliani, S., Frattini, A., Imberti, L., et al.: Partial V(D)J recombination activity leads to Omenn syndrome. Cell. **93**, 885–896 (1998)
60. Corneo, B., Moshous, D., Gungor, T., Wulffraat, N., Philippet, P., Le Deist, F.L., et al.: Identical mutations in RAG1 or RAG2 genes leading to defective V(D)J recombinase activity can cause either T-B-severe combined immune deficiency or Omenn syndrome. Blood. **97**, 2772–2776 (2001)
61. Ege, M., Ma, Y., Manfras, B., Kalwak, K., Lu, H., Lieber, M.R., et al.: Omenn syndrome due to ARTEMIS mutations. Blood. **105**, 4179–4186 (2005)
62. Giliani, S., Bonfim, C., de Saint, B.G., Lanzi, G., Brousse, N., Koliski, A., et al.: Omenn syndrome in an infant with IL7RA gene mutation. J. Pediatr. **148**, 272–274 (2006)
63. Roifman, C.M., Gu, Y., Cohen, A.: Mutations in the RNA component of RNase mitochondrial RNA processing might cause Omenn syndrome. J. Allergy Clin. Immunol. **117**, 897–903 (2006)
64. Joshi, A.Y., Ham, E.K., Shah, N.B., Dong, X., Khan, S.P., Abraham, R.S.: Atypical Omenn syndrome due to adenosine deaminase deficiency. Case Rep. Immunol. **2012**, 919241 (2012)
65. Allenspach, E., Rawlings, D.J., Scharenberg, A.M.: X-linked severe combined immunodeficiency. In: Pagon, R.A., Adam, M.P., Ardinger, H.H., Wallace, S.E., Amemiya, A., LJH, B., et al. (eds.) GeneReviews(R). University of Washington, Seattle. All rights reserved, Seattle (WA) (1993)
66. Thiel, C.T.: Cartilage-hair hypoplasia – anauxetic dysplasia spectrum disorders. In: Pagon, R.A., Adam, M.P., Ardinger, H.H., Wallace, S.E., Amemiya, A., LJH, B., et al. (eds.) GeneReviews(R). University of Washington, Seattle. All rights reserved, Seattle (WA) (1993)
67. Riley Jr., P., Weiner, D.S., Leighley, B., Jonah, D., Morton, D.H., Strauss, K.A., et al.: Cartilage hair hypoplasia: characteristics and orthopaedic manifestations. J. Child. Orthop. **9**, 145–152 (2015)
68. Hacihamdioglu, B., Hacihamdioglu, D., Delil, K.: 22q11 deletion syndrome: current perspective. Appl. Clin. Genet. **8**, 123–132 (2015)
69. Yu, H., Zhang, V.W., Stray-Pedersen, A., Hanson, I.C., Forbes, L.R., de la Morena, M.T., et al.: Rapid molecular diagnostics of severe primary immunodeficiency determined by using targeted next-generation sequencing. J. Allergy Clin. Immunol. **138**, 1142–1151 (2016.) e2
70. Itan, Y., Casanova, J.L.: Novel primary immunodeficiency candidate genes predicted by the human gene connectome. Front. Immunol. **6**, 142 (2015)
71. Chinen, J., Notarangelo, L.D., Shearer, W.T.: Advances in basic and clinical immunology in 2014. J. Allergy Clin. Immunol. **135**, 1132–1141 (2015)
72. Bigley, V., Haniffa, M., Doulatov, S., Wang, X.N., Dickinson, R., McGovern, N., et al.: The human syndrome of dendritic cell, monocyte, B and NK lymphoid deficiency. J. Exp. Med. **208**, 227–234 (2011)
73. Rodrigues, N.P., Janzen, V., Forkert, R., Dombkowski, D.M., Boyd, A.S., Orkin, S.H., et al.: Haploinsufficiency of GATA-2 perturbs adult hematopoietic stem-cell homeostasis. Blood. **106**, 477–484 (2005)
74. Patel, J.P., Puck, J.M., Srinivasan, R., Brown, C., Sunderam, U., Kundu, K., et al.: Nijmegen breakage syndrome detected by newborn screening for T cell receptor excision circles (TRECs). J. Clin. Immunol. **35**, 227–233 (2015)
75. Varon, R., Vissinga, C., Platzer, M., Cerosaletti, K.M., Chrzanowska, K.H., Saar, K., et al.: Nibrin, a novel DNA double-strand break repair protein, is mutated in Nijmegen breakage syndrome. Cell. **93**, 467–476 (1998)
76. Gregorek, H., Chrzanowska, K.H., Dzierzanowska-Fangrat, K., Wakulinska, A., Pietrucha, B., Zapasnik, A., et al.: Nijmegen breakage syndrome: long-term monitoring of viral and immunological biomarkers in peripheral blood before development of malignancy. Clin. Immunol. (Orlando, Fla). **135**, 440–447 (2010)

77. Puck, J.M.: Neonatal screening for severe combined immunodeficiency. Curr. Opin. Pediatr. **23**, 667–673 (2011)
78. Punwani, D., Wang, H., Chan, A.Y., Cowan, M.J., Mallott, J., Sunderam, U., et al.: Combined immunodeficiency due to MALT1 mutations, treated by hematopoietic cell transplantation. J. Clin. Immunol. **35**, 135–146 (2015)
79. Zhang, Y., Yu, X., Ichikawa, M., Lyons, J.J., Datta, S., Lamborn, I.T., et al.: Autosomal recessive phosphoglucomutase 3 (PGM3) mutations link glycosylation defects to atopy, immune deficiency, autoimmunity, and neurocognitive impairment. J. Allergy Clin. Immunol. **133**, 1400–9, 1409.e1–5 (2014)
80. Stray-Pedersen, A., Backe, P.H., Sorte, H.S., Morkrid, L., Chokshi, N.Y., Erichsen, H.C., et al.: PGM3 mutations cause a congenital disorder of glycosylation with severe immunodeficiency and skeletal dysplasia. Am. J. Hum. Genet. **95**, 96–107 (2014)
81. Willmann, K.L., Klaver, S., Dogu, F., Santos-Valente, E., Garncarz, W., Bilic, I., et al.: Biallelic loss-of-function mutation in NIK causes a primary immunodeficiency with multifaceted aberrant lymphoid immunity. Nat. Commun. **5**, 5360 (2014)
82. Martin, E., Palmic, N., Sanquer, S., Lenoir, C., Hauck, F., Mongellaz, C., et al.: CTP synthase 1 deficiency in humans reveals its central role in lymphocyte proliferation. Nature. **510**, 288–292 (2014)
83. Ostrander, D.B., O'Brien, D.J., Gorman, J.A., Carman, G.M.: Effect of CTP synthetase regulation by CTP on phospholipid synthesis in *Saccharomyces cerevisiae*. J. Biol. Chem. **273**, 18992–19001 (1998)
84. Fairbanks, L.D., Bofill, M., Ruckemann, K., Simmonds, H.A.: Importance of ribonucleotide availability to proliferating T-lymphocytes from healthy humans. Disproportionate expansion of pyrimidine pools and contrasting effects of de novo synthesis inhibitors. J. Biol. Chem. **270**, 29682–29689 (1995)
85. Dobbs, K., Dominguez Conde, C., Zhang, S.Y., Parolini, S., Audry, M., Chou, J., et al.: Inherited DOCK2 deficiency in patients with early-onset invasive infections. N. Engl. J. Med. **372**, 2409–2422 (2015)
86. Fukui, Y., Hashimoto, O., Sanui, T., Oono, T., Koga, H., Abe, M., et al.: Haematopoietic cell-specific CDM family protein DOCK2 is essential for lymphocyte migration. Nature. **412**, 826–831 (2001)

Next-Generation Sequencing Based Testing for Disorders of the Skeleton

Allison Tam, Victor Wei Zhang, Lee Jun Wong, and Sandesh C.S. Nagamani

Abstract While disorders that involve the skeleton are common, most forms of genetic skeletal disorders are typically rare and not encountered routinely in clinical practice. The presentations and etiologies of genetic forms of skeletal disorders are very heterogeneous; therefore, they can be challenging to diagnose. An accurate diagnosis is very important for counseling regarding the natural history and recurrence risks as well as for appropriate management. Detailed medical and family history, physical examination, radiological evaluations, laboratory, biochemical and molecular tests are all important components in the assessment of genetic skeletal disorders. Molecular testing using next-generation sequencing (NGS) techniques can help identify the pathogenic genetic variants and thus confirm the diagnoses of specific bone disorders, even in conditions which there are overlapping clinical, radiographic and histological features. As there are limitations and advantages in using whole exome sequencing versus targeted gene panels, the decision of which test to use, should be made based on a case-by-case basis.

Keywords Skeletal Disorders • Bone Development • Molecular Diagnosis • Genetics Next-Generation Sequencing

1 Introduction

Disorders that involve the skeleton are commonly encountered in clinical practice. These disorders can result from numerous causes including age-related processes (e.g. senile osteoporosis), hormonal imbalances (e.g. postmenopausal osteoporosis

A. Tam • V.W. Zhang • L.J. Wong
Department of Molecular and Human Genetics, Baylor College of Medicine, Houston, TX, USA

S.C.S. Nagamani, MD (✉)
Department of Molecular and Human Genetics, Baylor College of Medicine, Houston, TX, USA

Texas Children's Hospital, Houston, TX, USA
e-mail: Nagamani@bcm.edu

L.-J.C. Wong (ed.), *Next Generation Sequencing Based Clinical Molecular Diagnosis of Human Genetic Disorders*, DOI 10.1007/978-3-319-56418-0_7

and hyperparathyroidism-related bone loss), medications (e.g. corticosteroid-induced avascular necrosis), kidney and gastrointestinal disorders (e.g. renal osteodystrophy), and developmental anomalies of the bone (e.g. achondroplasia). Some bone disorders like osteoporosis have a high prevalence and have been estimated to affect over ten million individuals in the United States [1]. In contrast, *constitutional* errors of bone development, which typically manifest in childhood, are relatively rare conditions. These developmental disorders of bone have a collective incidence of 1 in 5000 births and can be categorized into: dysostoses (malformations of single skeletal elements), disruptions (malformations of bones due to non-skeletal causes), skeletal dysplasia (developmental disorders that involve bone and/or cartilage), and osteolyses (dissolution of preexisting bone) [2, 3]. Such disorders can present with short stature, abnormal patterning, altered size and structure of the bones, increased bone fragility, and secondary involvement of the nonskeletal tissues. Many disorders of the skeleton whether they are developmental or acquired, early or late-onset, have significant impact on the lives of affected individuals. An accurate diagnosis is important for counseling regarding the natural history and recurrence risks as well as for appropriate management. This chapter focuses on the diagnostic challenges in some genetic forms of skeletal disorders and the role of next-generation sequencing techniques in their diagnosis.

2 Genetic Forms of Skeletal Disorders

Genetic forms of skeletal disorders are heterogeneous in their presentations and etiologies. A "nomenclature" was developed in the 1970s in an attempt to classify these disorders, and these classifications have been updated and revised over the years [4–7]. The recognition of new phenotypes and the rapid advances in the molecular diagnostic techniques have led to significant increase in the number of disorders and identification of the causative genes. These have necessitated a more thorough evaluation of the nosology and classification of genetic skeletal disorders. The 2015 classification by the Nosology group of the International Skeletal Dysplasia Society identified over 430 conditions and categorized them into 42 groups based on molecular, biochemical, and/or radiographic criteria [8]. The conditions included those with primary bone involvement as well as overgrowth syndromes and lysosomal storage disorders with significant skeletal manifestations. While delving into further specifics of the classification are beyond the scope of this chapter, a review of the classification highlights the genetic and clinical heterogeneity of these disorders. Overall, there are 336 genes that have been identified to cause 436 disorders. Mutations in the same gene can give rise to phenotypically distinct disorders (e.g. metatrophic dysplasia and brachyolmia due to *TRPV4* mutations) or varying severity of the same disorder (e.g. *COL1A1* mutations in osteogenesis imperfecta types I [mild] vs. type II [perinatal lethal]), while mutations in different genes can give rise to disorders with overlapping clinical features (e.g. ciliopathies with major skeletal involvement). Mutations in genes encoding extracellular matrix proteins, transcription factors, signal transducers, enzymes, cellular transporters, chaperone proteins, intracellular binding proteins, RNA processing molecules, and ciliary proteins can present with skeletal involvement of varying severity and patterns.

3 Diagnostic Challenges in Genetic Skeletal Disorders

An accurate diagnosis of genetic skeletal disorders requires detailed medical and family history, physical examination, radiologic evaluations, as well as laboratory, biochemical, and molecular tests. Most forms of genetic skeletal disorders are typically rare and are not encountered in routine clinical practices. Hence, their diagnosis and treatment are often performed by centers with specialized expertise. Some of the pertinent questions that may help to narrow the diagnostic considerations include: (1) Is the bone involvement primary or a part of multisystem involvement (e.g. lysosomal storage disorders, overgrowth syndromes, or inflammatory osteoarthropathy)? (2) Is the involvement localized to a few bones (dysostoses) or is it generalized (typically skeletal dysplasia)? (3) Is there a particular pattern of bone involvement (e.g. ribs and vertebral bones involvement in spondylocostal dystotoses vs. vertebral bones and the ends of the long bones in spondyloephiphyseal dysplasia)? (4) Is there a particular part of bone involved - epiphyseal or ends of the bones (e.g. multiple epiphyseal dysplasia types 1–6) vs. diaphyseal or midsection of long bones (e.g. diaphyseal dysplasia) vs. metaphyseal or the part of the bone joining epiphyses to the diaphysis (e.g. metaphyseal dysplasia, Jansen type)? (5) If the long bones are involved, is the involvement predominantly the proximal (rhizomelic), middle (mesomelic), distal (acromelic), or combinations thereof (acromesomelic)? (6) Are there specific diagnostic clues on exam or X-rays (e.g. blue sclera, tooth abnormalities in type I collagen-related osteogenesis imperfecta or interosseous membrane calcification and exuberant callous formation in osteogenesis imperfecta type V)? (7) For disorders of increased bone fragility, are they associated with decreased (e.g. osteogenesis imperfecta) or increased bone mineral density (e.g. osteopetrosis)?

Systematic assessment based on the site, severity, and nature of involvement can lead to the diagnosis in many genetic disorders of the bone without the further need for confirmatory molecular testing (e.g. achondroplasia). However, many a time, the diagnosis is not apparent and further molecular tests may be necessary.

4 Molecular Diagnosis of Genetic Skeletal Disorders

The ability to identify the pathogenic genetic variants that cause specific bone disorders can be helpful in the diagnosis given overlapping clinical, radiographic and histological features in many conditions. Until recently, molecular diagnostic testing for skeletal dysplasias was limited to sequencing a single or a few select genes by the Sanger sequencing method. This approach is effective when the possibility of the provisional diagnosis being correct is high and the number of genes to be interrogated is few. However, in scenarios wherein there is genetic heterogeneity or the phenotype is not distinct enough to make a clinical diagnosis, interrogating numerous genes known to cause the phenotype would be a more time- and cost-effective strategy. For example, when the clinical and radiologic features are suggestive of a metaphyseal dysplasia, it would be more reasonable to investigate the seven genes that are known to cause eight conditions within this group at one time. Alternatively, when the majority of individuals with a particular disorder harbor pathogenic variants in one or few genes (e.g. *COL1A1* and *COL1A2* in osteogenesis imperfecta) and only a minority of affected individuals have mutations in one of the numerous other associated genes (e.g. *CRTAP*, *PPIB*, *LEPRE1*, *WNT1*, *FKBP10*, *SERPINF1* etc.), Sanger sequencing of the most commonly mutated genes followed by panel testing when required may be a reasonable approach.

5 Next-Generation Sequencing in Genetic Skeletal Disorders

Next-Generation Sequence (NGS) technologies have had a significant impact on the diagnosis of genetic disorders. Whole exome sequencing (WES) and targeted gene panels have been increasingly used in clinical practice. WES has the advantage of being able to sequence the entire coding portion of the genome and has been shown to have a diagnostic yield rate of 25% [9, 10]. Targeted gene panels focus on a set of genes known to cause particular phenotypes and typically have deeper coverage for the regions of interest. Currently, numerous gene panels are available for clinical diagnosis of a wide range of genetic skeletal disorders (genetic testing registry http://www.ncbi.nlm.nih.gov/gtr/ and GeneTests https://www.genetests.org). These range from large panels of over 150 genes for diagnosis of "many forms of skeletal dysplasia," to panels of over 50 genes for diagnosis of "disproportionate short stature," to assays that aim to assist in diagnoses of focused phenotypes like "osteogenesis imperfecta", "low bone mass", "osteopetrosis", high bone mass", and "Stickler syndrome", amongst others.

Table 1 A total of 34 genes responsible for disorders with high bone mass and low bone mass were utilized to create a next-generation sequencing based panel test. The total number of coding exons (CDS) and targeted bases are also shown

High Bone Mass Panel: *ANKH, CA2, CLCN7, CTSK, FAM123B, FAM20C, LEMD3, OSTM1, SOST, TCIRG1, TGFB1, TNFRSF11A, TNFRSF11B, TNFSF11, TYROBP* (15 genes) **Low Bone Mass Panel:** *ALPL, B4GALT7, COL1A1, COL1A2, COL3A1, COL5A1, COL5A2, CRTAP, FBN1, FKBP10, LEPRE1, PLOD2, PLOD3, PPIB, SERPINF1, SLC34A1, SLC39A13, SLC9A3R1, SP7* (19 genes)	
Number of CDS	602
Target size	98,962 bp (CDS ± 20 bp)
Enrichment	In solution capture library
Sequencing info	Illumina HiSeq 2000, 75 cycle, single-end

Table 2 Sequencing statistics for a total of 11 representative samples tested for disorders of high and low bone mass. All of the exons with low coverage (i.e., any base with coverage <20×) were "gap-filled" with Sanger sequencing

Sample ID	Mean coverage (bp)	Total reads per 100 bp	Minimal coverage*	# of CDS <10×	# of CDS <20×
#1	1148 ± 543	1572 ± 734	0/21×	12	14
#2	1201 ± 561	1656 ± 766	0/23×	11	13
#3	997 ± 430	1373 ± 591	0/22×	12	15
#4	969 ± 455	1328 ± 616	0/23×	14	20
#5	881 ± 402	1207 ± 545	0/24×	12	16
#6	1155 ± 568	1579 ± 765	0/21×	13	19
#7	1106 ± 546	1513 ± 737	0/23×	13	17
#8	1211 ± 586	1659 ± 792	0/20×	10	14
#9	523 ± 262	715 ± 356	0/21×	18	21
#10	945 ± 474	1290 ± 639	0/20×	13	16
#11	1072 ± 549	1465 ± 745	0/28×	13	19

* refers to the value of lowest coverage of all exons/ the coverage for 1st CDS >20×

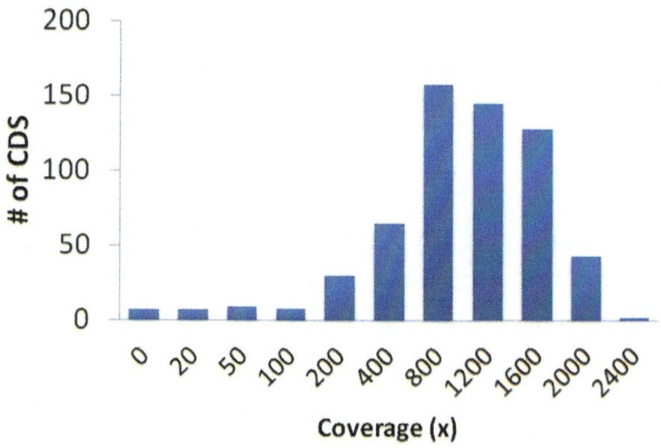

Fig. 1 Coverage depth for 602 coding exons in a representative panel used for diagnosis of disorders with high and low bone mass. Note that a vast majority of exons have deep coverage

Our previous experience on a diagnostic panel of disorders of low and high bone mass that included 34 related genes spanning 602 exons with complete coverage for coding exons from NGS and Sanger sequencing revealed 100% concordance while detecting previously identified pathogenic variants during the validation phase (Tables 1 and 2) [11]. The diagnostic utility of the panel was further underscored by the fact that a molecular diagnosis was achieved in four individuals (three with osteogenesis imperfecta and one with osteopetrosis) in whom, the genetic cause for the phenotype was not known.

As compared to exome or whole genome sequencing, panel testing offers advantages that include deeper coverage and fewer regions with insufficient coverage that could translate to decreased false negative rate (Fig. 1). In addition, regions with insufficient coverage, regions with high homologous sequences, and pseudogenes may be resolved by specifically designed PCR primers followed by

NGS. Panel-based testing is also typically more cost-efficient and is associated with fewer variants of uncertain significance (VUS) and incidental findings.

The increased use of panel-based tests has fueled the rapid growth and uptake of these diagnostic modalities in the clinic. The ability to interrogate multiple relevant genes in a single test is an attractive option for patients and physicians for whom such testing is associated with decreased costs and turn-around time, and increased diagnostic efficiency.

6 Other Diagnostic Evaluations

Biochemical tests may be useful for diagnosis in certain disorders. Some examples include urine oligosaccharides for mucopolysaccharidoses, low plasma alkaline phosphatase and elevated pyridoxal 5′-phosphate in hypophosphatasia, and abnormal sterol metabolites in chondrodysplasia punctata 2, X-linked [12, 13]. Skin biopsy and analysis of collagen secretion and amount are helpful in diagnosing osteogenesis imperfecta though this has currently been replaced by molecular diagnosis [14]. Tissue histology is typically not routinely performed but could be informative (e.g. osteogenesis imperfecta type IV). These additional modalities could be beneficial in confirmation of diagnosis when molecular testing reveals variants of uncertain significance [15].

7 Utility of an Accurate Diagnosis

(a) **For management**

The utility of making an accurate diagnosis cannot be overstated. Establishing a diagnosis provides psychological benefits and "closure" to families, enables access to the necessary support services, and guides the initiation of appropriate treatment and surveillance measures [16–18]. For example, a diagnosis of moderate-to-severe form of osteogenesis imperfecta may prompt the initiation of bisphosphonate therapy from infancy. Such therapy can be of utility in improving the bone mineral density [19–23]. Enzyme replacement therapies have been approved or being evaluated for some genetic skeletal disorders (e.g., Morquio A syndrome, hypophosphatasia) and their use is typically initiated after a definitive diagnosis [24, 25]. Many forms of genetic disorders of bone are associated with patterning defects (e.g. abnormal digits of the hand), scoliosis, or other bone malformations that may need surgical interventions. In addition, many disorders can be associated with extra-skeletal complications including neurologic (e.g., brain stem compression in achondroplasia and Morquio A syndrome), auditory (e.g., nerve entrapment in osteopetrosis and osteogenesis imperfecta), visual (e.g., optic nerve compression in osteopetrosis) and pulmonary systems (restrictive lung disease due to rib cage abnormalities). An appropriate diagnosis can thus be of significant use in initiating disease-specific surveillance measures.

(b) **For reproductive decisions**

Establishing a molecular diagnosis is important in counseling for recurrence risks and guiding reproductive decisions. Individuals with skeletal dysplasia consider the risk of transmitting the condition and the medical impact of the condition on a child as major concerns with respect to having children [26]. A not-so-infrequent-scenario is when the abnormalities of bone are detected prenatally during ultrasound examinations. Recognition of specific skeletal anomaly on ultrasound is extremely challenging and thus a definitive diagnosis is often dependent on molecular confirmation [27]. An accurate molecular diagnosis may be important for decisions regarding continuing the pregnancy or preparing to deliver the child at a tertiary care center. Panel testing could especially be of utility in such situation wherein a diagnosis may have to be reached in a short period of time. Many laboratories now offer panel-based testing for prenatal diagnosis of genetic skeletal disorders.

(c) **For evaluation of a heritable cause for fractures vs. non-accidental trauma**

Distinguishing fractures due to a genetic form of brittle bone disorder from acquired causes can have significant implications. Children with osteogenesis imperfecta can present with many fractures in various stages of healing. This is also the case in children who sustain non-accidental trauma (NAT) due to physical abuse. NAT is the leading cause of fractures in infancy and typically mandates reporting to appropriate authorities. Thus, differentiating a heritable form of bone disorder that predisposes to fracture from NAT can have medical, social, as well as legal consequences. Whereas often, the history, location and type of fractures, and other associated injuries may help in differentiating between osteogenesis imperfecta and NAT, this is not always the case. Hence, comprehensive molecular testing could be of significant utility in such scenarios.

8 Strengths and Limitations of Panel Testing in Clinic

Targeted panel tests typically provide deeper coverage than untargeted capture and sequencing of the exome or genome. The gaps in sequence due to the presence of pseudogenes or GC-rich regions are typically known and can be supplemented with Sanger sequencing of such regions to comprehensively interrogate the genes of interest. Some panels can be more affordable than whole exome sequencing. However, panel testing has limitations. The pace of discoveries in genetic disorders of bone typically makes any panel inadequate within a short span of time. Adding new genes and revalidation of such panels imposes burden of costs and time on the diagnostic laboratories. When bones are involved along with other organ manifestations, the differential diagnosis may be broad enough that whole exome sequencing could yield better results than targeted sequencing.

Acknowledgements The authors wish to acknowledge funding sources. SNSC is a recipient of the DDCF Clinical Scientist Development Award. This work was supported by the Doris Duke Charitable Foundation, Grant #2013095. The project described was supported by Baylor College of Medicine IDDRC Grant Number 1 U54 HD083092 from the Eunice Kennedy Shriver National Institute of Child Health & Human Development. The content is solely the responsibility of the authors and does not necessarily represent the official views of the Eunice Kennedy Shriver National Institute of Child Health & Human Development or the National Institutes of Health".

References

1. Wright, N.C., et al.: The recent prevalence of osteoporosis and low bone mass in the United States based on bone mineral density at the femoral neck or lumbar spine. J. Bone Miner. Res. **29**, 2520–2526 (2014)
2. Spranger, J., Brill, P., Nishimura, G., Superti-Furga, A., Unger, S.: Bone Dysplasias An Atlas of Genetic Disorders of Skeletal Development. Oxford University Press, New York (2012)
3. Krakow, D., Rimoin, D.L.: The skeletal dysplasias. Genet. Med. **12**, 327–341 (2010)
4. International nomenclature of constitutional diseases of bones. Ann. Radiol. (Paris). **13**, 455–464 (1970)
5. Nomenclature for the constitutional (intrinsic) diseases of bone. Radiology. **99**, 699–702 (1971)
6. A nomenclature for constitutional (intrinsic) diseases of bones. J. Pediatr. **78**, 177–179 (1971)
7. McKusick, V.A., Scott, C.I.: A nomenclature for constitutional disorders of bone. J. Bone Joint Surg. Am. **53**, 978–986 (1971)
8. Bonafe, L., et al.: Nosology and classification of genetic skeletal disorders: 2015 revision. Am. J. Med. Genet. A. **167**, 2869–2892 (2015)
9. Yang, Y., et al.: Clinical whole-exome sequencing for the diagnosis of mendelian disorders. N. Engl. J. Med. **369**, 1502–1511 (2013)
10. Yang, Y., et al.: Molecular findings among patients referred for clinical whole-exome sequencing. JAMA. **312**, 1870–1879 (2014)
11. Sule, G., et al.: Next-generation sequencing for disorders of low and high bone mineral density. Osteoporos. Int. **24**, 2253–2259 (2013)
12. Kelley, R.I., et al.: Abnormal sterol metabolism in patients with Conradi-Hunermann-Happle syndrome and sporadic lethal chondrodysplasia punctata. Am. J. Med. Genet. **83**, 213–219 (1999)
13. Cole, D.E., et al.: Increased serum pyridoxal-5′-phosphate in pseudohypophosphatasia. N. Engl. J. Med. **314**, 992–993 (1986)
14. Wenstrup, R.J., Willing, M.C., Starman, B.J., Byers, P.H.: Distinct biochemical phenotypes predict clinical severity in nonlethal variants of osteogenesis imperfecta. Am. J. Hum. Genet. **46**, 975–982 (1990)
15. Glorieux, F.H., et al.: Osteogenesis imperfecta type VI: a form of brittle bone disease with a mineralization defect. J. Bone Miner. Res. **17**, 30–38 (2002)
16. Dogba, M.J., et al.: The impact of severe osteogenesis imperfecta on the lives of young patients and their parents - a qualitative analysis. BMC Pediatr. **13**, 153 (2013)
17. Graungaard, A.H., Skov, L.: Why do we need a diagnosis? A qualitative study of parents' experiences, coping and needs, when the newborn child is severely disabled. Child Care Health Dev. **33**, 296–307 (2007)
18. Lenhard, W., Breitenbach, E., Ebert, H., Schindelhauer-Deutscher, H.J., Henn, W.: Psychological benefit of diagnostic certainty for mothers of children with disabilities: lessons from down syndrome. Am. J. Med. Genet. A. **133A**, 170–175 (2005)

19. Bishop, N., et al.: Risedronate in children with osteogenesis imperfecta: a randomised, double-blind, placebo-controlled trial. Lancet. **382**, 1424–1432 (2013)
20. Dwan, K., Phillipi, C.A., Steiner, R.D., Basel, D.: Bisphosphonate therapy for osteogenesis imperfecta. Cochrane Database Syst. Rev. **7**, CD005088 (2014)
21. Hald, J.D., Evangelou, E., Langdahl, B.L., Ralston, S.H.: Bisphosphonates for the prevention of fractures in osteogenesis imperfecta: meta-analysis of placebo-controlled trials. J. Bone Miner. Res. **30**, 929–933 (2015)
22. Phillipi, C.A., Remmington, T., Steiner, R.D.: Bisphosphonate therapy for osteogenesis imperfecta. Cochrane Database Syst. Rev. CD005088 (2008)
23. Rijks, E.B., et al.: Efficacy and safety of bisphosphonate therapy in children with osteogenesis imperfecta: a systematic review. Horm. Res. Paediatr. **84**(1), 26–42 (2015)
24. Hendriksz, C.J., et al.: Efficacy and safety of enzyme replacement therapy with BMN 110 (elosulfase alfa) for Morquio a syndrome (mucopolysaccharidosis IVA): a phase 3 randomised placebo-controlled study. J. Inherit. Metab. Dis. **37**, 979–990 (2014)
25. Whyte, M.P., et al.: Enzyme-replacement therapy in life-threatening hypophosphatasia. N. Engl. J. Med. **366**, 904–913 (2012)
26. Hunter, A.G.: Some psychosocial aspects of nonlethal chondrodysplasias: I assessment using a life-styles questionnaire. Am. J. Med. Genet. **78**, 1–8 (1998)
27. Krakow, D., Lachman, R.S., Rimoin, D.L.: Guidelines for the prenatal diagnosis of fetal skeletal dysplasias. Genet. Med. **11**, 127–133 (2009)

Diagnosing Hereditary Cancer Susceptibility Through Multigene Panel Testing

Holly LaDuca, Shuwei Li, A.J. Stuenkel, Virginia Speare, Jill S. Dolinsky, and Elizabeth C. Chao

Abstract Hereditary cancer diagnostics has considerably evolved with the clinical availability of multigene hereditary cancer panels. Over the past few years, multigene hereditary cancer panels have contributed to a growing number of diagnoses of hereditary cancer syndromes, including patients who would likely have been missed with a traditional testing approach. While panels are largely based on next generation sequencing (NGS), panel design is not always straightforward as there are a number of factors that need to be considered to correctly and reliably diagnose hereditary cancer syndromes. In this chapter, assay design and the interpretation/reporting of multigene panel results are reviewed from the perspective of a commercial genetic testing laboratory. Key observations in multigene panel cohorts are also presented, including the identification of atypical and expanding phenotypes, carriers of pathogenic variants in moderate penetrance genes, and individuals harboring pathogenic variants in multiple cancer susceptibility genes. Such observations have highlighted the need for data sharing and collaborative efforts, which is also discussed.

Keywords Hereditary cancer • Multigene panel • Moderate penetrance • Atypical phenotype • Data sharing

Abbreviations

ACMG	American College of Medical Genetics and Genomics
ASCO	American Society of Clinical Oncology
CAP	College of American Pathologists

H. LaDuca (✉) • S. Li • A.J. Stuenkel • V. Speare • J.S. Dolinsky
Department of Clinical Diagnostics, Ambry Genetics, Inc.,
15 Argonaut, Aliso Viejo, CA 92656, USA
e-mail: hladuca@ambrygen.com

E.C. Chao
Department of Clinical Diagnostics, Ambry Genetics, Inc.,
15 Argonaut, Aliso Viejo, CA 92656, USA

Department of Pediatrics, Division of Genetics and Genomics, University of California, Irvine, Irvine, CA, USA

L.-J.C. Wong (ed.), *Next Generation Sequencing Based Clinical Molecular Diagnosis of Human Genetic Disorders*, DOI 10.1007/978-3-319-56418-0_8

ClinGen	Clinical Genome Resource
CMMR-D	Constitutional mismatch repair deficiency syndrome
CRC	Colorectal cancer
ESP	Exome Sequencing Project
ExAC	Exome Aggregation Consortium
GTR	Genetic Testing Registry
IARC	International Agency for Research on Cancer
LOG	Log of likelihood ratio
NCCN	National Comprehensive Cancer Network
NGS	Next generation sequencing
PGL-PCC	Paraganglioma-pheochromocytoma

1 Introduction

The application of next generation sequencing (NGS) in hereditary cancer diagnostics led to the development of multigene hereditary cancer panel testing. Since becoming clinically available in March of 2012, multigene hereditary cancer panels have contributed to a growing number of diagnoses of hereditary cancer syndromes. For a number of patients, such diagnoses would have been missed with the traditional single gene approach to testing. Examples include the diagnosis of well-characterized cancer syndromes such as Li-Fraumeni syndrome in patients not meeting National Comprehensive Cancer Network (NCCN) *TP53* testing criteria [1], and the identification of pathogenic/likely pathogenic variants (herein collectively referred to as 'pathogenic variants') in moderate penetrance breast cancer genes such as *ATM* that were not routinely tested prior to the availability of multigene panels. Findings such as these have created much discussion in the cancer genetics community and have highlighted the need for data sharing and collaborative efforts to further characterize gene-specific cancer risks and management recommendations.

Though NGS-based testing has proven to be more efficient and cost-effective than a single-gene testing approach, assay design needs to be carefully considered and executed to yield a clinical-grade test. Furthermore, the interpretation and reporting of hereditary cancer testing results is a complex process, which relies on considerable laboratory expertise and sound policies. In this chapter, these components of multigene panel testing methods are reviewed from the perspective of a commercial genetic testing laboratory. In addition, updates are provided on clinical observations in multigene panel cohorts, testing and medical management guidelines, and data sharing and collaborative efforts.

2 Assay Design

There are a number of important considerations in NGS-based multigene panel design to ensure maximal clinical and analytical sensitivity. In the context of panels for genetically heterogeneous diseases such as cancer, clinical sensitivity, defined as the ability of a test to identify or diagnose a given disease, is largely dependent upon the selection of genes for the panel. Due to significant locus heterogeneity in the inheritance of susceptibility to common cancers, multigene cancer panels have demonstrated increased clinical sensitivity over single-gene counterparts. For example, in a study of 966 patients undergoing hereditary breast and ovarian cancer testing, Kapoor et al. reported that a pathogenic variant was identified in 7.5% of patients receiving multigene tests compared with 4% of patients receiving limited testing of *BRCA1/2* alone [2].

The American College of Medical Genetics and Genomics (ACMG) currently recommends limiting gene selection to those with sufficient evidence for a causative role in disease [3]. In the cancer genetics community there has been much debate over gene selection for hereditary cancer panels, particularly surrounding the inclusion of moderate penetrance genes in breast cancer susceptibility [4]. One guideline would suggest limiting genes to those where evidence supports a minimum of a twofold increase in hereditary risk for at least one of the cancers targeted by the panel [5]. Emerging consensus suggests that when the evidence is strong and supported by multiple studies, the gene ought to be included even if absolute risks fall into the moderate range, i.e. two to fourfold increased cancer risk. Examples include *CHEK2* and *ATM* where there is reproducible, and clear evidence [6–13] to support a role in increasing the risk of breast cancer, but information remains limited on the most appropriate medical management for patients with pathogenic variants in these genes [4, 14].

Once gene targets have been selected for inclusion on a hereditary cancer panel, the analytical sensitivity of the NGS assay, or ability of the assay to identify variants in the genes analyzed, must be evaluated in the context of the genes being analyzed. Key components of multigene panel methodology include (1) complete coverage of coding exons, flanking intronic regions, and relevant regulatory regions, (2) a robust bioinformatics pipeline for alignment and variant calling/filtering, and (3) the addition of companion diagnostics for alterations with limited detection via NGS. In addition, at minimum, confirmation of pathogenic and likely pathogenic variants identified on NGS with an orthogonal sequencing methodology is recommended in most cases, or when alterations do not meet conservative quality thresholds set by individual laboratories [15, 16].

Complete sequence analysis of coding exons, flanking intronic regions and relevant regulatory regions such as 5′ untranslated regions and other non-coding loci known to harbor pathogenic variants is necessary to achieve clinical-grade sensitivity. While target enrichment methods used with most multigene panels can yield on average a high depth of coverage (100–10,000X), factors such as GC-rich, repetitive and highly homologous regions pose challenges to sequence capture, alignment, variant-calling, and analysis. One example is in Lynch syndrome, a relatively common hereditary cancer syndrome, caused by pathogenic variants of genes in the DNA mismatch repair pathway including *PMS2*. For this reason, this gene is critical to any panel which evaluates hereditary risk of colorectal, endometrial, ovarian, uroepithelial, or a variety of other rarer cancer types. However, applying traditional capture methods to generate NGS data for exons 11–15 in *PMS2* is unreliable since these exons exhibit high sequence homology. Traditional or "out-of-the-box" capture methods rely on probes and primers in this region which may not be specific to the *PMS2* gene, and in this case, variants belonging to the untranslated pseudogene could be falsely identified and reported. Additionally, there are risks for false negative results in this scenario where a true gene variant might not be identified when the mutant allele is effectively diluted by the presence of an increased number of wild-type reads generated from both gene and pseudogene regions. One solution which has been previously described is to use gene-specific primers to amplify these exons using long-range PCR [17, 18] prior to sequencing by NGS. In this way, one can be certain that analyzed sequence is not from pseudogene regions. Additional approaches include custom bioinformatics based on the premise that one expects standard capture and NGS data to result from four rather than two alleles and concomitant adjustment of thresholds and genotype calculations [19]. For other regions recalcitrant to either unique capture or NGS analysis, or where coverage is insufficient for reliable heterozygous variant detection, custom sequence-specific PCR amplification followed by Sanger sequencing is typically performed.

Gross deletions and duplications are not routinely detected through standard NGS analysis protocols. As such, a companion methodology, such as targeted chromosomal microarray, should be included with NGS cancer panels to detect gross deletions and duplications. An analysis of >20,000 cancer panel cases revealed that 7% of pathogenic variants detected were gross deletions/duplications [20], demonstrating the importance of including deletion/duplication analysis in hereditary cancer diagnostics. Deletions and duplications can also be analyzed from NGS data using normalized depth of coverage and paired-end mapping [21, 22]. One might suggest at first glance that this method is more time and cost-effective than running a companion array, however, the sensitivity and specificity of this bioinformatics analysis are highly dependent of the quality of NGS data and the availability of a large pool of matched controls. Because these costs are difficult to quantify, it is challenging to directly compare these approaches and the efficiency of any one approach may vary between laboratories depending on test volume.

Since initially offering multigene hereditary cancer panels in 2012, our laboratory has made several gene additions/removals following both judicial and scientific advances. One of the first modifications to our panels involved the addition of *BRCA1/2* genes to relevant panels following the US Supreme Court ruling that naturally occurring DNA is not patent-eligible [23]. A subsequent change was the removal of *STK11* from our hereditary breast cancer panels as it was contributing to VUS burden without increasing diagnostic yield. Retrospective review of internal data from single- and multigene panel testing revealed that in the absence of other features of Peutz-Jeghers syndrome, breast cancer is not an indication for *STK11* testing. More recently, three colorectal cancer/polyposis susceptibility genes, *GREM1*, *POLE*, and *POLD1*, were added to our hereditary colorectal cancer and expanded cancer panels based on an increasing amount of evidence supporting their role in colorectal cancer/polyposis [24–26]. These examples highlight the need for laboratories to routinely re-evaluate multigene panel content based on internal data and medical literature.

3 Results Interpretation and Reporting

The interpretation of multigene cancer panel results should ideally occur at multiple levels, beginning at the laboratory level in the context of the specific gene variant(s) detected and associated disease(s) diagnosed. After results are reported, a secondary interpretation should occur at the clinical level in the context of the patient's phenotype leading to an appropriate plan for medical management based on results. Over the past few years, efforts have been made to standardize the interpretation and reporting of molecular results. In 2013, the Working Group of the ACMG Laboratory Quality Assurance Committee proposed standards for laboratories providing clinical next-generation sequencing (NGS) services [3]. Within these guidelines, authors include recommendations related to the reporting of NGS results, primarily focusing on standards for interpretation and classification of variants, the format in which findings are designated and presented, the inclusion of assay performance technical details, and reporting of incidental findings. More recently, the ACMG and the Association of Molecular Pathology published a joint guideline on the interpretation of sequence variants, aimed at assessing a variants' pathogenicity [15]. While these documents provide a necessary framework for variant interpretation and reporting, there are many complex case-, gene-, and alteration-specific factors impacting results interpretation and reporting that remain to be addressed. In this section, we provide examples of such factors encountered in our experience.

3.1 Laboratory Approach to Results Interpretation and Reporting

3.1.1 Variant Interpretation

Results interpretation begins with an assessment of each specific variant detected on testing. The variant interpretation criteria proposed by the ACMG are based on a five-tier variant classification algorithm using the following terms to indicate the likelihood of association with disease- pathogenic, likely pathogenic, variant of unknown significance, likely benign, and benign- where "likely" refers to a > 90% likelihood of a variant being disease-causing or benign [15]. These criteria utilize multiple lines of weighted evidence to classify variants, including but not limited to alteration type and location, functional studies, phenotype data, population frequency data, co-segregation data, and in silico prediction models. A similar five-tier variant classification algorithm has been developed and utilized by Ambry Genetics [5, 27] which incorporates lines of evidence assessing functional impact on the protein, fitness of the amino acid or nucleotide position, and pathogenicity of the variant. When assessing the pathogenicity of a variant, it is necessary to review all data related to co-occurrence, segregation, phenotype, and case-control analysis, whether it be from published literature, internal data, or a combination of both.

Table 1 Results interpretation based on variant classification

Alteration Type	Interpretation	Medical management	Family member testing
Pathogenic Mutation	Sufficient evidence to classify as pathogenic (capable of causing disease)	Appropriate changes in medical management (i.e. high risk surveillance) recommended	Targeted testing recommended
Variant, Likely Pathogenic (VLP)	Strong evidence in favor of pathogenicity	Appropriate changes in medical management (i.e. high risk surveillance) recommended	Targeted testing recommended
Variant, Unknown Significance (VUS)	Limited and/or conflicting evidence regarding pathogenicity	Based on personal and family clinical histories	Targeted testing of informative family members recommended to collect co-segregation data
Variant, Likely Benign (VLB)	Strong evidence against pathogenicity	Based on personal and family clinical histories	Not recommended
Benign	Very strong evidence against pathogenicity	Based on personal and family clinical histories	Not recommended

Under this algorithm, both pathogenic mutations and likely pathogenic variants are interpreted as clinically actionable. Interpretations for each result type are further described in Table 1. Of note, this algorithm is based on more stringent thresholds proposed by the International Agency for Research on Cancer (IARC), where "likely" refers to a > 95% confidence of a variant being disease-causing or benign [28].

Co-segregation analysis directly measures the association of the variant with disease phenotype in pedigrees. It is a robust method as it is not susceptible to uncertainties in allele frequency, population stratification or selection bias. This approach borrows the idea of genetic linkage analysis to compare the likelihood of observed genotype/phenotype data in pedigrees under the linkage model (assuming the variant is disease causing) with the likelihood under the neutral model (the variant segregates randomly with the disease). The difference between linkage and co-segregation analyses is that linkage analysis aims to identify a marker to track the gene of interest whereas co-segregation analysis is interested in the segregation of the variant itself [29]. As such, the likelihood ratio of co-segregation could be derived similarly as we compute the linkage LOD score (log of likelihood ratio), assuming the recombination fraction is zero and correcting for the fact that the proband is known to carry the variant. Conventionally, a LOD score of 3, which implies a 1000:1 chance that variant is disease causing versus neutral, represents strong evidence of disease association. The absence of co-segregation, on the other hand, provides evidence against pathogenicity.

Several challenges have emerged in interpreting alterations identified on multigene cancer panel testing, such as interpretation of co-segregation data for moderate penetrance genes and limited population frequency information for ethnic minority groups in the U.S. For most hereditary cancer conditions, even if one has the disease predisposing genotype, the individual will not always be affected, which is referred to as incomplete or moderate penetrance. For genetic diseases exhibiting complete penetrance under the autosomal dominant inheritance model, 10 informative meioses (for example a pedigree with one unaffected and one affected parent, five unaffected wild-type offspring and five affected offspring carrying the variant) would reach a likelihood ratio of 1023:1 (LOD score of 3). The power of the co-segregation method decreases dramatically for genes with incomplete penetrance since a high LOD score would hardly be obtained in reality without a large number of informative meioses. For example, for autosomal dominant diseases exhibiting 80% penetrance, the same size pedigree (two parents, four affected offspring and six unaffected offspring) would end up with a likelihood ratio of 69:1 (LOD score of 1.8, Fig. 1). If we assume a penetrance rate of 50%, the likelihood ratio would be 5:1 (LOD score of 0.7). Therefore, it is usually more difficult to confidently classify variants with incomplete penetrance solely based on co-segregation analysis.

Practically, multiple algorithms have been developed to quantify the likelihood of co-segregation; however each algorithm considers penetrance to a varying degree. Petersen et al. developed a Bayesian method to assess the causality of variant based on affected individuals [30]. To overcome its limitation of allowing different risks (such as age and gender), Thompson et al. introduced a more general model based on full pedigree likelihood, which could incorporate a constant incomplete penetrance parameter or age-dependent penetrance model by specifying liability classes with appropriate density or accumulative distribution function [31]. More recently, Mohammadi et al. developed an algorithm using precise age of onset information [32]. The LOD score is usually computed using statistical tools for linkage analysis in pedigrees with necessary corrections. Now in the age of whole exome/genome sequencing, another new application of the "old" likelihood ratio method is genome-wide linkage analysis, which can add extra evidence in addition to the commonly used filtering approaches in family based analysis [33].

Another challenge in variant interpretation has been limited variant frequency data in non-Caucasian populations. This substantially impacts the risk of an inconclusive result in genetic testing for a patient belonging to one or more ethnic minority groups in the U.S. In an assessment of VUS rates from Ambry Genetics' internal multigene panel data, African Americans, Hispanics, and Asians consistently exhibited higher VUS rates than Caucasians and Ashkenazi Jews (Table 2, unpublished internal data). While similar reports exist for various single-gene assays, the consequence is greater for multigene panel assays due to the increased number of genes analyzed and therefore increased number of alterations detected. A listing of the genes included on each multigene panel at our commercial laboratory can be found in Table 3.

For a number of years, the largest population frequency databases included 1000 Genomes and the Exome Sequencing Project (ESP) [34–36], comprised of data from 2577 and 6503 samples, respectively (Table 4). In 2014, the Exome Aggregation

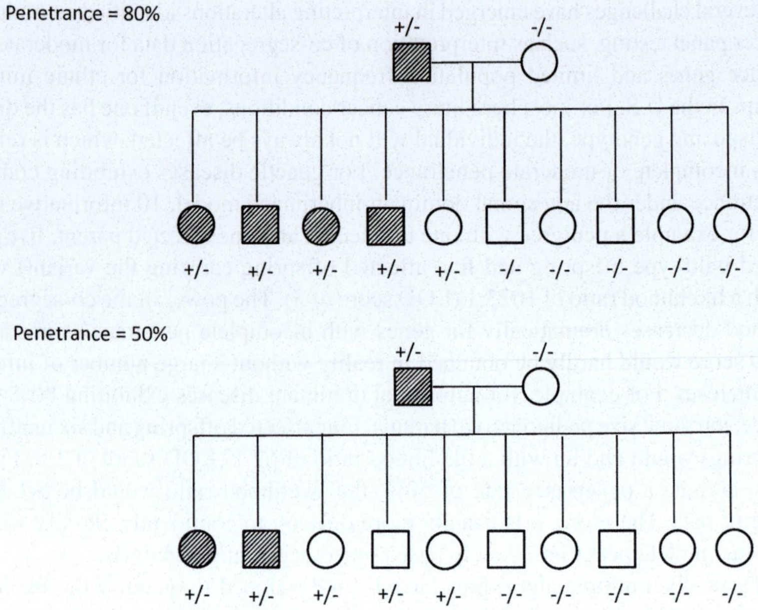

Fig. 1 Effect of disease penetrance on LOD score. For an autosomal dominant disease that is 80%-penetrant, co-segregation results would result in a likelihood ratio of 69:1 (LOD score of 1.8, top pedigree). For the same family, the likelihood ratio would be 5:1 (LOD score of 0.7) if the disease was 50%-penetrant (bottom pedigree)

Table 2 VUS rate by ethnicity and multigene panel type

Multigene panel test	# of Genes	% VUS rate by ethnicity Caucasian (n = 90,224)	African American/ Black (n = 8544)	Ashkenazi Jewish (n = 8609)	Asian (n = 5656)	Hispanic (n = 8055)
BRCAplus	6	4.2	8.7	2.3	13.1	7.8
GYNplus	13	10.4	16.6	8.6	22.4	12.6
BreastNext	17	20.5	33.6	17.3	39.6	24.8
OvaNext	24	25.9	40.0	22.2	48.1	32.8
ColoNext	17	14.8	22.7	12.4	32.4	25.1
CancerNext	32	29.5	45.5	25.8	52.1	35.3
PancNext	13	17.0	35.6	15.2	29.8	21.3
RenalNext	19	16.2	31.5	17.7	35.0	18.3
PGLNext	12	10.6	20.2	10.0	25.0	14.0
CancerNext-Expanded	49	38.2	54.7	29.7	65.1	46.7

Initially presented in: Panos, L. et al., American Society of Human Genetics, Baltimore, MD, 2015 [abstract]. Data has since been updated and is current through June 2016.

Table 3 Multigene panel gene content

	Multigene panel test name									
Gene	BRCAplus	GYNplus	PGLNext	PancNext	BreastNext	ColoNext	RenalNext	OvaNext	CancerNext	CancerNext-Expanded
APC				x		x			x	x
ATM				x	x			x	x	x
BAP1							x			x
BARD1					x			x	x	x
BRCA1	x	x		x	x			x	x	x
BRCA2	x	x		x	x			x	x	x
BRIP1		x			x			x	x	x
BMPR1A						x			x	x
CDH1	x				x	x		x	x	x
CDK4									x	x
CDKN2A				x					x	x
CHEK2					x	x		x	x	x
EPCAM		x		x		x	x	x	x	x
FH			x				x			x
FLCN							x			x
GREM1						x			x	x
MAX			x							x
MEN1			x							x
MET							x			x
MITF							x			x
MLH1		x		x		x	x	x	x	x
MRE11A					x			x	x	x
MSH2		x		x		x	x	x	x	x

(continued)

Table 3 (continued)

	Multigene panel test name									
MSH6		x		x		x	x	x	x	x
MUTYH					x	x		x	x	x
NBN					x			x	x	x
NF1			x		x			x	x	x
PALB2	x	x		x	x			x	x	x
PMS2		x		x		x	x	x	x	x
POLD1						x			x	x
POLE						x			x	x
PTEN	x	x			x	x	x	x	x	x
RAD50					x			x	x	x
RAD51C		x			x			x	x	x
RAD51D		x			x			x	x	x
RET			x							x
SDHA			x				x			x
SDHAF2			x							x
SDHB			x				x			x
SDHC			x				x			x
SDHD			x				x			x
SMAD4						x			x	x
SMARCA4								x	x	x
STK11				x		x		x	x	x
TMEM127			x							x
TP53	x	x		x	x	x	x	x	x	x
TSC1							x			x
TSC2							x			x
VHL			x				x			x
Gene Count	6	13	12	13	17	17	19	24	32	49

Table 4 Ethnic distribution of population frequency database samples

Database	Total Samples	European/European American	Ashkenazi Jewish	African/African American	Latino	East Asian	South Asian	Other
ESP	6503	4300	n/a	2203	n/a	n/a	n/a	n/a
1000 Genomes	2504	503	n/a	661	347	504	489	n/a
ExAC	60,706[a]	36,677[b]	n/a	5203	5789	4327	8256	454
GnomAD[c] – exomes	126,216	67,418	4935	8685	17,856	8674	15,450	3198
GnomAD[c] – genomes	15,136	9044	146	4257	381	798	n/a	510

[a]Includes 1851 samples from 1000 Genomes and 3936 samples from the ESP
[b]Includes 3307 Finnish patients and 33,370 Non-Finnish patients
[c]GnomAD is the second release of ExAC

Consortium (ExAC) released sequence data for over 60,000 unrelated samples, including those from the ESP, 1000 Genomes Project, and others [37]. This data will serve as a tremendous resource for variant assessment, particularly for individuals belonging to other ethnic groups. One limitation to the ExAC database specifically related to cancer is that the dataset includes 7601 samples from cancer patients. As such, the complete dataset cannot be reliably used as population data for cancer susceptibility genes. In March of 2016, ExAC released an update to the database to allow users to view frequency data excluding cancer samples that allows for ExAC to be reliably utilized in the interpretation of variants in cancer susceptibility genes [38]. Recently, this group released a larger dataset comprised of both exome and whole-genome sequencing data and is now referred to as the Genome Aggregation Database (GnomAD) [39].

3.1.2 Gene-Specific Considerations

Sparked by the emergence of lower-cost massively parallel sequencing techniques, tremendous progress in the characterization of cancer susceptibility genes has been realized in recent years. The application of NGS in high-risk patients/families has resulted in the identification of numerous genes with novel hereditary cancer implications, as well as associations with new tumor types for known predisposition genes. Such research success stories include the identification of *RAD51C* and *RAD51D* pathogenic variants as causes of hereditary ovarian cancer [40, 41] and the establishment of *MAX* as a paraganglioma-pheochromocytoma (PGL-PCC) predisposition gene [42]. Although critical for the advancement of the field overall, each of these discoveries further complicates the testing and results interpretation landscape as the lists of characterized susceptibility genes, associated tumor spectra, and penetrance continuously evolve. As such, the information provided by diagnostic laboratories on test reports is critical to avoid clinical misinterpretation and/or medical management decisions based on obsolete data.

First, laboratories must clearly indicate the primary tumor(s) correlated with each gene. For proper medical management, it is critical that an individual knows if they carry a pathogenic variant in a gene conferring increased colorectal cancer risk (e.g. *APC, MLH1,* and *POLD1*), versus a gene linked primarily to breast cancer risk (e.g. *BRCA1, ATM,* and *PALB2*). Often, an identified genetic change is expected to increase risk for multiple tumor types, such as (among others) elevated thyroid, endometrial, and breast cancer risks due to pathogenic variants in the *PTEN* gene. Laboratories must also investigate available evidence to determine which gene-specific cancer associations are sufficiently supported for report inclusion and which have been proposed based on limited data. As hereditary cancer predisposition is an area of aggressive research, laboratories are expected to have in place a mechanism for routine review of available evidence and modification of report content and/or policies when appropriate. In a recent example, data from a large collaborative effort revealed that pathogenic variants in the *PALB2* gene carry a cumulative female breast cancer risk (by age 70) of 33–58%, which is higher than

Table 5 Recessive phenotypes associated with biallelic pathogenic variants in cancer predisposition genes

Gene	Biallelic disease	References
ATM	Ataxia Telangiectasia	[46]
BRIP1	Fanconi Anemia (type J)	[47–49]
CHEK2	Stronger breast cancer phenotype	[50]
MRE11A	Ataxia Telangiectasia-like Disorder	[51]
NBN	Nijmegen Breakage Disorder	[52–54]
RAD51C	Fanconi Anemia (type O)	[55]
BRCA2	Fanconi Anemia (type D1)	[56]
PALB2	Fanconi Anemia (type N)	[57, 58]
MLH1, MSH2, MSH6, PMS2	Constitutional Mismatch Repair-Deficiency Syndrome (CMMR-D)	[59]
FH	Fumarase deficiency	[60, 61]
SDHA	Mitochondrial complex II deficiency (Leigh syndrome)	[62]

previously estimated [43]. Following this report, timely review and evaluation of the new penetrance data for possible results report inclusion is expected by clinical laboratories offering *PALB2* analysis.

Secondly, laboratories must emphasize that substantial differences exist in the magnitude of cancer risk conferred by each predisposition gene. For example, *CHEK2* and *APC* are two genes implicated in hereditary colorectal cancer (CRC). An individual carrying a germline pathogenic variant in the *CHEK2* gene has approximately twice the risk of developing CRC in his/her lifetime compared to the general population [44]. There are currently no recommendations for intensive gastrointestinal surveillance or consideration of prophylactic surgical interventions for carriers of *CHEK2* pathogenic variants. Comparatively, the *APC* gene is classically characterized by a lifetime CRC risk approaching 100% in the absence of screening and/or surgical interventions [45]. Both of these genes elevate lifetime CRC risk; however, without including the level of risk associated with each, clinicians would be missing vital information to guide medical management/recommendations and the patient decision-making process.

As a final point, laboratories must consider that many of the genes known to cause dominantly-inherited cancer predisposition can also result in a recessive phenotype when an individual carries two pathogenic variants in *trans* (Table 5). These recessive conditions are often severe with earlier-onset and more complex phenotypes compared to the associated dominantly-inherited condition. For example, monoallelic pathogenic variants in the *MLH1* gene underlie autosomal dominant Lynch syndrome, characterized by adult-onset gastrointestinal and gynecologic malignancies, whereas biallelic pathogenic variants in *MLH1* result in constitutional mismatch repair deficiency (CMMR-D) syndrome, a rare autosomal recessive disorder characterized by café-au-lait macules and childhood-onset hematologic and brain cancers [63]. When reporting results in these genes, a complete interpretation for an individual with one pathogenic variant would include a positive diagnosis of the dominant cancer syndrome as well as a carrier designation for the associated recessive condition. In addition, for patients found to carry a single pathogenic variant in a gene with a characterized recessive phenotype, carrier risks should be clearly explained on results reports to aid in post-test counseling/discussion and reproductive decision making.

The inclusion of gene-specific risk information, both in terms of expected phenotype (i.e. tumor spectrum), penetrance (i.e. lifetime risks), and when applicable, reproductive risks, is a critical aspect of a responsible laboratory reporting system for hereditary cancer testing. However, the approach to results interpretation is further improved when alteration-specific information is incorporated.

3.1.3 Allele-Specific Considerations

As research in the field of hereditary cancer testing progresses, it is apparent that not all disease-causing alleles in the same gene confer comparable levels of risk. In some cases, specific alleles have been associated with significantly increased risk compared to what is generally accepted for the respective gene. In *ATM*, a gene associated with moderate risk for both female breast and pancreatic cancers, nearly all pathogenic variants are estimated to approximately double lifetime breast cancer risk, but a small group of pathogenic missense variants impacting the 3′ functional domains have been correlated with significantly higher risks [64–66]. The mechanism of pathogenicity for these higher-risk *ATM* alleles (dominant negative) differs and thereby confers an increased level of breast cancer risk for females comparable to those seen with *BRCA2* pathogenic variants. For women receiving a positive *ATM* report, awareness of these allele-specific risk differences can alter disease screening, prevention, and reproductive choices. Conversely, there are many documented examples of lower-risk alleles in classically highly-penetrant genes. In *BRCA1*, a high-risk breast and ovarian cancer susceptibility gene, a recurrent arginine to tryptophan amino acid substitution at codon 1699 (p.R1699W) has been reported as a typical pathogenic variant expected to confer risks of approximately 65% for female breast cancer and 40% for ovarian cancer by age 70 [67]. Interestingly, a different amino acid change impacting the same residue (arginine to glutamine, p.R1699Q) has been correlated with a significantly lower level of risk: 24% for breast <u>or</u> ovarian cancer by the same age of 70 [67]. Regardless of the scenario, the inclusion of allele-specific risk information on results reports, when available, provides more accurate and personalized information to patients and their families.

3.1.4 Case-Specific Considerations

The laboratory results interpretation process would not be complete without the review of provided clinical information for each case. Clinical data can be powerful in highlighting cases requiring additional analysis, confirmation, or report modification. When reviewing results from multigene cancer tests, it's not uncommon to encounter molecular results where there is a suspicion of discrepancy with previous testing data or reported clinical details. In these instances, the clinical and molecular data must be reconciled before results can be issued. Often, follow-up with the ordering healthcare provider and/or other diagnostic laboratories is required to complete the phenotypic and previous testing picture. Once resolved, any difference between current and previous testing results can be clearly addressed on the new clinical report.

One of the more complex scenarios to arise as a result of NGS panel adoption is the more frequent detection of "low-level" alterations. These alterations are present in the analyzed DNA, but at a level significantly lower (i.e. <25%) than the ratio of mutant to wild-type alleles expected for a heterozygous germline alteration. When identified in a cancer predisposition gene a low-level finding is most commonly the result of either (1) a *de novo* alteration, which is mosaic in the patient analyzed, (2) a somatic alteration detected due to tumor contamination in the provided specimen, most commonly believed to be of hematologic origin or (3) a technical artifact of preferential allele-specific amplification or capture. Application of a secondary laboratory methodology for confirmation paired with a thorough investigation of the patient's clinical history will often suggest a likely cause for low-level findings. However, as additional testing is almost always required for confirmation of the underlying etiology, a detailed explanation along with clinical significance should be included on results reports and communicated to the ordering clinician to avoid misinterpretation and ensure appropriate follow-up.

3.2 Challenges of Clinical Result Interpretation

When a clinical laboratory is not directly affiliated with the clinic or physician who saw the patient, interpreting a genomic variant with clinical context becomes more challenging. Laboratorians must then rely solely on the clinical information provided with the test order or invest significant resources in re-contacting the clinician and obtaining additional information. Although the ACMG, the College of American Pathologists (CAP) and other professional guidelines strongly support the provision of clinical history along with test requisition [68], compliance remains limited. In our experience, nearly 11% provide little to no clinical information; this number grows even more when tests are requisitioned through reference or send-out laboratories where the test order becomes one more step removed from the patient encounter.

Poor match of a specific patient's result to their clinical presentation can be a first clue to a lab error, including an inaccurate result due to a limitation of the methodology such as allele drop out where the presence of a rare polymorphism may affect the allelic balance in sequence data. Recurrent mismatch of genotype and phenotype data might suggest that our current understanding of a particular condition or a gene's role in that condition may be limited in scope. This can be common due to the ascertainment bias that often plagues initial reports of a gene involved in a particular heritable condition. It could also suggest additional phenotypic features resulting from perturbation of that gene. It is important for the laboratorian to not only be aware of this information but to store it in a query-able and retrievable format for future analysis and development of testing as well as future research of penetrance and expressivity.

As noted above, the presence of an active malignancy or cellular dysplasia affecting the hematopoetic lineages can interfere with genetic testing, most notably by the identification of somatic alterations in blood which can often be mistaken as being germline. One notable example is that of an otherwise healthy 72-year-old female with breast cancer in which biallelic (*trans*) truncating *ATM* variants were identified through multigene panel testing. This result is consistent with a diagnosis of Ataxia-Telangiectasia (A-T), a severe, neurodegenerative condition typically leaving an individual wheelchair bound by age 10. Although later-onset cases have been reported, none have been diagnosed in the seventh decade of life [69–72]. Clinical correlation made this genetic result unlikely to be germline and prompted further hematologic work-up in this patient to identify the source of these alterations which were likely somatic and confined to dysplastic or malignant cells present in the tested blood sample. Appropriate clinical correlation prevented the misdiagnosis of this patient with A-T and the associated elevated risks of breast and pancreatic risks due to a germline pathogenic variant in this gene.

4 Observations from Multigene Panel Cohorts

Over the past few years, laboratories and clinics have been accumulating genotype and phenotype data from patients undergoing multigene panel testing. A number of groups have published or presented on retrospective multigene panel cohorts, and several common observations have been noted including atypical/expanding phenotypes for well-characterized conditions and the identification and characterization of patients with pathogenic variants in moderate penetrance genes. In addition, reports of identifying individuals with pathogenic variants in multiple genes have emerged.

4.1 Diagnosis of Hereditary Cancer Predisposition in Patients Not Meeting Testing Criteria/Expanding Phenotypes

As multigene hereditary cancer testing panel data abounds, the spectra of patients who test positive for pathogenic variants in genes associated with known genetic syndromes is expanding, with reports of patients with pathogenic variants in well-characterized genes continuing to emerge. For example, there have now been multiple reports of patients with pathogenic *TP53* variants who do not meet clinical

Table 6 Pathogenic variant carriers not meeting gene-specific testing criteria from a multigene panel cohort (unpublished data, presented at scientific meetings[a–d])

Gene	Hereditary syndrome	Results [% positive (total cohort tested)]	Clinical testing criteria	Pathogenic variant carriers not meeting clinical testing criteria
TP53[a]	Li-Fraumeni	0.31% (n = 22,226)	NCCN [1]	31/66 (47%)
CDH1[b]	Hereditary diffuse gastric cancer	0.07% (n = 19,218)	International Gastric Cancer Linkage Consortium (IGCLC) [82]	10/13 (77%)[e]
PTEN[c]	Cowden/ *PTEN* Hamartoma Tumor	0.15% (n = 14,897)	NCCN [1]	17/23 (74%)
NF1[d]	Neurofibromatosis	0.39% (n = 2597)	NIH diagnostic criteria for NF1 [83]	4/10 (40%)

[a]Rana, H. et al., The American Society of Human Genetics Annual Meeting, Baltimore, MD, 2015 [abstract]

[b]Sturgeon D, et al. The Collaborative Group of the Americas on Inherited Colorectal Cancer (CGA) 18th Annual Meeting, New Orleans, LA, 2014 [presentation]

[c]Weltmer et al. The American Society of Human Genetics Annual Meeting, San Diego, CA, 2014 [presentation]

[d]Summerour P, et al. The National Society of Genetic Counselor's Annual Meeting, New Orleans, LA, 2014 [abstract]

[e]Five of these individuals were classified as "IGCLC-, partial phenotype" (gastric cancer >40 or lobular breast cancer present but IGCLC criteria not met) and 5 did not meet criteria at all

criteria for Li-Fraumeni syndrome, including early-onset colorectal cancer patients and those suspicious for hereditary breast/ovarian cancer [73–77]. In addition, there have been several reports of pathogenic *CDH1* variants being identified in families without diffuse gastric cancer [74, 78–80], as well as pathogenic mismatch repair gene variants in families not meeting diagnostic/testing criteria for Lynch syndrome [73, 76, 78, 81].

Review of internal cases referred for a range of hereditary cancer multigene panel tests has yielded similar observations (Table 6). It should be noted that the data represented in Table 6 is unpublished and therefore warrants further study. Furthermore, for data collected on *PTEN* and *NF1* pathogenic variant carriers, a possible bias exists in the information collected by clinicians at the time of testing, whereby they are not asking details of syndromic features that may have been carefully assessed if the syndrome had been initially suspected. Regardless, it highlights the point that individuals at increased risk for cancer and other medical conditions associated with these genes are going undiagnosed in the absence of multigene panel testing. Collectively, this data demonstrates the role of commercial laboratory in further characterizing the phenotypic spectra of various hereditary cancer syndromes and contributing to the development of testing guidelines which encompass a larger number of genes and address the wide range of phenotypes being observed in pathogenic variant carriers.

4.2 *The Role of Moderate Penetrance Genes in Hereditary Cancer Susceptibility*

The utilization of multigene panel testing has also led to increased identification of patients harboring pathogenic variants in moderate penetrance breast/ovarian cancer genes [5, 73–76, 79–81, 84]. The 'moderate penetrance' terminology is generally applied to genes conferring a two to fourfold increase in breast cancer risk, most of which are part of the Fanconi anemia/homologous repair pathway. In a study of 360 unselected ovarian cancer patients undergoing multigene panel testing (inclusive of *BRCA1/2*), 17 patients (4.7%) were identified to carry a pathogenic variant in a moderate penetrance gene (20% of total pathogenic variants detected; 73.9% of non-*BRCA1/2* pathogenic variants) [76]. In a separate study of *BRCA1/2*-negative breast and ovarian cancer patients undergoing multigene panel testing, 49 patients (5.4%) carried a pathogenic variant in a moderate penetrance gene (73.1% of total pathogenic variants detected) [81]. In larger multigene cancer panel cohorts not selected based on a breast/ovarian indication, the likelihood of detecting a pathogenic variant in a moderate penetrance gene can exceed that of detecting pathogenic variants in high penetrance genes [20].

Phenotype data collected by clinical diagnostic laboratories can help further characterize the role of moderate penetrance genes in hereditary cancer susceptibility. For example, while multiple large studies have not supported an increased risk for breast cancer with *RAD51C/D* [41, 85–90], retrospective review of 33 *RAD51C/D* pathogenic variant carriers ascertained via multigene panel testing revealed that 10 pathogenic variants (30.3%) occurred in breast cancer families with no reported ovarian cancer (unpublished internal data). Advantages of laboratory-selected phenotype and genotype data include access to larger numbers of pathogenic variant carriers with a wide range of testing indications. Limitations include selection bias based on referral of patients with clinical histories suggestive of hereditary cancer susceptibility and lack of control cohorts for comparison. Therefore, published laboratory-based multigene panel data has, for the most part, been limited to validation studies and retrospective descriptive analyses.

In a recent report on panel testing for hereditary breast cancer, Easton and colleagues reviewed the evidence surrounding breast cancer risk for genes on clinically-available multigene panel tests [4]. Authors found sufficient evidence to support increased breast cancer risk for several moderate penetrance genes including *NF1*, *PALB2*, *ATM*, *CHEK2*, and *NBN*. For *RAD51C*, *RAD51D*, and *BRIP1* authors found sufficient evidence to support an increased risk of ovarian cancer but not breast cancer. Conclusions from this review of evidence highlighted the need for large studies (population and family based) to help determine risk estimates, and supported the continued collection of multigene panel data. Further elucidation of the role of moderate penetrance genes in cancer susceptibility will involve collaborations between clinicians, and laboratorians, and researchers.

4.3 Patients Carrying Multiple Pathogenic Variants

Since a number of different genes are analyzed on multigene panel testing, the possibility exists of detecting alterations in more than one cancer susceptibility gene [5, 73, 74, 76, 80, 81]. Retrospective review of 54,658 multigene panel cases revealed that 95 (0.17%) patients carried pathogenic variants in more than one gene (unpublished internal data). The vast majority of these cases carried pathogenic variants in two genes (n = 94), and the remaining patient carried a pathogenic variant in a high penetrance gene along with pathogenic variants in three different moderate penetrance genes. For patients harboring a mono-allelic *MUTYH* pathogenic variant in addition to a pathogenic variant in a different gene, the *MUTYH* pathogenic variants was not considered as a second pathogenic variant, since *MUTYH*-associated polyposis is an autosomal recessive disorder. A number of combinations were observed in this cohort, including multiple pathogenic variants in high risk genes (e.g. *BRCA2* and *TP53*), multiple pathogenic variants in moderate penetrance genes (e.g. *ATM* and *CHEK2*), and multiple pathogenic variants in a combination of high and moderate penetrance genes (e.g. *TP53* and *CHEK2*) (Fig. 2, Table 7). An additional 29 patients carried a pathogenic variant in a high or moderate penetrance gene in combination with a moderate risk allele such as p.I1307K in *APC* or p.I157T in *CHEK2*. In cases with multiple pathogenic variants, it is difficult to assess cancer risks, as any combined effects of genes are unknown. As additional cases are accumulated, comparative studies between multiple and single pathogenic variant cases will need to be performed.

4.4 Diagnosing Hereditary Cancer Predisposition Based on NGS of Tumor DNA

The primary indication for NGS of tumor tissue is the identification of therapeutic targets. Incidentally, germline pathogenic variants may be detected through tumor testing as these are present in all tissues and analysis of tumor tissue alone cannot determine the origin of a pathogenic variant. Incidental detection of a germline pathogenic variant has clear implications for both the patient and family beyond the utility for treatment planning. Therefore, the potential for detection of a germline pathogenic variant ought to be part of patient-provider shared decision making and informed consent process [91]. Still, relatively little is known about how often a germline pathogenic variant is detected in the course of the genomic profiling of tumors. Of further concern can be qualitative differences in the bioinformatics, variant assessment, and risk reporting by a laboratory focused solely on tumor and not germline genetic testing [92]. Our commercial laboratory has recently seen an increase in the number of requests for germline genetic testing based on a genomic profiling result, including requests for multigene panel tests to confirm suspected germline pathogenic variant and assess for additional germline alterations that may

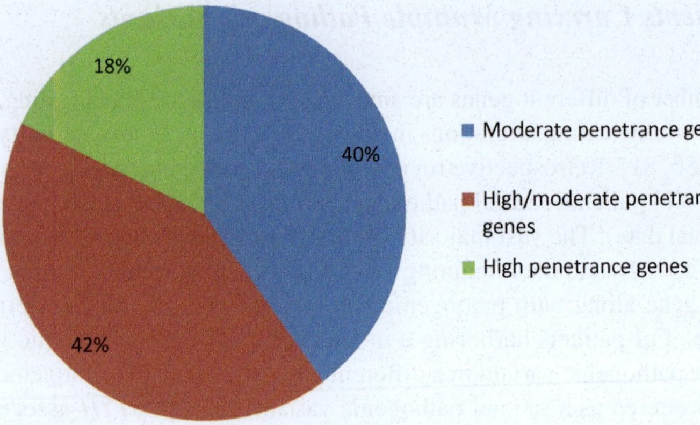

Fig. 2 Distribution of cases with multiple pathogenic variants. General results categories are summarized for 95 patients with pathogenic variants in more than one gene (unpublished internal data). Most patients either carried pathogenic variants in multiple moderate penetrance genes (n = 38) or in a combination of high and moderate penetrance genes (n = 40), with a smaller portion carrying pathogenic variants in two different high penetrance genes (n = 17). Specific pathogenic variants combinations are shown in Table 7

Table 7 Gene combinations observed in cases with multiple pathogenic variants

Two high penetrance genes	High/moderate penetrance genes	Two moderate penetrance genes
BRCA1/TP53	*BRCA1/ATM* (3)	*ATM/CHEK2* (7)
BRCA1/PMS2	*BRCA1/CHEK2* (5)	*ATM/NBN*
BRCA1/BRCA2	*BRCA1/MRE11A*	*ATM/PALB2* (2)
BRCA2/CDKN2A	*BRCA1/NBN* (2)	*ATM/BRIP1*
BRCA2/MSH6 (2)	*BRCA1/NF1*	*ATM/MRE11A*
BRCA2/PMS2	*BRCA1/RAD50*	*ATM/RAD51D*
BRCA2/SMAD4	*BRCA2/ATM* (7)	*ATM/BARD1*
BRCA2/TP53 (3)	*BRCA2/BARD1*	*ATM/RAD50* (2)
FLCN/SDHB	*BRCA2/BRIP1*	*CHEK2/BARD1* (2)
TP53/CDH1	*BRCA2/CHEK2/NBN/RAD50*	*CHEK2/BRIP1* (5)
CDKN2A/PTEN	*BRCA2/RAD51D*	*CHEK2/PALB2* (3)
MSH2/PTEN	*BRCA2/NF1*	*CHEK2/RAD50* (3)
MSH6/PMS2 (2)	*CDKN2A/BRIP1*	*CHEK2/NBN* (3)
	MLH1/CHEK2	*CHEK2/NF1*
	MSH6/ATM (2)	*PALB2/BARD1*
	MSH6/CHEK2	*PALB2/MRE11A*
	MUTYH biallelic/*CHEK2*	*PALB2/NF1*
	PMS2/RAD51C	*PALB2/RAD50*
	PMS2/CHEK2	*NBN/RAD51D*
	PTEN/ATM	
	TP53/BRIP1	
	TP53/CHEK2 (3)	
	TP53/MRE11A	
	TP53/PALB2	

Gene combinations observed in more than one patient are indicated in parentheses contribute to the patient's history. A retrospective review of laboratory cases revealed that 13% (n = 74) of pathogenic variants in 17 different cancer risk genes identified via tumor profiling were of germline origin [93]. Preliminary data from others has shown that potentially pathogenic germline variants in cancer risk genes are found in 3% to 7% of tumor profiling results [94]. It is not currently known, but is crucial to determine what percentage of these had personal or family histories consistent with the germline diagnosis. This information will be critical in establishing guidelines for analysis of germline DNA following a positive result in tumor tissue. As genomic profiling continues to provide valuable data for personalized and precise management of cancer patients, clinicians should prepare patients for the possibility of the incidental detection of inherited risk for cancer through education and counseling.

5 Testing Guidelines

Sequencing multiple genes simultaneously and returning complex results back to clinicians and ultimately to patients and their families has fundamentally changed how hereditary cancer risk is assessed and managed. Analyzing single genes in an "a la carte" approach may be sufficient to make a diagnosis when the patient's clinical history is definitive and the differential diagnosis is short. However, histories are often complex and multigene testing enables probing the entire genetic differential diagnosis with a single test. This changing paradigm has created challenges for oncologists, geneticists and other cancer care providers who must gauge the clinical value of multigene tests based on what is, for now, limited evidence of clinical utility [4]. Given the challenges that providers face in assessing the utility of multigene tests, there is a role for large groups such as the National Comprehensive Cancer Network (NCCN), comprehensive databases such as the Genetic Testing Registry (GTR) and professional organizations, such as the ACMG and the American Society of Clinical Oncology (ASCO) to provide expert opinion and well informed management guidelines.

5.1 NCCN

The NCCN is an alliance of cancer centers that was founded in 1995 with the mission to improve cancer care through providing clinical practice guidelines based on evidence and expert opinion. The guidelines encompass aspects of cancer care including screening, diagnosis, therapy and survivorship. In 2015 an overview of multigene testing was added within the guidelines for Genetic/Familial High-Risk Assessment: Breast and Ovarian [1]. Included in the content for breast and ovarian cancer prevention are recommendations for intervention with breast cancer MRI screening for carriers of pathogenic variants in *ATM*, *BRCA1*, *BRCA2*, *CDH1*,

CHEK2, *PALB2*, *PTEN*, *STK11* and *TP53*. Presumably pathogenic variants in these genes are frequently detected by a multigene test rather than costly sequential testing. The addition of *ATM, PALB2* and *CHEK2* to this list is notable since guidelines for management based on these genes had not been previously published.

5.2 GTR

The GTR is an additional resource where providers may identify multigene panel tests and the laboratories performing this type of testing. The registry is supported by the National Institutes of Health in the USA but content is contributed to by laboratories worldwide [95]. The registry is a comprehensive repository of genetic testing information that is presented in a standardized format via a publically available website. Laboratories submit required information to the database and the information is manually curated. Statements about the purpose of the test are selected by contributors from a standardized set of indications. Contributors who provide statements about clinical validity and clinical utility must include pertinent citations. A search of the GTR for "cancer panel" yielded 424 tests offered from 49 labs [96]. The GTR does not itself provide guidelines for medical management based on genetic testing information but the information provided may assist providers in selecting appropriate testing for their patients and understanding the potential benefits and limitations of testing.

5.3 ACMG

The ACMG continually addresses issues in the clinical application of NGS by publishing guidelines on topics such as the interpretation of sequence variants and the reporting of incidental findings [15, 97]. The ACMG has stated strongly that determining a genetic etiology for disease is a demonstration of the clinical utility of genetic and genomic testing [98]. Genetic/genomic testing is a broad term and multigene panels for identification of hereditary cancer risk have not been specifically evaluated by the ACMG.

5.4 ASCO

Another important resource, the ASCO is a leader in providing education in cancer genetics. In 1996 an ASCO policy statement regarding genetic testing was published and later updated in 2003 [99, 100]. The most recent statement in 2010 was published before the widespread introduction of NGS and multigene testing to the oncology marketplace [101]. Although ASCO has not provided specific practice guidelines for the use of multigene testing in the context of cancer risk assessment, the organization remains committed to the incorporation of genetic data into clinical care as demonstrated by an expert statement regarding the collection and use of cancer family histories in oncology care [102].

Commentary on the value and use of multigene testing has been provided by multiple experts [103–105]. A recurrent theme in the commentary surrounds the need for involvement of a provider trained in genetics in the testing process [4]. Clearly there is an opportunity to develop guiding principles for the responsible use of multigene testing in clinical care. Contributions to evidence-focused research from diverse stakeholders including laboratories, clinicians, professional groups, and patients will be needed to fully understand the value of this new paradigm in oncology care.

6 Data Sharing and Consortium Variant Classification/ Interpretation Efforts

NGS has allowed for the accumulation of massive amounts of genetic data which has proven to be very useful in the interpretation of individual genetic testing results. For example, large public databases with exome sequencing data generalizable to the population and partitioned by ethnicity, such as the Exome Variant Server [35], the 1000 genomes project [36] and the ExAC database [37] have vastly improved our ability to interpret variants as benign polymorphisms based on high frequency in general population cohorts. However, rare sequence variants remain difficult to classify.

By pooling data across genetics laboratories, we increase our chances of accumulating enough data on a given variant. Recently, large scale resources for sharing variant data have emerged, including ClinVar, an open resource for clinicians, researchers and labs seeking variant information [106]. ClinVar is a core component of the NIH-funded Clinical Genome Resource (ClinGen). These data sharing initiatives are helping to identify differences in variant classifications between laboratories. In addition, consensus guidelines are now available to standardize variant classification, set by groups such as the ACMG and the IARC [15, 28]. Working in silos in competition with one another is not efficient for laboratories to further our understanding of genomic variation and of its impacts on patient care. Rather, through resources such as ClinGen, laboratories now have great opportunities to work together with expert consensus groups to interpret alterations based on pooled data with considerations for interpretation guidelines. This will lead to the interpretation of variants to the best of our collective ability, allowing for the most significant impact on patient possible [107].

Data sharing initiatives are also supported by groups such as the National Society of Genetic Counselors, which has a position statement encouraging the sharing of data collected through clinical genetic and genomic testing [108]. Data sharing is even more powerful when coupled with the continued work of gene specific expert consortia which pool resources to classify variants, such as efforts in *BRCA1* and *BRCA2* classification by the Evidence-Based Network for the Interpretation of Germline Mutation Alleles (ENIGMA) [109] and efforts in the classification of mismatch repair gene alterations by the International Society for Gastrointestinal Hereditary Tumours Incorporated (InSiGHT) [110]. In recent years, with the massive amounts of data accumulating at commercial laboratories, commercial entities such as Ambry Genetics, Inc. have been invited to become an integral part of these consortia, providing large volumes of valuable data on unique rare variants. Altogether, responsible data-sharing is necessary for comprehensive, accurate and consistent variant classification by all laboratories and clinicians interpreting genetic test results, which translates into the most optimal utility of genetic testing in patient care [108, 111].

7 Conclusions

In conclusion, multigene hereditary cancer panel testing is a complex process that is continuously evolving as new susceptibility genes are identified and existing susceptibility genes are further characterized. As demonstrated in this chapter, there are many necessary considerations in developing multigene panel methodology to maximize the detection of pathogenic variants. While guidelines from professional organizations such as the ACMG have been helpful in standardizing NGS-based testing methods and results interpretation, molecular laboratories must address gene, allele, and case-specific factors when developing analytic techniques and reporting policies in order to provide clinicians and patients with the most accurate, clinically relevant, and personalized results possible.

Complexities of multigene panel testing are not limited to the laboratory setting; they carry over to clinic as well. Multigene panels have led to the identification of atypical and expanding phenotypes as well as the identification of patients with pathogenic variants in moderate penetrance genes. Collectively, multigene panel data on genes such as *TP53, CDH1, PTEN* and *NF1* that are associated with what are believed to be well-characterized syndromes bring into questioning the utility of following predefined testing criteria for cancer predisposition genes, as multigene testing will identify additional individuals with hereditary susceptibility to cancer that may have otherwise gone undiagnosed [5, 78]. It remains to be determined that the clinical utility or whether identical management guidelines are appropriate in the setting of diagnosed patients without supportive histories. Identification of a growing number of patients carrying pathogenic variants in moderate penetrance breast/ovarian genes has led to controversy over inclusion of those genes into panels as well as appropriate medical management for these patients. Analysis of genotypic-phenotypic cancer panel data through data sharing and other collaborative efforts is needed to further characterize hereditary cancer susceptibility genes, define testing criteria, and integrate management guidelines into specific patient and family specific care plans.

References

1. The NCCN Clinical Practice Guidelines in Oncology™ Genetic/Familial High-Risk Assessment: Breast and Ovarian V2.2015. In: National Comprehensive Cancer Network, Inc. http://www.nccn.org/ (2015)
2. Kapoor, N.S., Curcio, L.D., Blakemore, C., Bremner, A.K., McFarland, R., West, J.G., Banks, K.C.: Multi-Gene panel testing detects equal rates of pathogenic BRCA1/2 mutations and has a higher diagnostic yield compared to limited BRCA1/2 analysis alone in patients at risk for hereditary breast cancer. Ann. Surg. Oncol. **22**(10), 3282–3288 (2015)
3. Rehm, H.L., Bale, S.J., Bayrak-Toydemir, P., Berg, J.S., Brown, K.K., Deignan, J.L., et al.: ACMG clinical laboratory standards for next-generation sequencing. Genet. Med. Off. J. Am. Coll. Med. Genet. **15**(9), 733–747 (2013). doi:10.1038/gim.2013.92
4. Easton, D.F., Pharoah, P.D., Antoniou, A.C., Tischkowitz, M., Tavtigian, S.V., Nathanson, K.L., et al.: Gene-panel sequencing and the prediction of breast-cancer risk. N. Engl. J. Med. **372**(23), 2243–2257 (2015). doi:10.1056/NEJMsr1501341
5. Laduca, H., Stuenkel, A.J., Dolinsky, J.S., Keiles, S., Tandy, S., Pesaran, T., et al.: Utilization of multigene panels in hereditary cancer predisposition testing: analysis of more than 2,000 patients. Genet. Med. Off. J. Am. Coll. Med. Genet. (2014). doi:10.1038/gim.2014.40
6. Chek Breast Cancer Case-Control Consortium: CHEK2*1100delC and susceptibility to breast cancer: a collaborative analysis involving 10,860 breast cancer cases and 9,065 controls from 10 studies. Am. J. Hum. Genet. **74**(6), 1175–1182 (2004). doi:10.1086/421251
7. Janin, N., Andrieu, N., Ossian, K., Lauge, A., Croquette, M.F., Griscelli, C., et al.: Breast cancer risk in ataxia telangiectasia (AT) heterozygotes: haplotype study in French AT families. Br. J. Cancer. **80**(7), 1042–1045 (1999). doi:10.1038/sj.bjc.6690460
8. Kilpivaara, O., Vahteristo, P., Falck, J., Syrjakoski, K., Eerola, H., Easton, D., et al.: CHEK2 variant I157T may be associated with increased breast cancer risk. Int. J. Cancer J. Int. du Cancer. **111**(4), 543–547 (2004). doi:10.1002/ijc.20299
9. Meijers-Heijboer, H., van den Ouweland, A., Klijn, J., Wasielewski, M., de Snoo, A., Oldenburg, R., et al.: Low-penetrance susceptibility to breast cancer due to CHEK2(*)1100delC in noncarriers of BRCA1 or BRCA2 mutations. Nat. Genet. **31**(1), 55–59 (2002). doi:10.1038/ng879
10. Olsen, J.H., Hahnemann, J.M., Borresen-Dale, A.L., Tretli, S., Kleinerman, R., Sankila, R., et al.: Breast and other cancers in 1445 blood relatives of 75 Nordic patients with ataxia telangiectasia. Br. J. Cancer. **93**(2), 260–265 (2005). doi:10.1038/sj.bjc.6602658
11. Renwick, A., Thompson, D., Seal, S., Kelly, P., Chagtai, T., Ahmed, M., et al.: ATM mutations that cause ataxia-telangiectasia are breast cancer susceptibility alleles. Nat. Genet. **38**(8), 873–875 (2006). doi:10.1038/ng1837
12. Thompson, D., Duedal, S., Kirner, J., McGuffog, L., Last, J., Reiman, A., et al.: Cancer risks and mortality in heterozygous ATM mutation carriers. J. Natl. Cancer Inst. **97**(11), 813–822 (2005). doi:10.1093/jnci/dji141
13. Weischer, M., Heerfordt, I.M., Bojesen, S.E., Eigentler, T., Garbe, C., Rocken, M., et al.: CHEK2*1100delC and risk of malignant melanoma: Danish and German studies and meta-analysis. J. Invest. Dermatol. **132**(2), 299–303 (2012). doi:10.1038/jid.2011.303
14. Mauer, C.B., Pirzadeh-Miller, S.M., Robinson, L.D., Euhus, D.M.: The integration of next-generation sequencing panels in the clinical cancer genetics practice: an institutional experience. Genet. Med. Off. J. Am. Coll. Med. Genet. (2013). doi:10.1038/gim.2013.160
15. Richards, S., Aziz, N., Bale, S., Bick, D., Das, S., Gastier-Foster, J., et al.: Standards and guidelines for the interpretation of sequence variants: a joint consensus recommendation of the American College of Medical Genetics and Genomics and the Association for Molecular Pathology. Genet. Med. Off. J. Am. Coll. Med. Genet. **17**(5), 405–424 (2015). doi:10.1038/gim.2015.30
16. Mu, W., Lu, H.M., Chen, J., Li, S., Elliott, A.M.: Sanger confirmation is required to achieve optimal sensitivity and specificity in next-generation sequencing panel testing. J. Mol. Diagn. **18**(6), 923–932 (2016). doi:10.1016/j.jmoldx.2016.07.006

17. Vaughn, C.P., Robles, J., Swensen, J.J., Miller, C.E., Lyon, E., Mao, R., et al.: Clinical analysis of PMS2: mutation detection and avoidance of pseudogenes. Hum. Mutat. **31**(5), 588–593 (2010). doi:10.1002/humu.21230
18. Li, J., Dai, H., Feng, Y., Tang, J., Chen, S., Tian, X., et al.: A comprehensive strategy for accurate mutation detection of the highly homologous PMS2. J. Mol. Diagn. **17**(5), 545–553 (2015). doi:10.1016/j.jmoldx.2015.04.001
19. Invitae. Sequencing exons 12-15 of PMS2 using next-generation sequencing (NGS). http://marketing.invitae.com/acton/attachment/7098/f-0139/1/-/-/-/-/WP103-1_PMS2%20Sequencing%20NGS%20Validation%20Summary.pdf (2015). Accessed July 2015
20. Wu, W., Choudhry, H. (eds.): Next generation sequencing in cancer research, volume 2. Springer International Publishing, Switzerland (2015)
21. Duan, J., Zhang, J.G., Deng, H.W., Wang, Y.P.: Comparative studies of copy number variation detection methods for next-generation sequencing technologies. PLoS One. **8**(3), e59128 (2013). doi:10.1371/journal.pone.0059128
22. Feng, Y., Chen, D., Wang, G.L., Zhang, V.W., Wong, L.J.: Improved molecular diagnosis by the detection of exonic deletions with target gene capture and deep sequencing. Genet. Med. Off. J. Am. Coll. Med. Genet. **17**(2), 99–107 (2015). doi:10.1038/gim.2014.80
23. Association for Molecular Pathology et al. v. Myriad Genetics Inc., et al 569 U. S. ____ (2013), Supreme Court of the United States (June 13, 2013)
24. Jaeger, E., Leedham, S., Lewis, A., Segditsas, S., Becker, M., Cuadrado, P.R., et al.: Hereditary mixed polyposis syndrome is caused by a 40-kb upstream duplication that leads to increased and ectopic expression of the BMP antagonist GREM1. Nat. Genet. **44**(6), 699–703 (2012). doi:10.1038/ng.2263
25. Palles, C., Cazier, J.B., Howarth, K.M., Domingo, E., Jones, A.M., Broderick, P., et al.: Germline mutations affecting the proofreading domains of POLE and POLD1 predispose to colorectal adenomas and carcinomas. Nat. Genet. **45**(2), 136–144 (2013). doi:10.1038/ng.2503
26. Valle, L., Hernandez-Illan, E., Bellido, F., Aiza, G., Castillejo, A., Castillejo, M.I., et al.: New insights into POLE and POLD1 germline mutations in familial colorectal cancer and polyposis. Hum. Mol. Genet. **23**(13), 3506–3512 (2014). doi:10.1093/hmg/ddu058
27. Pesaran, T., Karam, R., Huether, R., Li, S., Farber-Katz, S., Chamberlin, A., et al.: Beyond DNA: an integrated and functional approach for classifying germline variants in breast cancer genes. Int. J. Breast Cancer. **2016**, 2469523 (2016). doi:10.1155/2016/2469523
28. Plon, S.E., Eccles, D.M., Easton, D., Foulkes, W.D., Genuardi, M., Greenblatt, M.S., et al.: Sequence variant classification and reporting: recommendations for improving the interpretation of cancer susceptibility genetic test results. Hum. Mutat. **29**(11), 1282–1291 (2008). doi:10.1002/humu.20880
29. Goldgar, D.E., Easton, D.F., Byrnes, G.B., Spurdle, A.B., Iversen, E.S., Greenblatt, M.S., et al.: Genetic evidence and integration of various data sources for classifying uncertain variants into a single model. Hum. Mutat. **29**(11), 1265–1272 (2008). doi:10.1002/humu.20897
30. Petersen, G.M., Parmigiani, G., Thomas, D.: Missense mutations in disease genes: a Bayesian approach to evaluate causality. Am. J. Hum. Genet. **62**(6), 1516–1524 (1998). doi:10.1086/301871
31. Thompson, D., Easton, D.F., Goldgar, D.E.: A full-likelihood method for the evaluation of causality of sequence variants from family data. Am. J. Hum. Genet. **73**(3), 652–655 (2003). doi:10.1086/378100
32. Mohammadi, L., Vreeswijk, M.P., Oldenburg, R., van den Ouweland, A., Oosterwijk, J.C., van der Hout, A.H., et al.: A simple method for co-segregation analysis to evaluate the pathogenicity of unclassified variants; BRCA1 and BRCA2 as an example. BMC Cancer. **9**, 211 (2009). doi:10.1186/1471-2407-9-211
33. Ott, J., Wang, J., Leal, S.M.: Genetic linkage analysis in the age of whole-genome sequencing. Nat. Rev. Genet. **16**(5), 275–284 (2015). doi:10.1038/nrg3908
34. 1000 Genomes Project [database on the Internet]. Available from: http://www.1000genomes.org/. Accessed 11 Nov 2016

35. Exome Variant Server, NHLBI GO Exome Sequencing Project (ESP) [database on the Internet] 2013. Available from: http://evs.gs.washington.edu/EVS/. Accessed 11 Nov 2016
36. Abecasis, G.R., Auton, A., Brooks, L.D., DePristo, M.A., Durbin, R.M., Handsaker, R.E., et al.: An integrated map of genetic variation from 1,092 human genomes. Nature. **491**(7422), 56–65 (2012). doi:10.1038/nature11632
37. Lek, M., Karczewski, K.J., Minikel, E.V., Samocha, K.E., Banks, E., Fennell, T., et al.: Analysis of protein-coding genetic variation in 60,706 humans. Nature. **536**(7616), 285–291 (2016). doi:10.1038/nature19057
38. Exome Aggregation Consortium (ExAC) [database on the Internet]. Available from: http://exac.broadinstitute.org. Accessed 11 Nov 2016
39. Genome Aggregation Database (GnomAD) [database on the Internet] 2016. Available from: http://gnomad.broadinstitute.org/. Accessed 11 Nov 2016
40. Loveday, C., Turnbull, C., Ramsay, E., Hughes, D., Ruark, E., Frankum, J.R., et al.: Germline mutations in RAD51D confer susceptibility to ovarian cancer. Nat. Genet. **43**(9), 879–882 (2011). doi:10.1038/ng.893
41. Meindl, A., Hellebrand, H., Wiek, C., Erven, V., Wappenschmidt, B., Niederacher, D., et al.: Germline mutations in breast and ovarian cancer pedigrees establish RAD51C as a human cancer susceptibility gene. Nat. Genet. **42**(5), 410–414 (2010). doi:10.1038/ng.569
42. Comino-Mendez, I., Gracia-Aznarez, F.J., Schiavi, F., Landa, I., Leandro-Garcia, L.J., Leton, R., et al.: Exome sequencing identifies MAX mutations as a cause of hereditary pheochromocytoma. Nat. Genet. **43**(7), 663–667 (2011). doi:10.1038/ng.861
43. Antoniou, A.C., Foulkes, W.D., Tischkowitz, M.: Breast-cancer risk in families with mutations in PALB2. N. Engl. J. Med. **371**(17), 1651–1652 (2014). doi:10.1056/NEJMc1410673
44. Xiang, H.P., Geng, X.P., Ge, W.W., Li, H.: Meta-analysis of CHEK2 1100delC variant and colorectal cancer susceptibility. Eur. J. Cancer. **47**(17), 2546–2551 (2011). doi:10.1016/j.ejca.2011.03.025
45. Jasperson, K.W.: Genetic testing by cancer site: colon (polyposis syndromes). Cancer J. **18**(4), 328–333 (2012). doi:10.1097/PPO.0b013e3182609300
46. Savitsky, K., Bar-Shira, A., Gilad, S., Rotman, G., Ziv, Y., Vanagaite, L., et al.: A single ataxia telangiectasia gene with a product similar to PI-3 kinase. Science. **268**(5218), 1749–1753 (1995)
47. Levitus, M., Waisfisz, Q., Godthelp, B.C., de Vries, Y., Hussain, S., Wiegant, W.W., et al.: The DNA helicase BRIP1 is defective in Fanconi anemia complementation group J. Nat. Genet. **37**(9), 934–935 (2005). doi:10.1038/ng1625
48. Levran, O., Attwooll, C., Henry, R.T., Milton, K.L., Neveling, K., Rio, P., et al.: The BRCA1-interacting helicase BRIP1 is deficient in Fanconi anemia. Nat. Genet. **37**(9), 931–933 (2005). doi:10.1038/ng1624
49. Litman, R., Peng, M., Jin, Z., Zhang, F., Zhang, J., Powell, S., et al.: BACH1 is critical for homologous recombination and appears to be the Fanconi anemia gene product FANCJ. Cancer Cell. **8**(3), 255–265 (2005). doi:10.1016/j.ccr.2005.08.004
50. Adank, M.A., Jonker, M.A., Kluijt, I., van Mil, S.E., Oldenburg, R.A., Mooi, W.J., et al.: CHEK2*1100delC homozygosity is associated with a high breast cancer risk in women. J. Med. Genet. **48**(12), 860–863 (2011). doi:10.1136/jmedgenet-2011-100380
51. Stewart, G.S., Maser, R.S., Stankovic, T., Bressan, D.A., Kaplan, M.I., Jaspers, N.G., et al.: The DNA double-strand break repair gene hMRE11 is mutated in individuals with an ataxia-telangiectasia-like disorder. Cell. **99**(6), 577–587 (1999)
52. Carney, J.P., Maser, R.S., Olivares, H., Davis, E.M., Le Beau, M., Yates 3rd, J.R., et al.: The hMre11/hRad50 protein complex and Nijmegen breakage syndrome: linkage of double-strand break repair to the cellular DNA damage response. Cell. **93**(3), 477–486 (1998)
53. Matsuura, S., Tauchi, H., Nakamura, A., Kondo, N., Sakamoto, S., Endo, S., et al.: Positional cloning of the gene for Nijmegen breakage syndrome. Nat. Genet. **19**(2), 179–181 (1998). doi:10.1038/549
54. Varon, R., Vissinga, C., Platzer, M., Cerosaletti, K.M., Chrzanowska, K.H., Saar, K., et al.: Nibrin, a novel DNA double-strand break repair protein, is mutated in Nijmegen breakage syndrome. Cell. **93**(3), 467–476 (1998)

55. Vaz, F., Hanenberg, H., Schuster, B., Barker, K., Wiek, C., Erven, V., et al.: Mutation of the RAD51C gene in a Fanconi anemia-like disorder. Nat. Genet. **42**(5), 406–409 (2010). doi:10.1038/ng.570
56. Howlett, N.G., Taniguchi, T., Olson, S., Cox, B., Waisfisz, Q., De Die-Smulders, C., et al.: Biallelic inactivation of BRCA2 in Fanconi anemia. Science. **297**(5581), 606–609 (2002). doi:10.1126/science.1073834
57. Reid, S., Schindler, D., Hanenberg, H., Barker, K., Hanks, S., Kalb, R., et al.: Biallelic mutations in PALB2 cause Fanconi anemia subtype FA-N and predispose to childhood cancer. Nat. Genet. **39**(2), 162–164 (2007). doi:10.1038/ng1947
58. Xia, B., Dorsman, J.C., Ameziane, N., de Vries, Y., Rooimans, M.A., Sheng, Q., et al.: Fanconi anemia is associated with a defect in the BRCA2 partner PALB2. Nat. Genet. **39**(2), 159–161 (2007). doi:10.1038/ng1942
59. Wimmer, K., Etzler, J.: Constitutional mismatch repair-deficiency syndrome: have we so far seen only the tip of an iceberg? Hum. Genet. **124**(2), 105–122 (2008). doi:10.1007/s00439-008-0542-4
60. Bourgeron, T., Chretien, D., Poggi-Bach, J., Doonan, S., Rabier, D., Letouze, P., et al.: Mutation of the fumarase gene in two siblings with progressive encephalopathy and fumarase deficiency. J. Clin. Invest. **93**(6), 2514–2518 (1994). doi:10.1172/JCI117261
61. Zinn, A.B., Kerr, D.S., Hoppel, C.L.: Fumarase deficiency: a new cause of mitochondrial encephalomyopathy. N. Engl. J. Med. **315**(8), 469–475 (1986). doi:10.1056/NEJM198608213150801
62. Bourgeron, T., Rustin, P., Chretien, D., Birch-Machin, M., Bourgeois, M., Viegas-Pequignot, E., et al.: Mutation of a nuclear succinate dehydrogenase gene results in mitochondrial respiratory chain deficiency. Nat. Genet. **11**(2), 144–149 (1995). doi:10.1038/ng1095-144
63. Felton, K.E., Gilchrist, D.M., Andrew, S.E.: Constitutive deficiency in DNA mismatch repair. Clin. Genet. **71**(6), 483–498 (2007). doi:10.1111/j.1399-0004.2007.00803.x
64. Gatti, R.A., Tward, A., Concannon, P.: Cancer risk in ATM heterozygotes: a model of phenotypic and mechanistic differences between missense and truncating mutations. Mol. Genet. Metab. **68**(4), 419–423 (1999). doi:10.1006/mgme.1999.2942
65. Goldgar, D.E., Healey, S., Dowty, J.G., Da Silva, L., Chen, X., Spurdle, A.B., et al.: Rare variants in the ATM gene and risk of breast cancer. Breast Cancer Res. BCR. **13**(4), R73 (2011). doi:10.1186/bcr2919
66. Tavtigian, S.V., Oefner, P.J., Babikyan, D., Hartmann, A., Healey, S., Le Calvez-Kelm, F., et al.: Rare, evolutionarily unlikely missense substitutions in ATM confer increased risk of breast cancer. Am. J. Hum. Genet. **85**(4), 427–446 (2009). doi:10.1016/j.ajhg.2009.08.018
67. Antoniou, A., Pharoah, P.D., Narod, S., Risch, H.A., Eyfjord, J.E., Hopper, J.L., et al.: Average risks of breast and ovarian cancer associated with BRCA1 or BRCA2 mutations detected in case series unselected for family history: a combined analysis of 22 studies. Am. J. Hum. Genet. **72**(5), 1117–1130 (2003). doi:10.1086/375033
68. American College of Medical Genetics and Genomics. Standards and guidelines for clinical genetics laboratories. www.acmg.net (2008)
69. Alterman, N., Fattal-Valevski, A., Moyal, L., Crawford TO, Lederman, H.M., Ziv, Y., et al.: Ataxia-telangiectasia: mild neurological presentation despite null ATM mutation and severe cellular phenotype. Am. J. Med. Genet. A. **143A**(16), 1827–1834 (2007). doi:10.1002/ajmg.a.31853
70. Mitui, M., Nahas, S.A., Du, L.T., Yang, Z., Lai, C.H., Nakamura, K., et al.: Functional and computational assessment of missense variants in the ataxia-telangiectasia mutated (ATM) gene: mutations with increased cancer risk. Hum. Mutat. **30**(1), 12–21 (2009). doi:10.1002/humu.20805
71. Saviozzi, S., Saluto, A., Taylor, A.M., Last, J.I., Trebini, F., Paradiso, M.C., et al.: A late onset variant of ataxia-telangiectasia with a compound heterozygous genotype, A8030G/7481insA. J. Med. Genet. **39**(1), 57–61 (2002)
72. Trimis, G.G., Athanassaki, C.K., Kanariou, M.M., Giannoulia-Karantana, A.A.: Unusual absence of neurologic symptoms in a six-year old girl with ataxia-telangiectasia. J. Postgrad. Med. **50**(4), 270–271 (2004)

73. Eliade, M., Skrzypski, J., Baurand, A., Jacquot, C., Bertolone, G., Loustalot, C., et al.: The transfer of multigene panel testing for hereditary breast and ovarian cancer to healthcare: what are the implications for the management of patients and families? Oncotarget. (2016). doi:10.18632/oncotarget.12699
74. Susswein, L.R., Marshall, M.L., Nusbaum, R., Vogel Postula, K.J., Weissman, S.M., Yackowski, L., et al.: Pathogenic and likely pathogenic variant prevalence among the first 10,000 patients referred for next-generation cancer panel testing. Genet. Med. Off. J. Am. Coll. Med. Genet. **18**(8), 823–832 (2016). doi:10.1038/gim.2015.166
75. Thompson, E.R., Rowley, S.M., Li, N., McInerny, S., Devereux, L., Wong-Brown, M.W., et al.: Panel testing for familial breast cancer: calibrating the tension between research and clinical care. J. Clin. Oncol. Off. J. Am. Soc. Clin. Oncol. **34**(13), 1455–1459 (2016). doi:10.1200/JCO.2015.63.7454
76. Walsh, T., Casadei, S., Lee, M.K., Pennil, C.C., Nord, A.S., Thornton, A.M., et al.: Mutations in 12 genes for inherited ovarian, fallopian tube, and peritoneal carcinoma identified by massively parallel sequencing. Proc. Natl. Acad. Sci. U. S. A. **108**(44), 18032–18037 (2011). doi:10.1073/pnas.1115052108
77. Yurgelun, M.B., Masciari, S., Joshi, V.A., Mercado, R.C., Lindor, N.M., Gallinger, S., et al.: Germline TP53 mutations in patients with early-onset colorectal cancer in the Colon cancer family registry. JAMA Oncol. **1**(2), 214–221 (2015). doi:10.1001/jamaoncol.2015.0197
78. Cragun, D., Radford, C., Dolinsky, J., Caldwell, M., Chao, E., Pal, T.: Panel-based testing for inherited colorectal cancer: a descriptive study of clinical testing performed by a U.S. laboratory. Clin. Genet. (2014). doi:10.1111/cge.12359
79. Kurian, A.W., Hare, E.E., Mills, M.A., Kingham, K.E., McPherson, L., Whittemore, A.S., et al.: Clinical evaluation of a multiple-gene sequencing panel for hereditary cancer risk assessment. J. Clin. Oncol. Off. J. Am. Soc. Clin. Oncol. **32**(19), 2001–2009 (2014). doi:10.1200/JCO.2013.53.6607
80. Tung, N., Battelli, C., Allen, B., Kaldate, R., Bhatnagar, S., Bowles, K., et al.: Frequency of mutations in individuals with breast cancer referred for BRCA1 and BRCA2 testing using next-generation sequencing with a 25-gene panel. Cancer. **121**(1), 25–33 (2015). doi:10.1002/cncr.29010
81. Minion, L.E., Dolinsky, J.S., Chase, D.M., Dunlop, C.L., Chao, E.C., Monk, B.J.: Hereditary predisposition to ovarian cancer, looking beyond BRCA1/BRCA2. Gynecol. Oncol. **137**(1), 86–92 (2015). doi:10.1016/j.ygyno.2015.01.537
82. Fitzgerald, R.C., Hardwick, R., Huntsman, D., Carneiro, F., Guilford, P., Blair, V., et al.: Hereditary diffuse gastric cancer: updated consensus guidelines for clinical management and directions for future research. J. Med. Genet. **47**(7), 436–444 (2010). doi:10.1136/jmg.2009.074237
83. National Institutes of Health Consensus Development Conference Statement: neurofibromatosis. Bethesda, Md., USA, July 13–15, 1987. Neurofibromatosis. **1**(3), 172–178 (1988)
84. Castera, L., Krieger, S., Rousselin, A., Legros, A., Baumann, J.J., Bruet, O., et al.: Next-generation sequencing for the diagnosis of hereditary breast and ovarian cancer using genomic capture targeting multiple candidate genes. Eur. J. Hum. Genet. EJHG. (2014). doi:10.1038/ejhg.2014.16
85. Akbari, M.R., Tonin, P., Foulkes, W.D., Ghadirian, P., Tischkowitz, M., Narod, S.A.: RAD51C germline mutations in breast and ovarian cancer patients. Breast Cancer Res. BCR. **12**(4), 404 (2010). doi:10.1186/bcr2619
86. Osorio, A., Endt, D., Fernandez, F., Eirich, K., de la Hoya, M., Schmutzler, R., et al.: Predominance of pathogenic missense variants in the RAD51C gene occurring in breast and ovarian cancer families. Hum. Mol. Genet. **21**(13), 2889–2898 (2012). doi:10.1093/hmg/dds115
87. Pelttari, L.M., Heikkinen, T., Thompson, D., Kallioniemi, A., Schleutker, J., Holli, K., et al.: RAD51C is a susceptibility gene for ovarian cancer. Hum. Mol. Genet. **20**(16), 3278–3288 (2011). doi:10.1093/hmg/ddr229

88. Romero, A., Perez-Segura, P., Tosar, A., Garcia-Saenz, J.A., Diaz-Rubio, E., Caldes, T., et al.: A HRM-based screening method detects RAD51C germ-line deleterious mutations in Spanish breast and ovarian cancer families. Breast Cancer Res. Treat. **129**(3), 939–946 (2011). doi:10.1007/s10549-011-1543-x
89. Vuorela, M., Pylkas, K., Hartikainen, J.M., Sundfeldt, K., Lindblom, A., von Wachenfeldt, W.A., et al.: Further evidence for the contribution of the RAD51C gene in hereditary breast and ovarian cancer susceptibility. Breast Cancer Res. Treat. **130**(3), 1003–1010 (2011). doi:10.1007/s10549-011-1677-x
90. Thompson, E.R., Boyle, S.E., Johnson, J., Ryland, G.L., Sawyer, S., Choong, D.Y., et al.: Analysis of RAD51C germline mutations in high-risk breast and ovarian cancer families and ovarian cancer patients. Hum. Mutat. **33**(1), 95–99 (2012). doi:10.1002/humu.21625
91. Catenacci, D.V., Amico, A.L., Nielsen, S.M., Geynisman, D.M., Rambo, B., Carey, G.B., et al.: Tumor genome analysis includes germline genome: are we ready for surprises? Int. J. Cancer J. Int. du Cancer. **136**(7), 1559–1567 (2015). doi:10.1002/ijc.29128
92. Varga, E., Chao, E.C., Yeager, N.D.: The importance of proper bioinformatics analysis and clinical interpretation of tumor genomic profiling: a case study of undifferentiated sarcoma and a constitutional pathogenic BRCA2 mutation and an MLH1 variant of uncertain significance. Familial Cancer. (2015). doi:10.1007/s10689-015-9790-3
93. Speare, V., Dolinsky, J.S., LaDuca, H., Horton, C., Panos, L., Mason, C., Dalton, E., Chao, E.: Germline testing in hereditary cancer genes subsequent to the identification of mutations in tumor specimens. J. Clin. Oncol. Off. J. Am. Soc. Clin. Oncol. **33**(15_suppl), abstr 1527 (2015)
94. Hall, M.J., Beryl Daly, M., Ross, E.A., Boyd, J., Sanford, E.M., Sun, J., Stephens, P., Liss, D., Chen, S., Miller, V.I., Yelensky, R., Giri, V.N.: Germline variants in cancer risk genes detected by NGS-based comprehensive tumor genomic profiling (CGP). J. Clin. Oncol. Off. J. Am. Soc. Clin. Oncol. **33**(15_suppl), abstr 11084 (2015)
95. Rubinstein, W.S., Maglott, D.R., Lee, J.M., Kattman, B.L., Malheiro, A.J., Ovetsky, M., et al.: The NIH genetic testing registry: a new, centralized database of genetic tests to enable access to comprehensive information and improve transparency. Nucleic Acids Res. **41**(Database issue), D925–D935 (2013). doi:10.1093/nar/gks1173
96. GTR: The Genetic Testing Registry [database on the Internet] 2012. Available from: http://www.ncbi.nlm.nih.gov/gtr/. Accessed 11 Nov 2016
97. ACMG Board of Directors: ACMG policy statement: updated recommendations regarding analysis and reporting of secondary findings in clinical genome-scale sequencing. Genet. Med. Off. J. Am. Coll. Med. Genet. **17**(1), 68–69 (2015). doi:10.1038/gim.2014.151
98. ACMG Board of Directors: Clinical utility of genetic and genomic services: a position statement of the American College of Medical Genetics and Genomics. Genet. Med. Off. J. Am. Coll. Med. Genet. **17**(6), 505–507 (2015). doi:10.1038/gim.2015.41
99. Statement of the American Society of Clinical Oncology: genetic testing for cancer susceptibility, adopted on February 20, 1996. J. Clin. Oncol. Off. J. Am. Soc. Clin. Oncol. **14**(5), 1730–1736 (1996); discussion 7–40
100. American Society of Clinical O: American Society of Clinical Oncology policy statement update: genetic testing for cancer susceptibility. J. Clin. Oncol. Off. J. Am. Soc. Clin. Oncol. **21**(12), 2397–2406 (2003). doi:10.1200/JCO.2003.03.189
101. Robson, M.E., Storm, C.D., Weitzel, J., Wollins, D.S., Offit, K., American Society of Clinical O: American Society of Clinical Oncology policy statement update: genetic and genomic testing for cancer susceptibility. J. Clin. Oncol. Off. J. Am. Soc. Clin. Oncol. **28**(5), 893–901 (2010). doi:10.1200/JCO.2009.27.0660
102. Lu, K.H., Wood, M.E., Daniels, M., Burke, C., Ford, J., Kauff, N.D., et al.: American Society of Clinical Oncology Expert Statement: collection and use of a cancer family history for oncology providers. J. Clin. Oncol. Off. J. Am. Soc. Clin. Oncol. **32**(8), 833–840 (2014). doi:10.1200/JCO.2013.50.9257
103. Bradbury, A.R., Patrick-Miller, L., Domchek, S.: Multiplex genetic testing: reconsidering utility and informed consent in the era of next-generation sequencing. Genet. Med. Off. J. Am. Coll. Med. Genet. **17**(2), 97–98 (2015). doi:10.1038/gim.2014.85

104. Robson, M.: Multigene panel testing: planning the next generation of research studies in clinical cancer genetics. J. Clin. Oncol. Off. J. Am. Soc. Clin. Oncol. **32**(19), 1987–1989 (2014). doi:10.1200/JCO.2014.56.0474
105. Kurian, A.W., Ford, J.M.: Multigene panel testing in oncology practice: how should we respond? JAMA Oncol. **1**(3), 277–278 (2015). doi:10.1001/jamaoncol.2015.28
106. Landrum, M.J., Lee, J.M., Riley, G.R., Jang, W., Rubinstein, W.S., Church, D.M., et al.: ClinVar: public archive of relationships among sequence variation and human phenotype. Nucleic Acids Res. **42**(Database issue), D980–D985 (2014). doi:10.1093/nar/gkt1113
107. Rehm, H.L., Berg, J.S., Brooks, L.D., Bustamante, C.D., Evans, J.P., Landrum, M.J., et al.: ClinGen – the clinical genome resource. N. Engl. J. Med. **372**(23), 2235–2242 (2015). doi:10.1056/NEJMsr1406261
108. National Society of Genetic Counselors. Clinical Data Sharing. Position Statements. www.nsgc.org (2015)
109. Spurdle, A.B., Healey, S., Devereau, A., Hogervorst, F.B., Monteiro, A.N., Nathanson, K.L., et al.: ENIGMA – evidence-based network for the interpretation of germline mutant alleles: an international initiative to evaluate risk and clinical significance associated with sequence variation in BRCA1 and BRCA2 genes. Hum. Mutat. **33**(1), 2–7 (2012). doi:10.1002/humu.21628
110. Ou, J., Niessen, R.C., Vonk, J., Westers, H., Hofstra, R.M., Sijmons, R.H.: A database to support the interpretation of human mismatch repair gene variants. Hum. Mutat. **29**(11), 1337–1341 (2008). doi:10.1002/humu.20907
111. Karam, R., Pesaran, T., Chao, E.: ClinGen and genetic testing. N. Engl. J. Med. **373**, 1376–1379 (2015). doi:10.1056/NEJMc1508700

Application of Next-Generation Sequencing in Noonan Spectrum Disorders

Hui Mei, Hongzheng Dai, Jinglan Zhang, Lee-Jun Wong, and Pilar L. Magoulas

Abstract Noonan spectrum disorders (NSDs) are a group of disorders with aberrant signal transduction in the RAS/mitogen-activated protein kinase (MAPK) pathway, and therefore they are also known as RASopathies. All NSDs known to date are caused by germline dominant mutations in genes encoding proteins participating in the RAS-MAPK pathway. The molecular characterization thus far explains approximately 80% of individuals affected with a NSD. Pathogenic variants in the *PTPN11* gene cause about 50% of all NSD cases. The other 17 genes account for an additional 20–30% NDS cases. High genetic heterogeneity in NSDs and their considerable overlap in clinical presentations had made the diagnosis of these disorders expensive and time consuming in a gene by gene approach. In this chapter, we provide a brief overview of clinical features of Noonan syndrome and closely related conditions, the molecular mechanisms underlying pathogenesis, and the advantages and challenges in implementing next generation sequencing (NGS) in clinical laboratories for the molecular diagnosis of NSDs.

Keywords RAS/MAPK • Signal transduction pathway • RASopathies • Noonan syndrome • Cardiofaciocutaneous syndrome • Costello syndrome • Noonan syndrome with multiple lentigines (LEOPARD syndrome) • NGS • Prenatal testing • Whole genome amplification

H. Mei, PhD (✉) • H. Dai • J. Zhang • L.-J. Wong
Department of Molecular and Human Genetics, Baylor College of Medicine,
Houston 77030, TX, USA
e-mail: hui.mei@bcm.edu

P.L. Magoulas (✉)
Department of Molecular and Human Genetics, Baylor College of Medicine,
Houston 77030, TX, USA

Division of Genetic Counseling, Texas Children's Hospital,
6701 Fannin St. suite 1560, Houston, TX 77030, USA
e-mail: plmagoul@texaschildrens.org

L.-J.C. Wong (ed.), *Next Generation Sequencing Based Clinical Molecular Diagnosis of Human Genetic Disorders*, DOI 10.1007/978-3-319-56418-0_9

1 Introduction

1.1 RAS/MAPK Pathway Biology

The RAS/mitogen-activated protein kinase (MAPK) pathway is a key signal transduction pathway that plays an essential role in cell determination, proliferation, differentiation, and survival. RAS proteins are members of the small guanosine-binding protein family. The RAS subfamily includes HRAS, NRAS, and KRAS. They act as signal switch molecules that transmit extracellular signals to activate downstream effectors (Fig. 1). Activation of the RAS/MAPK pathway is initiated by growth factors binding to the transmembrane tyrosine kinase receptors, which undergo dimerization, autophosphorylation, and subsequent activation. The activated receptors interact with a set of specific adaptor proteins, including growth factor receptor-bound protein (GRB2), CBL, and SHP2. GRB2 recruits son of sevenless (SOS), a guanine nucleotide exchange factor (GEF). SOS1 is the major GEF that activates the RAS proteins by facilitating a conformational switch that is dependent on the exchange of GDP for GTP. Activated RAS propagates the signaling cascade by activating the effector MAPKKK (RAF). There are three RAF serine/threonine kinases (ARAF, BRAF, and RAF1). Phosphorylated RAF then activates the MAPKKs -MAP2K1 and/or MAP2K2, which in turn activates the MAPK proteins, ERK1 and ERK2. The substrates of ERK1 and ERK2 include nuclear components, transcription factors, membrane proteins, and protein kinases that in turn control vital cellular functions.

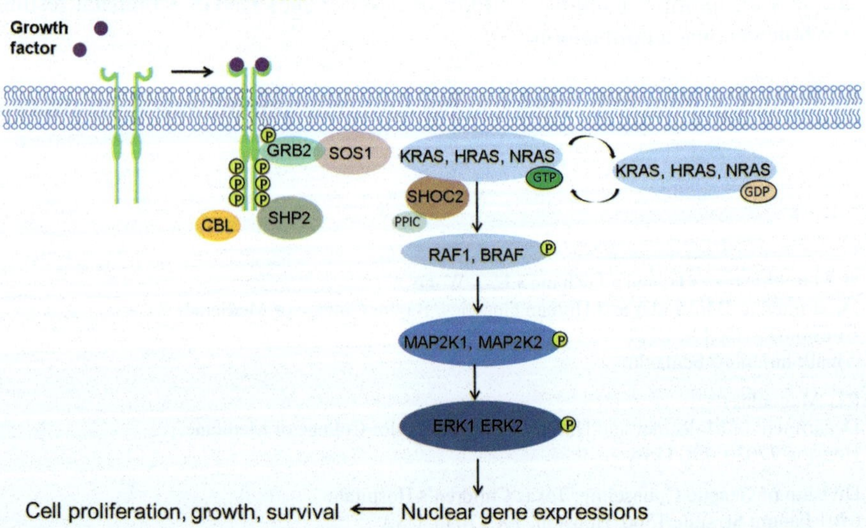

Fig. 1 The RAS/ MAPK signal transduction pathway

1.2 The RASopathies

The RASopathies are a group of genetically heterogeneous developmental disorders caused by defects in genes involved in the RAS/MAPK signaling pathway [43]. These disorders include Noonan syndrome (NS), Costello syndrome, Noonan syndrome with multiple lentigines (NSML; formerly called LEOPARD syndrome), cardio-facio-cutaneous syndrome (CFC), capillary malformation-arteriovenous malformation syndrome, Legius syndrome, and neurofibromatosis type 1 (NF1). While these individual disorders are rare, as a group, the prevalence of these disorders is between 1 in 1000 and 1 in 2500 live births. Although each RASopathy exhibits distinctive phenotypic features, the common dysregulation of RAS/MAPK signaling likely underlies the overlapping clinical manifestations, including dysmorphic craniofacial features, cardiac malformations, skin abnormalities, musculoskeletal and ocular abnormalities, varying degrees of intellectual disability, and increased cancer risk (Fig. 2; Table 1).

NF1 was the first disease gene identified in the RAS/MAPK pathway [8, 65, 67]. Recently, a number of additional genes causative for the RASopathies have been identified, including *PTPN11, SOS1, RAF1, KRAS, NRAS, HRAS, SHOC2, RIT1, BRAF, MAP2K1,MAP2K2, CBL, SOS2, LZTR1, RRAS, RASA2, and A2ML1* (Table 2a) [3–5, 9, 10, 12, 13, 15, 19, 30, 35, 36, 41, 44, 46–48, 53, 58, 61, 66, 68]. Table 2b summarizes several RASopathy genes, their respective phenotypes, and percentage of that gene accounts for all NSD cases. Discovery of specific RASopathy genes made it possible to develop genetic tests that facilitate the postnatal diagnosis of symptomatic individuals with characteristic features of a RASopathy as well as prenatal diagnosis of fetuses with normal karyotypes that present with specific ultrasound findings [4, 27, 37].

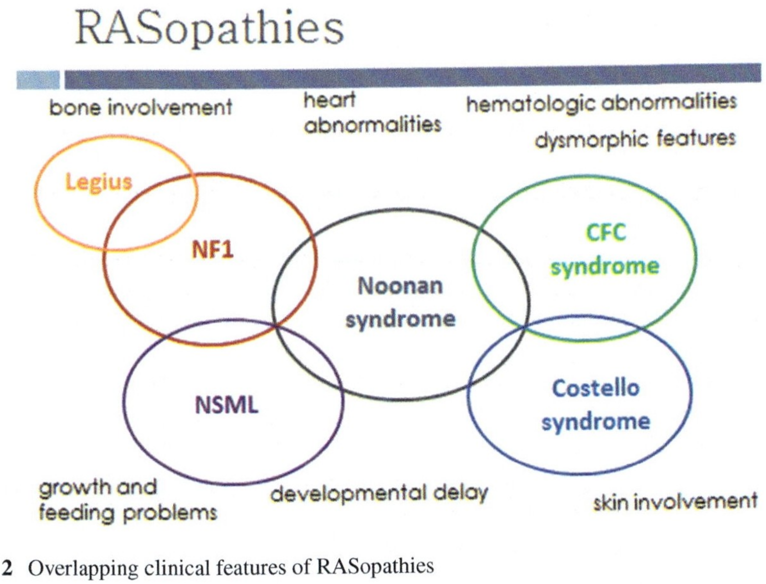

Fig. 2 Overlapping clinical features of RASopathies

Table 1 Summary of RASopathies

Syndromes	RAS/MA PK gene	Characteristic clinical features
Neurofibromatosis I	*NF1*	Multiple café-au-lait spots; axillary and inguinal freckling; multiple
Noonan syndrome	*PTPN11, SOS1, RAF1, KRAS, NRAS, SHOC2, RIT1, CBL*	Short stature; broad or webbed neck; characterisitic facies, including a borad forehead, hypertelorism, down-slanting palpebral fissures, ptosis, and low-set posteriorly rotated ears; congenital heart defect; coagulation defects; ocular abnormalities; normal intelligence to mild intellectual disability; predispostion to cancer
Noonan syndrome with multiple lentigines	*PTPN11, RAF1*	Similar to Noonan syndrome; but with multiple skin lentigines present as dispersed flat, black-brown macules
Costello syndrome	*HRAS*	Failure to thrive; short stature; Noonan facies but more coarse, including
Cardiofaciocutaneous syndrome	*BRAF1,* *MAP2K1,* *MAP2k2,KR AS*	Noonan facies; congenital heart defects; cutaneous abnormalities, including xerosis, hyperkeratosis, ichthyosis, keratosis pilaris, eczema; developmental delay or intellectual disability;
Leguis syndrome	*SPRED1*	Multiple café-au-lait spots; intertriginous freckling; lipoma; macrocephaly;
Capillary malformation-arteriovenous malformation	*RASA1*	Multiple, small capillary malformations; arteriovenous malformations and/or arteriovenous fistulas

2 Noonan Spectrum Disorders

2.1 Noonan Syndrome

Noonan syndrome (NS) is an autosomal dominant disorder characterized by short stature, congenital heart defect, and developmental delay of variable degree. Short stature is present in over 70% of the NS patients; the heart defects are one of the key features presented in approximately 50–80% of the NS patients, including primarily pulmonary valve stenosis and hypertrophic cardiomyopathy (Allanson and Roberts GeneReviews; http://www-ncbi-nlm-nih-gov.ezproxyhost.library.tmc.edu/books/NBK1124/). Other important findings include broad or webbed neck, unusual chest shape with superior pectus carinatus and inferior pectus excavatum, cryptorchidism, and bleeding disorders. Patients with NS exhibit distinctive craniofacial features, including a broad forehead, hypertelorism, down-slanting palpebral fissures, and low-set, posteriorly rotated ears. In addition, individuals with NS have an increased risk of hematological malignancy, including juvenile myelomonocytic leukemia (JMML) [60]. While the postnatal presentation of NS is well known, the recurrent prenatal findings of NS have also been noted, including polyhydramnios, increased nuchal translucency (NT), cystic hygroma, pleural effusions, hydrops, and cardiac defects [27, 37].

Table 2a Selected genes involved in RASopathies

RAS/MAPK pathway gene	Chromosome location	OMIM	NM#	Number of exons	Protein	Protein function	Inheritance
NF1	17q11.2	162,200	NM_000267.3	57	Neurofibromin	RASGAP	AD
PTPN11	12q24.1	176,876	NM_002834.3	16	Tyrosine-protein phosphatase non-receptor type 11	Phosphatase	AD
SOS1	2p22.1	182,530	NM_005633.3	23	Son of sevenless homolog1	RAS GEF	AD
RAF1	3p25.1	164,760	NM_002880.3	17	RAF proto-oncogene serine/threonine-protein kinase	Kinase	AD
BRAF	7q34	164,757	NM_004333.4	18	V-RAF murine sarcoma viral oncogene homolog B1	Kinase	AD
KRAS	12p12.1	190,070	NM_004985.3	5	GTPase KRas	GTPase	AD
NRAS	1p13.2	164,790	NM_002524.3	7	GTPase NRas	GTPase	AD
HRAS	11p15.5	190,020	NM_005343.2	6	GTPase HRas	GTPase	AD
CBL	11q23.3	165,360	NM_005188.3	16	CBL	E3 ubiquitin ligase	AD
SHOC2	10q25.2	602,775	NM_007373.3	9	SHOC2	Scaffolding	AD
RIT1	1q22	609,591	NM_006912.5	6	RIC-like protein without CAAX motif 1	GTPase	AD
MAP2K1	15q22.31	176,872	NM_002755.3	11	Dual specificity mitogen-activated protein kinase kinase 1	Kinase	AD
MAP2K2	19p13.3	601,263	NM_030662.3	11	Dual specificity mitogen-activated protein kinase kinase 2	kinase	AD
SPRED1	15q14	609,291	NM_152594.2	7	Sprouty-related EVH1 domian-containing protein 1	SPROUTY-related EVH1 domain-containing protein 1	AD

NS is caused by activating pathogenic variants in genes in the RAS/MAPK signaling pathway (Table 1). The heterozygous missense pathogenic variants in these genes lead to the constitutive activation or enhanced activity of proteins encoded by these genes, resulting in increased signaling through the RAS/MAPK pathway.

2.1.1 PTPN11

Gain-of-function missense pathogenic variants in the *PTPN11* are the most common cause of NS. Approximately 50% of the patients with a clinical diagnosis of NS harbor pathogenic variants in *PTPN11* (Allanson and Roberts GeneReviews; http://www-ncbi-nlm-nih-gov.ezproxyhost.library.tmc.edu/books/NBK1124/; [62]). The *PTPN11* gene is located at 12q24.1 and consists of 16 exons (Table 2a and 2b, Fig. 3). The protein product of *PTPN11*, SHP2, is a non-receptor protein tyrosine phosphatase (PTP) [62, 69]. It is composed of N- and C-terminal SH2 domains and a single catalytic PTP domain (Fig. 3). Interactions between the N-terminal SH2 domain and the PTP domain are involved in switching the protein between its inactive and active conformation. In the inactive state, the N-terminal SH2 domain directly binds the PTP domain and inhibits its catalytic activity by blocking the access of the substrate to the catalytic site. Once the N-terminal SH2 domain binds the phosphotyrosine peptide, conformational change results in the active state. The majority of NS-causing pathogenic variants in *PTPN11* cluster in and around the interacting residues of the N-terminal SH2 domain and the PTP domain. Pathogenic variants in this region disrupt the stability of the catalytically inactive form of SHP2, causing constitutive or prolonged activation of the protein (gain of function). Most pathogenic variants indentified in Noonan syndrome were missense gain of function pathogenic variants (Table 3). Four pathogenic variants affecting residues involved in the N-terminal SH2/PTP interaction, Y63C, Q79R, N308D, and N308S, were identified in approximately 40% of Noonan patients with

Table 2b Summary of RASopathy genes by phenotype and diagnostic yield

Gene	Phenotype(s)	Percentage of cases
PTPN11	NS	50%
SOS1	NS	10–13%
RIT1	NS	9%
RAF1	NS	3–7%
KRAS	NS, CFC	<5%, <5%
NRAS	NS	Unknown (<1%)
LZTR1	NS	Unknown
SOS2	NS	Unknown
SHOC2	Noonan-like	Unknown
CBL	Noonan-like	Unknown (<1%)
BRAF	CFC	75–80%
MEK1	CFC	10–15%
MEK2	CFC	
HRAS	CS	80–90%

PTPN11 pathogenic variants (Table 3) [2]. Pathogenic variants affecting residues not only involved in N-terminal SH2/PTP interaction but also involved in controlling the catalytic activity, substrate specificity, or the flexibility of the linker stretch between N-terminal SH2 and C-terminal SH2, have all been identified in NS patients. Two pathogenic variants affecting residues binding the phosphopeptide in SH2 domain, Thr42 (N-terminal SH2) and Glu139 (C-terminal SH2), were identified in approximately 6% of Noonan patients with *PTPN11* pathogenic variants [2]. Some germ line *PTPN11* pathogenic variants overlap with somatically acquired *PTPN11* pathogenic variants associated with JMML. Some *PTPN11* pathogenic variants are distinctively associated with Noonan syndrome or cancer [2]. Although most of the *PTPN11* pathogenic variants arise as *de novo* events, PTPN11 pathogenic variants have been detected in 30–75% of patients with familial NS [59, 69].

2.1.2 SOS1

The second most common cause of NS is missense pathogenic variant in the *SOS1* gene, accounting for approximately 15% of the NS cases [28, 46, 47, 61, 70]. The *SOS1* gene is located at 2p22.1 and consistent of 23 exons. *SOS1* encodes the guanine nucleotide exchange factors for RAS and acts as a positive regulator of RAS by stimulating the guanine nucleotide exchange. N-terminal of SOS1 is an autoinhibition regulatory domain including tandem histone-like folds (HF), a Dbl-homology domain (DH) and a pleckstrin-homology domain (PH). C-terminal of SOS1 is the catalytic domain including the RAS exchanger motif (REM), CDC25 domain, and a tail providing docking sites for adaptor proteins required for receptor anchoring. The N-terminal inhibits the SOS1's GEF activity by blocking the GDP-RAS allosteric binding site resided in the C-terminal. Once SOS1 is recruited to the membrane, the N-terminal autoinhibition is relieved, allowing the RAS binding to the allosteric site, which in turn, promotes the RAS binding to the catalytic site through the conformational change of the CDC25 domain.

The most common pathogenic variant, c.2536G > A (p.Q846K), is located in the CDC25 domain, which accounts for approximately 12% of the *SOS1* pathogenic variants (Table 4) [28, 46, 47, 61, 70]. Other SOS1 pathogenic variants also tend to cluster in specific regions. Pathogenic variants at three residues located in the PH-REM linker (Ser 548, Leu550, and Arg552) account for approximately 40% of SOS1 pathogenic variants (Table 4). The second pathogenic variant cluster is located in the PH domain, which accounts for approximately 20% of *SOS1* pathogenic variant. The last pathogenic variant cluster is resided in the interaction region of DH domain and REM domain (16% of all pathogenic variants). Most SOS1 pathogenic variants arise as *de novo* events. Rarely, familial cases have been reported. In these cases, the parents who transmitted pathogenic variants had similar clinical features as the affected children [46, 47, 70].

Individuals with *SOS1* pathogenic variants have typical features of Noonan syndrome, however some are reported to have more ectodermal manifestations, including sparse eyebrows and skin abnormalities, similar to what is more typically seen

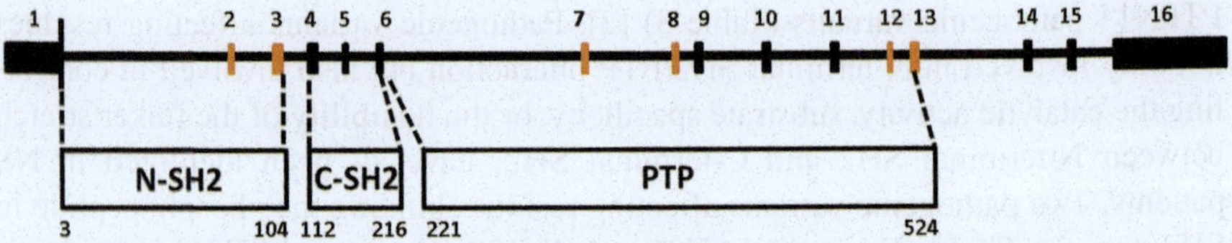

Fig. 3 Genomic organization and function domains of PTPN11 gene. Coding exons are shown as filled boxes. The peptide structure shows functional domains with their amino acid boundaries. *The hot spots for the pathogenic variants* are shown as orange filled boxes. (Adapted from [62, 69])

Table 3 Selected recurrent PTPN11 pathogenic variants in Noonan Syndrome. The four most common PTPN11 pathogenic variants are highlighted in red

DNA Nucleotide change	Protein amino acid change	Exon	Function domain
c.124A>G	p.T42A	2	N-SH2
c.179G>C	p.G60A	3	N-SH2
c.181G>A	p.D61N	3	N-SH2
c.182A>G	p.D61G	3	N-SH2
c.184T>G	p.Y62D	3	N-SH2
c.188A>G	**p.Y63C**	**3**	**N-SH2**
c.214G>A	p.A72S	3	N-SH2
C.215C>G	p.A72G	3	N-SH2
c.218C>T	p.T73I	3	N-SH2
c.228G>C	p.E76D	3	N-SH2
c.236A>G	**p.Q79R**	**3**	**N-SH2**
c.317A>C	p.D106A	3	Linker
c.417G>C,T	p.E139D	4	C-SH2
c.836A>G	p.Y279C	7	PTP
c.922A>G	**p.N308D**	**8**	**PTP**
C.923A>G	**p.N308S**	**8**	**PTP**
c.1403C>T	p.T468M	12	PTP
c.1510A>G	p.M504V	13	PTP

in CFC syndrome, when compared to other individuals with NS. Growth may also be less affected in *SOS1*-related NS when compared to other genetic causes of NS (such as *PTPN11*-related NS) [28].

2.1.3 KRAS and NRAS

Pathogenic variants in RAS genes have been reported in a small portion of NS patients, including *KRAS* and *NRAS*. RAS contains a G domain and a C terminal membrane targeting region. The G domain directly binds to GDP or GTP, and two switch motifs within the G domain, G2 (switch I) or G3 (switch II), are the main parts that facilitate the activation of RAF by GTP [64]. The Thr35 that binds directly to GTP is located in G2 (switch I) domain. The conformation change of these two switch motifs mediates the switch between ON-state of GTP bound RAS and OFF-state of GDP bound RAS. In the active state, RAS has high affinity of downstream effectors, which in turn, stimulates downstream signaling pathways. RAS also has an intrinsic GTPase activity that needs the binding of GTPase activating proteins (GAPs) to hydrolyze a bound GTP molecule into GDP. The G3 play a crucial role in hydrolysis of GTP to GDP. The balance between SOS1 and GAP activity determines the guanine nucleotide status of RAS, thus regulating RAS activity.

The *KRAS* gene is located at 12p12.1 and encodes two isoforms; the KRASA expressed in a tissue-specific and developmentally restricted manner, and a ubiquitously expressed KRASB. The *KRAS* pathogenic variants have been identified in approximately 2% NSD patients [7, 53, 70]. The KRAS pathogenic variants increase the signaling of the RAS/MAPK pathway through three distinct mechanisms: reducing the RAS GTPase activity; interfering with the guanine nucleotide binding of KRAS; or disrupting the KRAS autoinhibition by membrane sequestration of its effector-binding site [31]. In fact, the most common pathogenic variant in *KRAS*, D153V in exon 6 of *KARSB*, activates signaling by perturbing membrane orientation to unleash autoinhibition. Somatic KRAS pathogenic variants are frequently detected in lung, colon, and pancreatic cancers. However, the most common somatic pathogenic variants at codons 12, 13, and 61 have not been indentified as germline pathogenic variants. Pathogenic variants in *KRAS* indentified in other RASopathy disorders are discussed later.

A few pathogenic variants in *NRAS* have been found in patients with NS [10, 16, 49]. The *NRAS* gene is located at 11p13.2 and consistent of seven exons. *NRAS* pathogenic variants are located within or near the switch I and switch II regions and thought to activate the MAPK pathway by accumulating in the GTP-bound conformation or to reducing the GTPase function.

2.1.4 RIT1

Recently, whole exome sequencing studies have identified pathogenic variants in a new RAS like gene, *RIT1*, in RASopathies patients with no detectable pathogenic variants in known Noonan-related genes [3]. The RIT1 gene is located at 1q22 and consists of six exons. RIT1 shares approximately 50% sequence identity with RAS at amino acid level, and it has an additional N-terminal extension. Similar to *NRAS* pathogenic variants, *RIT1* pathogenic variants are clustered in the G3 (switch II) GTPase activity domain. A few *RIT1* pathogenic variants also have been identified in G1 and G2 domains. Overall, *RIT1* pathogenic variants are identified in 4–9% NS patients.

Table 4 Selected recurrent SOS1 pathogenic variants in Noonan Syndrome

DNA nucleotide change	Protein amino acid change	Exon	Function domain
c.322G>A	P.E108K	4	HF
c.806T>C	P.M269 T	6	DH
c.806T>G	p.M269R	6	DH
c.1642A>C	p.S548R	10	PH-REM linker
c.1649T>C	p.L550R	10	PH-REM linker
c.1654A>G	p.R552G	10	PH-REM linker
c.1655G>A	p.R552K	10	PH-REM linker
c.1655G>T	p.R552M	10	PH-REM linker
C.1656G>C	p.R552S	10	PH-REM linker
c.210T>C	p.Y702H	14	REM
c.2536G>A	p.E846K	16	CDC25

2.1.5 RAF1 and BRAF

Pathogenic variants in *RAF1* have been identified in 3–17% NS patients. The *RAF1* gene is located at 3p25.2 and consistent of 17 exons. *RAF1* encodes a serine/threonine kinase that is one of the direct downstream RAS effectors. RAF1 share three conserved cysteine-rich regions, CR1, CR2, and CR3 with the other two RAFs, ARAF and BRAF [32]. CR1 domain (CRD) and most of the RAS binding domain (RBD) bind to RAS-GTP. CR2 is rich in serine and threonine residues. CR3 is the kinase domain containing the highly conserved glycine-rich G-loop GXGXXG motif. Pathogenic variants in RAF1 identified in NS patients are clustered in CR2 and CR3 domains. CR2 domain contains an inhibitory phosphorylation site (Ser 259). The dephosphorylation of Ser259, which is required for RAF1 translocation to the cell membrane and its catalytic activity, has been shown to be the primary pathogenic mechanism in the activation of RAF1. This pathogenic variant accounts for approximately 70% of all the identified *RAF1* pathogenic variants in NS patients. The other four pathogenic variants that reside within the C-terminal kinase domain, Asp 486, Thr491, Ser612, and Leu613, account for almost all the remaining 30% of *RAF1* pathogenic variants. Individuals with *RAF1*-related NS have a significant risk of hypertrophic cardiomyopathy, which often presents in the newborn period as severe cardiac involvement.

Pathogenic variants in *BRAF* have been reported in a few NSD patients, the majority pathogenic variants of *BRAF* have been identified in patients with CFC syndrome. The detailed discussion of BRAF is under the CFC syndrome section.

2.1.6 SHOC2

A recurrent pathogenic variant, p.S2G in the *SHOC2* gene, has been identified in NS patient with a unique phenotypic feature of loose anagen hair [12]. The *SHOC2* gene is located at 10q25.2 and consists of nine exons. The *SHOC2* gene encodes a protein composed almost entirely of leucine-rich repeats (LRR) with a lysine-rich sequence at the N-terminus. The LRR of SHOC2 functions as a scaffold linking

RAS to the downstream effector, RAF1. SHOC2 binds to RAS-GTP and promotes the catalytic subunit of protein phosphatase 1 (PP1C) translocation to the cell membrane. This facilitates PP1C dephosphorylation at residue Ser259 of RAF1. As mentioned previously, the Ser259 is the major hot spot for NS-causing pathogenic variant in RAF1. The recurrent pathogenic variant p.S2G in SHOC2 is proposed to promote an aberrant protein N-myristoylation that results in constitutive membrane targeting of SHOC2, leading to prolonged PP1C-mediated RAF1 dephosphorylation at Ser259 and consequently increased MAPK pathway activation [12, 24].

2.1.7 CBL

Pathogenic variants in a tumor-suppressor gene, *CBL*, have been reported as a rare cause of NS [30]. The *CBL* gene is located at 11q23.3, consisting of 16 exons. The CBL gene encodes an E3 ubiquitin ligase that negatively regulates the downstream signaling of receptor tyrosine kinases (RTKS). CBL contains an N-terminal tyrosine kinase-binding (TKB) domain, a short linker and a C-terminal zinc-binding RING-finger domain mediating the E3 ubiquitin ligase activity. CBL catalyzes the ubiquitination of activated RTKs to switch off signaling via receptor degradation or recycling [52]. Pathogenic variants in CBL identified in NS patients are clustered within the RING finger domain or the adjacent linker region.

2.2 Costello Syndrome

Costello syndrome (CS) is one of the rare RASopathies characterized by failure to thrive in infancy, short stature, developmental delay, coarse facial features, ectodermal abnormalities, hypotonia, and cardiac abnormalities. Clinical features of CS overlap with other RASopathy disorders. Relative or absolute macrocephaly is typical in CS patient and the characteristic facial appearances include coarse face (full cheeks, wide mouth with full lips and broad nasal base and full nasal tip). Ectodermic abnormalities in CS patient include soft skin, curly or spare, fine hair, deep palmar and plantar creases. The majority of CS individuals have cardiac abnormalities including hypertrophic cardiomyopathy, valve abnormalities (usually valvar pulmonic stenosis), and arrhythmia. Individuals with CS have an approximately 15% lifetime risk to develop malignant tumors, including rhabdomyosarcoma, ganglioneuroblastoma, and bladder carcinoma.

Only a single gene, *HRAS*, has been reported to cause CS. The *HRAS* gene is located at 11p15.5 and consists of six exons. The most common *HRAS* pathogenic variants affect residues Gly12 and Gly13 that are found in approximately 80–90% CS individuals. Especially, pathogenic variant p.G12S is found in more than 80% CS individuals and also found in cancers. The second most common pathogenic variant causes CS is p.G12A, unlike p.G12S, this pathogenic variant is not commonly found in cancers. These pathogenic variants disrupt guanine nucleotide binding and cause a reduction of GTP hydrolysis, resulting in HRAS remaining in the active state. Interestingly, the two most common residues mutated in CS are also the most frequently mutated positions in cancers. Other HRAS pathogenic variants are also observed in CS individuals with very low frequency. Up to date, there is only one documented case of germline mosaicism reported in CS [56]. The p.G12S pathogenic variant identified in the affected son was present in 7–8% cells of the father.

2.3 *Cardio-Facio-Cutaneous (CFC) Syndrome*

CFC is another rare RASopathy syndrome characterized by cardiac abnormalities, distinctive craniofacial appearance, and cutaneous abnormalities. CFCS has considerable clinical features overlap with NS and CS. Individuals with CFC have distinct facial features similar to NS, including macrocephaly, a high forehead, bitemporal narrowing, and facial dysmorphia that is coarser compared to NS. The ectodermal findings include dry and hyperkeratotic skin, ichthyosis, eczema, sparse, curly hair, and sparse eyebrows and eyelashes. Cardiac abnormalities also occur in the majority of individuals with CFC; the most common being pulmonic stenosis, septal defects and hypertrophic cardiomyopathy. Neurologic and/or cognitive deficits are present in nearly all individuals with CFC, ranging from mild to severe. Neoplasia, mostly acute lymphoblastic leukemia (ALL), has been reported in some individuals.

Four genes have been identified to be associated with CFC syndrome, *BRAF, MAP2K1, MAP2K2*, and *KRAS*. Pathogenic variants in *BRAF* are found in approximately 75% of individuals with CFC. The *BRAF* gene is located at 7q34 and consists of 18 exons. The *BRAF* gene is also known as a proto-oncogene. Somatic pathogenic variants in this gene are frequently found in various cancers, including malignant melanoma, thyroid, colorectal, ovary and lung cancers. However, the CFC-associated pathogenic variants only partially overlap with the cancer-associated pathogenic variants in *BRAF*. Unlike the cancer-associated pathogenic variants, which are clustered in the glycine-rich loop and activation segment, the majority of CFC-associated pathogenic variants are clustered in the cysteine-rich domain (CR1) in exon 6 and in the protein kinase domain (CR3). The pathogenic variant Q257R in the CR1 domain is the most common *BRAF* pathogenic variant identified in CFC individuals, followed by E501G, G469E and N581D. The functional analyses of *BRAF* pathogenic variant proteins have demonstrated that some *BRAF* pathogenic variants have increased kinase activity and some BRAF pathogenic variants have exhibited impaired kinase activity [36, 48]. However, further *in vivo* studies in Zebrafish have demonstrated that both kinase-active and kinase-impaired *BRAF* pathogenic variants result in similar phenotypic dysregulation of MAPK signaling [1]. The functional difference between the kinase-active and kinase-impaired *BRAF* pathogenic variants remains unknown.

Pathogenic variants in *MAP2K1* and *MAP2K2* are identified in the approximately 25% of the pathogenic variant positive CFC individuals [17, 33, 34, 40, 48, 54]. *MAP2K1* and *MAP2K2* are threonine/tyrosine kinases (MEK1 and MEK2) and both isoforms have the equally ability to phosphorylate and activate ERK substrates (ERK1 and ERK2) [71]. Functional studies of MEK CFC mutant proteins by examining the ERK phosphorylation have showed that all pathogenic variants are more active than wild-type MEK [48].

In addition, pathogenic variants in *KRAS* have been identified in 2–3% CFC individuals (Rauen [42], gene reviews).

2.4 Noonan Syndrome with Multiple Lentigines (NSML)

NSML (formerly referred to as LEOPARD syndrome) is a rare autosomal dominant disorder that is an allelic NS. NSML is characterized by the craniofacial features of NS as well as multiple lentigines, electrocardiogram (ECG) conduction abnormalities, ocular hypertelorism, pulmonary valve stenosis, abnormal genitalia, growth retardation, and sensorineural deafness [21, 50]. Multiple lentigines present as flat, black-brown macules, mostly on face, neck and upper part of the trunk with sparing of mucosa (Gelb and Tartagila [20], gene reviews). In general lentigines appear at the age of 4–5 years and increase to the thousands by puberty. Heart defects are observed in approximately 85% of NSML individuals, including ECG anomalies, HCM and pulmonary valve stenosis. Growth retardation is observed in approximately 50% of NSML individuals with a final height in most individuals less the 25th percentile for age. Pathogenic variants in the PTPN11 gene have been identified in approximately 90% NSML individuals. However, unlike the PTPN11 pathogenic variants associated with NS, the most common PTPN11 pathogenic variants associated with NSML are clustered in the catalytic PTP domain, causing a loss of function of SHP2 catalytic activity [18, 25]. In addition to PTPN11, pathogenic variants in RAF1 have been reported in about 5% of NSML individuals. Rarely NSML individuals have pathogenic variants in the BRAF and MAP2K1 genes [26, 38, 50].

3 Molecular Diagnosis of NSDs by NGS

As of 2016, 18 genes have been discovered to play a role in the pathogenesis of RASopathies, including *A2ML1, BRAF, CBL, HRAS, KRAS, LZTR1, MAP2K1, MAP2K2, NF1, NRAS, PTPN11, RAF1, RASA2, RIT1, SHOC2, SPRED1, SOS1,* and *SOS2* [3, 5, 8–10, 12, 30, 35, 36, 41, 44, 46–48, 53, 58, 61, 65–68]. The genes associated with NSDs over 30 kb coding sequences in total in terms of coding region and 20 bp adjacent intronic sequences which make standard stepwise molecule testing of NSDs expensive by Sanger sequencing. Therefore, thorough clinical evaluation and preliminary differential diagnosis based on presenting symptoms for suspected NSDs are often the prerequisite to warrant a high diagnostic yield. However, the clinical delineation of NSDs can be difficult, as these disorders present with wide variabilities in the affected organs as discussed above. Prioritization of genes to be tested is primarily dependent on the distinct phenotypes seen in the disorders and the disease prevalence. The molecular diagnosis can be achieved for the most of NSD cases by taking this gene-by-gene approach. Some s cannot be identified in primary screening for prevalent mutations in genes such as PTPN11, RAF1 or SOS1. A follow-up sequencing test should be considered for other less common NSDs.

All of known NSD pathogenic variants are single nucleotide changes and no deletions or duplications involving *PTPN11, KRAS, SOS1, RAF1, BRAF*, or *MAP2K1* have been reported to cause NSDs. Therefore, sequencing for these genes is considered to be sufficient the molecular testing for NSDs. However, unlike other NSDs, NF1 is caused by loss of function pathogenic variants in NF1 and exonic copy number variation in this gene represents an important fraction of NF1 pathogenic variants. A different and more comprehensive test strategy should be considered if NF1 is included in the NSD panel. Recent report on *MAP2K2* suggests deletion could also be a novel mechanism for the etiology of RASopathy [39].

When designing an NGS panel for NSD genes by a capture approach, the paralogues sequences in three genes, *PTPN11, MAP2K and MAP2K2*, and the high GC contents in *BRAF* and *HRAS* genes should be considered in order to achieve 100% coverage of the coding regions. For example, higher probe density in the high-GC content exon 1 of the PTPN11 gene might improve the capture efficiency and reduce gap-filling by Sanger sequencing. Any probe that may capture the paralogous sequences such as exons in the NF1 gene should be avoided, so the enriched DNA fragments by hybridization will be less prone to ambiguous mapping after sequencing Amplicon-based NGS for 12 NSD genes has been clinically validated and demonstrated satisfactory detection rate [29]. As any other amplicon based NGS tests, redundant primer pairs should be included to avoid allele drop-out caused by rare SNPs. In addition, the amplicon size has to be optimized for sequencers and to generate overlapping reads in long exons to ensure full coverage. Recent studies have reported the pathogenic variant detection rates from 19% to 68% in patients who had sequencing test for RASopathies [6, 11, 29]. We have designed a capture based 12 gene NSD NGS panel, including the *BRAF, CBL, HRAS, KRAS, MAP2K1, MAP2K2, NRAS, PTPN11, RAF1, RIT1, SHOC2*, and *SOS1* genes. For the first 73 patients evaluated by our laboratory, 38 pathogenic or likely pathogenic variants (52% of the cases) were detected (data not published). It is worth noting that the recently discovered RIT1 gene accounts for about 10% of the diagnosed patients in our cohort.

3.1 NGS Based Prenatal Diagnosis

It has been estimated that Noonan spectrum syndrome has high prevalence as 1 in 1000–2500 live birth previously [3]. Most of those typical presentations of postnatal Noonan spectrum disorders are not identified in the first or second trimester of pregnancy. Although indications of Noonan spectrum disorders related prenatal abnormalities have been observed as mention above, such associations are only suggestive. Precise molecular diagnosis then can play a critical role to provide early medical intervention, especially for patients with structural cardiac defects identified usually after delivery.

3.1.1 Whole Genome Amplification for Prenatal DNA

The quantity of DNA isolated from chorionic villi sampling (CVS) or amniotic fluid is usually not sufficient for NGS based targeted gene enrichment. For example, it has been estimatated that about 100 ng DNA per ml of direct amniotic fluid at early gestational age. Thus, it is often a challenge to obtain sufficient amount of DNA for NGS analysis of multiple target genes, which usually requires at least one microgram of DNA for target gene enrichment. Thus, unbiased whole genome amplification (WGA) is necessary to generate sufficient DNA. Efforts have been reported in various samples with low DNA input, such as cancer tissues from micro dissect [47] and cells derived from *in vitro* fertilization for preimplementation examination [14].

WGA can be achieved by different mechanisms. In principle they can be categorized into PCR-based and Non-PCR based. PCR based WGA utilizes PCR amplification with modifications of primers to incorporate randomized nucleotides for PCR reaction working on whole genome as uniformly as possible. Some of those PCR based WGA can be regarded as modified library preparation protocols for NGS. Non-PCR based WGA methods are isothermal amplification using Phi29 DNA polymerase (MDA) [57], T4 replisome [51] or T7 RNA polymerase(LinDA or LADS) [23, 55]. Performance of each WGA methods has been evaluated under different platforms: MDA or GenomePlex in aCGH [63], in amplicon-NGS [45], MDA in whole exome sequencing [22].

3.1.2 Validation of Capture/NGS Based Prenatal Diagnosis

A custom-designed capture library targeted to 12 genes *(BRAF, CBL, HRAS, KRAS, MAP2K1, MAP2K2, NRAS, PTPN11, RAT1, RIT1, SHOC2, SOS1*), was designed to capture a total of 135 CDS of these genes with a total target region of 24.7 kb. DNA extracted from varies kind of validation samples, such as blood, cultured chorionic villus sampling (CVS), cultured amniocentesis (AMNIO) and tissues were amplified using MDA based method Illustra GenomiPhi V2 DNA Amplification Kit (GE Healthcare, Waukesha, Wisconsin) with 100 ng of DNA input for WGA. Amplified DNA products were used for library construction and then for NimbleGen based capture NGS. WGA performance was accessed first on representativeness. As shown in Table 5 an average 95.3% of total captured CDS (ranged from 84.4% to 100%) have relative fold change within 0.5 ~ 1.5 compared to non-WGA samples, which is acceptable for evenness of amplification efficiency. So the amplification performance in terms of representativeness is consistently good for most regions of genes in Noonan Spectrum test.

Accuracy of whole genome amplification was accessed by the pathogenic variant calling. All negative samples and positive pathogenic variants were correctly identified through double blinded test and listed in Table 5. Sample 4, i.e. DNA from Sample 1 mixed with 5% of DNA from Sample 2, was used to mimic the maternal contamination in fetus sample, which is not a rare event during clinical sample collection for direct AMNIO or CVS. NGS results of variants calling indicated that WGA process we used here would not exaggerate the contamination so maternal

Table 5 Summary of Noonan NGS panel validation

Validation ssample	Sample type	Percentage of CDS with relative fold change within 0.5~1.5	Variant	Concordant with Sanger sequencing
1	Prenatal(tissue)/trio	88.10%	neg	Yes
2	Postnatal(blood)/trio	99.30%	neg	Yes
3	Postnatal(blood)/trio	99.30%	neg	Yes
4	Mixed of sample 1 and sample 2	94.80%	neg	Yes
5	Prenatal(CCVS)	97.10%	neg	Yes
6	Prenatal(CCVS)	91.10%	neg	Yes
7	Prenatal(CCVS)	84.40%	PTPN11:c.215C>G	Yes
8	Prenatal(CAMNIO)	100%	neg	Yes
9	Prenatal(CCVS)	99.20%	neg	Yes
10	Postnatal(blood)	93.30%	SOS1:c.1656G>C	Yes
11	Postnatal(blood)	98.50%	RAF1:c.781C>A	Yes
12	Postnatal(blood)	99.30%	HRAS:c.34G>A	Yes

alleles could be distinguishable for the low (~5%) heterozygosity, which would be expected from the validation results on evenness of amplification efficiency as mentioned above. To further rule distinguish possible maternal contamination and possible chimerical pathogenic variants in prenatal samples, we always perform Trio-analysis (simultaneously run prenatal sample and parental samples) by which maternal alleles can be identified from variants calls in prenatal samples.

3.1.3 Insufficiently Covered Regions and Pseudogene

At the average coverage depth of 500X, there are 3 recurrent insufficiently (<20X) covered exons (*PTPN11*-exon1, *BRAF*-exon1, and *MAP2K2*-exon1) that require specific amplification followed by sequencing. These insufficiently covered exons have been consistently observed in both WGA and nonWGA analyses, and are similar to what have been previously reported ([29]).

Pseudogene interference has been observed in *PTPN11*-exon 6, *MAP2K2*-exon 6. *PTPN11* exon6 has several highly homologous regions across the genome (with identities ranging from 91.0% to 97.5%). Capture based enrichment methods would easily bring these pseudogene sequences into the final sequencing libraries. Routine bioinformatic analytical pipeline is usually not designed to distinguish homologous or pseudogene sequences from active genes. Thus, sequences in these regions with problematic alignment or without effective specific capture probes should be carefully reviewed, and re-aligned if necessary. The pseudogene sequences can be distinguished by their recurrent appearance and *in cis* alignment pattern. Nevertheless, it is necessary to use active gene specific primers for Sanger sequencing to rule out ambiguous NGS calls.

3.2 Trio Analysis for the Detection of De Novo Findings

As mentioned above, trio analysis can be used to distinguish between possible maternal allele from contamination and possible chimerical low heterozygocity allele in fetus DNA. Moreover, observance of parental alleles could provide valuable information on pathogenicity of a variant detected in prenatal samples because of autosomal dominant pattern caused by pathogenic alleles in RASopathies. From 100 trio prenatal cased analyzed we identified four *de novo* novel pathogenic/likely pathogenic variants and one probably *de novo* pathogenic variant (paternal DNA unavailable), eight inherited novel variants with unknown significance and one inherited reported variant with unknown significance (Table 6). For novel VUS and reported VUS, it is interesting to notice that these VUS were observed paternally or maternally, which could provide additional information for variant classification if clinical presentations of parents are available. On the other hand, some cases appeared to be *de novo* at the time of test request. However, the molecular results indicated that one of the parents carried the same pathogenic variant. Thus, due to variable penetrance and expressivity, careful clinical evaluation of the parents should be a general practice before and after prenatal diagnosis. The success of trio analysis with WGA suggests possible application of non-invasive prenatal genetic testing (NIPT) for RASopathies in the near future.

Table 6 Variants identified in 100 prenatal cases

	Gene	Variant	Inherited or De novo	Variant classification	Sample type
1	PTPN11	c.215C>T (p.A72V)	De novo	Likely pathogenic	CCVS
2	PTPN11	c.1505OT(p.S502L)	Not inherited from mother	Pathogenic	CCVS
3	RIT1	c.268A>G (p.M90V)	De novo	Likely pathogenic	CAMNIO
4	RIT1	c.170OG (p.A57G)	De novo	Likely pathogenic	CAMNIO
5	SOS1	c.508A>G (p.K170E)	De novo	Pathogenic	CAMNIO
6	SOS1	c.1051OG (p.L351V	Inherited from mother	VUS	CCVS
7	SOS1	c.911G>C (p.R304P)	Inherited from father	VUS	CCVS
8	MAP2K2	c.1112G>A (p.R371Q)	Inherited from mother	VUS	CAMNIO
9	RAF1	c.124_125GOAT,p.A421	Inherited from father	VUS	CAMNIO
10	CBL	c.1324C>A(p.L442M)	Inherited from father	VUS	CCVS
11	CBL	c.2635G>A (p.V879l)	Inherited from father	VUS	CCVS
12	RIT1	c.634C>T (p.R212W)	Inherited from mother	VUS	CCVS
13	SHOC2	c.1594A>G (p.S532G)	Inherited from father	VUS	CCVS

References

1. Anastasaki, C., Estep, A.L., Marais, R., Rauen, K.A., Patton, E.E.: Kinase-activating and kinase-impaired cardio-facio-cutaneous syndrome alleles have activity during zebrafish development and are sensitive to small molecule inhibitors. Hum. Mol. Genet. **18**(14), 2543–2554 (2009)
2. Aoki, Y., Niihori, T., Narumi, Y., Kure, S., Matsubara, Y.: The RAS/MAPK syndromes: novel roles of the RAS pathway in human genetic disorders. Hum. Mutat. **29**(8), 992–100 (2008)
3. Aoki, Y., Niihori, T., Banjo, T., Okamoto, N., Mizuno, S., Kurosawa, K., Ogata, T., Takada, F., Yano, M., Ando, T., Hoshika, T., Barnett, C., Ohashi, H., Kawame, H., Hasegawa, T., Okutani, T., Nagashima, T., Hasegawa, S., Funayama, R., Nagashima, T., Nakayama, K., Inoue, S., Watanabe, Y., Ogura, T., Matsubara, Y.: Gain-of-function mutations in RIT1 cause Noonan syndrome a RAS/MAPK pathway syndrome. Am. J. Hum. Genet. **93**, 173–180 (2013)
4. Bakker, M., Pajkrt, E., Mathijssen, I.B., Bilardo, C.M.: Targeted ultrasound examination and DNA testing for Noonan syndrome, in fetuses with increased nuchal translycency and normal karyotype. Prenat. Diagn. **31**, 833–840 (2011)
5. Bertola, D.R., Yamamoto, G.L., Almeida, T.F., Buscarilli, M., Jorge, A.A., Malaquias, A.C., Kim, C.A., Takahashi, V.N., Passos-Bueno, M.R., Pereira, A.C.: Further evidence of the importance of RIT1 in Noonan syndrome. Am J Hum Genet A. **93**, 173–180 (2014)
6. Bhoj, E.J., Yu, Z., Guan, Q., Ahrens-Nicklas, R., Cao, K., Luo, M., Tischler, T., Deardorff, M., Zackai, E., Santani, A.: Phenotypic predicators and final diagnoses in patients referred for RASopathy testing by targeted next-generation sequencing. Genet. Med. (2016). doi:10.1038/gim.2016.169
7. Carta, C., Pantaleoni, F., Bocchinfuso, G., Stella, L., Vasta, I., Sarkozy, A., Digilio, C., Palleschi, A., Pizzuti, A., Grammatico, P., Zampino, G., Dallapiccola, B., Gelb, B.D., Tartaglia, M.: Germline missense mutations affecting KRAS isoform B are associated with a severe Noonan syndrome phenotype. Am. J. Hum. Genet. **79**, 129–135 (2006)
8. Cawthon, R.M., O'Connell, P., Buchberg, A.M., Viskochil, D., Weiss, R.B., Culver, M., Stevens, J., Jenkins, N.A., Copeland, N.G., White, R.: Identification and characterization of transcripts from the neurofibromatosis 1 region: the sequence and genomic structure ofEVI2 and mapping of other transcripts. Genomics. **7**, 555–565 (1990)
9. Chen, P.C., Yin, J., Yu, H.W., Yuan, T., Fernandez, M., Yung, C.K., Trinh, Q.M., Peltekova, V.D., Reid, J.G., Tworog-Dube, E., Morgan, M.B., Muzny, D.M., Stein, L., McPherson, J.D., Roberts, A.E., Gibbs, R.A., Neel, B.G., Kucherlapati, R.: Next-generation sequencing identifies rare variants associated with Noonan syndrome. Proc. Natl. Acad. Sci. U. S. A. **111**(31), 11473–11478 (2014)
10. Cirstea, I.C., Kutsche, K., Dvorsky, R., Gremer, L., Carta, C., Horn, D., Roberts, A.E., Lepri, F., Merbitz-Zahradnik, T., König, R., Kratz, C.P., Pantaleoni, F., Dentici, M.L., Joshi, V.A., Kucherlapati, R.S., Mazzanti, L., Mundlos, S., Patton, M.A., Silengo, M.C., Rossi, C., Zampino, G., Digilio, C., Stuppia, L., Seemanova, E., Pennacchio, L.A., Gelb, B.D., Dallapiccola, B., Wittinghofer, A., Ahmadian, M.R., Tartaglia, M., Zenker, M.: A restricted spectrum of NRAS mutations causes Noonan syndrome. Nat. Genet. **42**, 27–29 (2010)
11. Čizmárová, M., Hlinková, K., Bertok, S., Kotnik, P., Duba, H.C., Bertalan, R., Poločková, K., Košťálová, Ľ., Pribilincová, Z., Hlavatá, A., Kovács, L., Ilenčíková, D.: New mutations associated with Rasopathies in the central European population and Genotyp-Phenotyp correlations. Ann. Hum. Genet. **80**(1), 50–62 (2016)
12. Cordeddu, V., Di Schiavi, E., Pennacchio, L.A., Ma'ayan, A., Sarkozy, A., et al.: Mutation of SHOC2 promotes aberrant protein N-myristoylation and causes Noonan-like syndrome with loose anagen hair. Nat. Genet. **41**, 1022–1026 (2009)
13. Cordeddu, V., Yin, J.C., Gunnarsson, C., Virtanen, C., Drunat, S., Lepri, F., De Luca, A., Rossi, C., Ciolfi, A., Pugh, T.J., Bruselles, A., Priest, J.R., Pennacchio, L.A., Lu, Z., Danesh, A., Quevedo, R., Hamid, A., Martinelli, S., Pantaleoni, F., Gnazzo, M., Daniele, P., Lissewski, C., Bocchinfuso, G., Stella, L., Odent, S., Philip, N., Faivre, L., Vlckova, M., Seemanova, E., Digilio, C., Zenker, M., Zampino, G., Verloes, A., Dallapiccola, B., Roberts, A.E., Cavé, H., Gelb, B.D., Neel, B.G., Tartaglia, M.: Activating mutations affecting the Dbl homology domain of SOS2 cause Noonan Syndrome. Hum. Mutat. **36**(11), 1080–1087 (2015)

The Applications and Challenges of Next-Generation Sequencing in Diagnosing Neuromuscular Disorders

Lijun Wang, Arunkanth Ankala, Hamoud Al Khallaf, Xinjian Wang, Mikhail Martchenko, Biao Dong, Ammar Husami, Kejian Zhang, and C. Alexander Valencia

Abstract Inherited neuromuscular disorders (NMDs) form a group of highly heterogeneous diseases with a relatively high incidence of 1 in 3000. NMDs affect the peripheral nervous and muscular systems, resulting in gross motor disability. Disease subtype diagnosis is complicated by the high clinical and genetic heterogeneities of the disease and consequently more than 50% of the cases remain molecularly uncharacterized. Traditional gene-by-gene approach is quite exhaustive and after a few negative tests the quest for diagnosis is often given up without establishing diagnosis. However in the recent years, the clinical applications of next-generation sequencing (NGS)-based comprehensive approach such as multi-gene panels, and exome sequencing have allowed for rapid diagnosis. Additionally, its applications in research settings have allowed for identification of new disease-causing genes and variants which translated into an improved clinical diagnostic yield. Here, we discuss the application of NGS technology in NMDs as a diagnostic and research tool. We conclude that such an application will tremendously broaden

L. Wang • X. Wang • A. Husami
Division of Human Genetics, Cincinnati Children's Hospital Medical Center, Cincinnati, OH, USA

A. Ankala • H. Al Khallaf
Department of Human Genetics, Emory University School of Medicine, Atlanta, GA, USA

M. Martchenko
Keck Graduate Institute, Claremont, CA, USA

B. Dong
State Key Laboratory of Biotherapy, West China Hospital, Sichuan University, Chengdu, People's Republic of China

K. Zhang • C. Alexander Valencia (✉)
Division of Human Genetics, Cincinnati Children's Hospital Medical Center, Cincinnati, OH, USA

Department of Pediatrics, University of Cincinnati College of Medicine, 3333 Burnet Ave. R1-1028, MLC 7016, Cincinnati, OH 45229, USA
e-mail: alexander.valencia@cchmc.org

L.-J.C. Wong (ed.), *Next Generation Sequencing Based Clinical Molecular Diagnosis of Human Genetic Disorders*, DOI 10.1007/978-3-319-56418-0_10

our knowledge of NMDs; the outcome of which includes rapid and accurate diagnosis that would result in an earlier and more effective intervention.

Keywords Neuromuscular disorder • Exome sequencing • Muscular dystrophy • Next-generation • Sequencing • Inherited

1 Introduction

NMDs are a group of genetic diseases that affect the muscles (e.g. myopathies and muscular dystrophies), nerves (e.g. Charcot–Marie–Tooth neuropathies), and neuromuscular junctions (e.g. myasthenic syndromes) result in long term incapacity and considerable burden to the patients, their families and health care systems. This group is composed of more than 200 monogenic disorders with a total incidence that is greater than 1 in 3000 [1]. For about half of the cases, the molecular cause has not been identified. An extensive clinical evaluation with complementary gene-by-gene testing is often required to reach a definitive diagnosis. Due to genetic heterogeneity and lack of segregation in sporadic cases, reaching a diagnosis is challenging, lengthy and costly. The genetic heterogeneity can be demonstrated by the number of genes involved in specific subgroups of inherited NMDs, namely, hereditary sensorimotor neuropathies (HSMN; 50 genes), Charcot–Marie–Tooth diseases (CMT; 30 genes) and congenital muscular dystrophies (CMD, 12 genes) [2, 3]. In other instances, some NMD-related genes are very large and have not been sequenced completely because it is costly and labor intensive by Sanger. For the patient, this gene-by-gene approach not only increases the number of tests that are required, but also delays the diagnosis and results in unnecessary investigations and treatments [4]. NGS interrogates multiple genes in parallel and can be extended to the whole exome. In the field of NMDs, this approach has been mainly used in a research setting to identify novel disease genes. Its diagnostic potential was soon realized and put into clinical use by creating different NMD panels and by examining the exome of NMD-related genes. In this review, we describe the complicated nature of this group of disorders and the challenges faced in their diagnosis. We discuss why molecular testing is considered the gold standard for diagnosing them and how utilizing NGS, the most recent molecular testing technique, has allowed that in a timely and more cost-effective manner. We also quote various recent reports of targeted multigene panel and exome studies to bring into perspective the true potential of NGS in the field of NMDs, not only as a diagnostic method, but also as a powerful gene discovery tool. Finally, the limitations of this technology and some proposed solutions to circumvent them are briefly discussed.

2 Heterogeneity of NMDs: Genetic and Phenotypic Overlap

NMDs are well known for their clinical and genetic heterogeneity. Muscular dystrophies which form the majority of inherited NMDs are a good example of the heterogeneous nature of these disorders. Muscular dystrophies share clinical, genetic, and pathological characteristics. Major clinical characteristics of the disease group include muscle degeneration and wasting, progressive muscle weakness, hypotonia, and although at very variable levels, elevated serum creatine kinase levels [5]. Very often cardiac involvement might also be present, accounting for higher morbidity and mortality. There are over 80 different genetically defined types of muscular dystrophies categorized into different subgroups based on the age of onset, the specific muscles involved, and common characteristic clinical features [6, 7]. However, there is a high genetic and phenotypic overlap among these subgroups. Congenital muscular dystrophies (CMDs) and limb-girdle muscular dystrophies (LGMDs) are the two major subgroups, the genetic heterogeneity of which has been expanding rapidly in recent years, with more and more genes being implicated [8]. Lack of pathognomonic signs or specific biochemical markers and the presence of high phenotypic overlap with other forms of NMDs often make clinical diagnosis difficult and molecular confirmation expensive.

3 Diagnostic Challenges Associated with NMDs

The diagnosis of NMDs involves the clinician, the pathologist and the molecular genetics laboratory. Even today, with the increasing literature on the characterization of the different types of the NMDs and published algorithms and guidelines on how to differentiate between them, a skillful clinician finds it difficult to tell what specific type of NMDs a patient has [9, 10]. That is due to some inherent nature of these disorders, including the lack of pathognomonic signs and/or symptoms. The fact that a large number of NMD patients present with non-specific clinical and histopathological features has been demonstrated in cases with congenital myopathies that are usually sub-classified into nemaline, core or centronuclear myopathies where no specific signs have been found in almost half of them, whereas other patients have diverse but overlapping clinical and histopathological manifestations [11].

Their clinical and genetic heterogeneities are another two inherent features of NMDs that complicate the picture even more. An example of how clinical heterogeneity can hinder the precise diagnosis of NMDs is when the clinician, based on the clinical picture, the mode of inheritance and the histopathology result, considers the diagnosis of autosomal recessive LGMD or nemaline myopathy. Knowing that there are around 15 known different genes that cause autosomal recessive LGMD and seven known different genes that cause nemaline myopathy, it is hard to tell which autosomal recessive LGMD or nemaline myopathy the patient has.

The clinical heterogeneity, where mutations in one gene cause different diseases, is another feather of NMDs that makes it even more difficult for the clinician to reach an accurate diagnosis. For example, mutations in the slow-skeletal beta cardiac myosin gene (*MYH7*) cause several disorders including: hypertrophic and dilated cardiomyopathy, Laing distal myopathy, both dominant and recessive myosin storage myopathy, minicore myopathy, scapuloperoneal myopathy, congenital fiber type disproportion (CFTD), and mixtures of the different phenotypes [12–20]. Caveolin 3 gene is another example of this phenomenon. Mutations in this gene cause four different muscle disease phenotypes: limb-girdle muscular dystrophy, distal myopathy, rippling muscle disease and hyperCKemia [21]. An overlap of these symptoms can present in one patient, or diverse clinical phenotypes with different severities can be caused by the same mutation in different individuals or even different family members [22]. This clinical heterogeneity makes it difficult for the clinician to suggest a candidate gene for molecular testing.

Another challenge is the unidentified genes in different NMDs. For instance, amyotrophic lateral sclerosis (ALS), a neurodegenerative disorder characterized by the loss of motor neurons in brain and spinal cord can be caused by mutations in several genes, including *SOD1*, *SETX*, *ALS2*, *TARDBP*, *FUS*, *ANG* and *C9orf72* [23–32]. However, for a significant number of patients who have this disorder, have no known mutation in any of these genes, suggesting that this disorder can be caused by other genes yet to be discovered. So, it is easy to imagine the dilemma a clinician would be facing in reporting the result to a patient thought to have ALS when the molecular testing for all the aforementioned ALS genes did not discover a mutation. In fact, despite the successful identification of so many NMD genes, many patients and families for whom the disease-causing gene is still unknown are estimated to be at least 40% of neuromuscular disorder patients [33, 34]. Thereby, these families do not have an accurate estimate of the recurrent risk of those disorders in future pregnancies [35].

Another challenge in diagnosing NMDs is that non-molecular investigations like muscle imaging with magnetic resonance imaging (MRI), computed tomography (CT) or ultrasound are not always helpful. These investigations have been utilized to differentiate muscle diseases from each other, since the pattern of affected muscles can be different in different muscle diseases [36–39]. However, the benefit of these investigations is limited to NMDs affecting mainly the muscles, besides the severity and the types of the muscles involved cannot accurately predict the responsible gene. These investigations at their best can only narrow down the list of the responsible genes. Even procedures like electromyography (EMG), muscle and nerve biopsy, in addition to being invasive, they are not always helpful in differentiating some NMDs subtypes. EMG can help in differentiating myopathies from motor neuropathies, anterior horn cell and neuromuscular junction disorders. This technique assesses several components of muscle electrical activity: the muscle's spontaneous activity, its response to the insertion of a probe, the character of the muscle's individual motor unit action potentials and the rapidity with which additional motor units are recruited in response to an electrical signal. However, one of the problems of differentiating myogenic from neurogenic diseases is that slow, chronic myogenic disorders can ultimately affect the innervating nerve leading to a mixed myogenic/ neurogenic picture, which can be hard to differentiate [34].

Another example of the limited utility of these investigations is the fact that different congenital myopathies can share pathological findings. It has become clear that many individuals with a genetically confirmed CMD have only non-specific histopathological features. It is also known that mutations in the same gene can cause different muscle pathologies. For example, mutations in α-skeletal actin can result in actin accumulations, nemaline myopathy, intranuclear rod myopathy, zebra body myopathy, CFTD, and cap disease [40–47]. The same genetic mutation can lead to different histopathological features in different members of the same family or in the same individual at different ages. That has been proved for ryanodine receptor gene (RYR1) mutations and has also been reproduced in a mouse model [48]. Therefore, the correlation between pathology result and the genetic cause is not always clear.

From what has been mentioned so far, it is obvious why molecular testing is considered the gold standard in diagnosing NMDs. Molecular diagnosis of these disorders has its own challenges. The genetic heterogeneity of these disorders is not only a challenge to the clinician, as discussed earlier; it is a challenge to molecular diagnosis as well. This high degree of genetic heterogeneity becomes very problematic when using conventional molecular diagnostic tool like Sanger sequencing, for example. Taking CMT neuropathies as an example, it would be time-consuming and costly to test the several different implicated genes (more than 30 genes) in diagnostic laboratories. Some of these genes are only Sanger sequenced in a very few specialized laboratories around the world, which explains why some samples need to be sent to several laboratories to cover all candidate genes. Some NMDs genes are the largest human genes such as *DMD* spanning more than 2.3 Mb with 79 exons, *TTN* with 363 exons with an open reading frame spanning more than 100 Kb, *NEB* with 183 exons and *RYR1* with 106 exons [49–52]. The implications of their size become clear when considering the cost of identifying their mutations by Sanger sequencing. The cost is very high since by Sanger sequencing only a small number of closely spaced gene exons can be sequenced at a time, which is why that can be costly and time consuming. Due to their large size, these genes might not be fully tested or only the mutation hot spot regions may be analyzed which might miss some less common and rare disease causing mutations in these genes.

Since different types of mutations have been detected in NMDs patients, diagnosing NMDs require the implementation of several molecular diagnostic techniques which can be laborious, costly, and time consuming. The following are some examples of the different mutation types that have been discovered in NMDs patients and the different testing platforms implemented to diagnose them. The first example is the types of mutations seen in the *DMD* gene. It is known that 60–65% of patients with Duchenne and Becker muscular dystrophies (DMD and BMD) have deletions in the *DMD* gene, 5–15% have duplications, and the rest have point mutations or small insertions–deletions, which means that sequencing alone is not enough to cover all types of mutations in this gene and a need for another molecular testing platform e.g. comparative genomic hybridization or Multiplex Ligation Probe Amplification (MLPA) is needed to achieve that [53–55]. Spinal muscular atrophy is an example of how different molecular testing platforms can be used in a complementary way to detect its most common mutation, where restriction enzyme digestion is used following PCR to discriminate between homozygous deletion of the centromeric and telomeric copies of the *SMN* gene [56]. To diagnose Facioscapulohumeral muscular dystrophy 1 (FSHD1), Southern blotting is used to size the macrosatellite repeat D4Z4 on chromosome 4q35 [57]. Repeat PCR is another example of another molecular testing platform utilized in NMDs diagnosis.

This technique has been useful in achieving rapid diagnosis in some NMDs like myotonic dystrophy types 1 (DM1) and 2 (DM2) [58–60]. It has also been proved useful in detecting the hexanucleotide repeat, the single commonest known cause of familial and sporadic ALS [31, 32].

An important obstacle in NMDs diagnosis that cannot be overlooked is the fact that not all NMD genes have been identified. Therefore, even after screening all known implicated genes, some NMD patients will remain without molecular diagnosis. This obstacle cannot be overcome by NGS through customized panels, since only mutations in the included genes will be discovered and mutations in other disease causing genes, yet to be discovered, will be missed. As will be discussed later, this obstacle can only be overcome by WES.

It is clear why molecular testing using NGS technology is considered an efficient diagnostic approach for these disorders. In the following parts of this review, we will briefly describe the technical aspect of this technology, discuss some literature reports of its successful implementation as a diagnostic and research tool, address some of its limitations and share some conclusions we have reached.

4 The Concept of NGS

Next-generation sequencing technologies, by sequencing hundreds of thousands to millions of DNA templates in parallel, resulted in higher throughput (Gb scale) and lowered sequencing cost [61, 62]. Generally, next-generation sequencing is composed of four steps; DNA isolation, target sequences enrichment, sequencing by next-generation platform and bioinformatics analysis. During the analysis, fragment sequences are aligned and variant calls are obtained and prioritized by applying various filters to identify the potentially causative gene variants. Before sequencing, a significant effort has focused on the development of 'target-enrichment' methods, in which genomic regions are selectively captured. To enrich for regions of interest that range in size from hundreds of kb to the whole exome, genomic enrichment steps, both traditional and novel, are being incorporated into overall experimental designs [63]. Traditional overlapping long-range PCR amplicons (approximately 5–10 kb) can only be used for up to several 100 kb. More recently, enrichment based on hybridization of fragmented genomic DNA to oligonucleotide capture probes has been successfully achieved by several groups [64–67]. Capture probes can be immobilized on a solid surface (Roche NimbleGen, Agilent Technologies, and Febit) or used in solution (Agilent, Roche NimbleGen). Another enrichment approach that relies on the use of molecular inversion probes (MIPs) was initially developed for multiplex target detection and SNP genotyping [68, 69]. In principle, single stranded oligonucleotides, consisting of a common linker flanked by target-specific sequences, anneal to their target sequence and become circularized by a ligase [65]. An alternative enrichment approach, developed by RainDance Technologies, individual pairs of PCR primers for the genomic regions of interest are segregated in water in emulsion droplets and then pooled to create a primer library [70]. Separately, emulsion droplets containing genomic DNA and PCR reagents are prepared. Following the merging of the droplets, DNA is amplified by the PCR and subsequently processed for NGS.

Massively parallel sequencing is one common feature shared by almost all current NGS platforms, following clonally amplified single DNA molecules, separated in a defined microchamber (called Flow Cells, FlowChips or PicoTiter plate) [63]. One exception to this is Pacific Biosciences' which uses single-molecule sequencing technology without clonal amplification [71]. In contrast, Sanger sequencing has orders of magnitude lower throughput by sequencing products produced in individual sequencing reactions. NGS is first carried out by fragmenting the genomic DNA into small pieces, usually in the range of 300–500 bps [72]. Then, platform-specific adapters are ligated to the ends of the DNA segments, permitting for their attachment and sequencing. In the NGS execution, sequencing results are obtained by reading optical signals during repeated cycles from either polymerase-mediated fluorescent nucleotide extensions of four different colors (e.g., Illumina's HiSeq system), or from iterative cycles of fluorescently-labeled oligonucleotide ligation (e.g., ABI SOLiD system), or by the principle of pyrosequencing (e.g., Roche 454 system). Nonoptical DNA sequencing by detecting the hydrogen protons generated by template-directed DNA polymerase synthesis on semiconductor-sensing ion chips has recently been developed as well [73–75]. In such a massively parallel sequencing process, NGS platforms produce up to 600 gigabases of nucleotide sequence from a single instrument run (e.g., Illumina's HiSeq 2000) [76]. The sequenced fragments are called "reads," which could be 25–100 bps from one or both ends. The massive capacity of NGS allows the sequencing of many randomly overlapping DNA fragments; therefore, each nucleotide in targeted regions may be included in many reads, allowing repeated analysis which provides depth of coverage. Increased depth of coverage usually improves sequencing accuracy, because a consensus voting algorithm is used in determining the final nucleotide calls [77].

5 Clinical Utility of NGS-Based Analysis for NMD Testing

From what has been discussed so far, it is clear why NGS-based analysis is suitable for testing of diseases with high genetic and allelic heterogeneity such as NMDs [4]. It also allows sequencing large genes involved in NMDs such as *DMD* and *TTN* making it a more suitable technique for such disorders. In certain occasions, NGS had expanded the clinical spectrum of some NMDs related genes. Expanding the clinical spectrum of *TTN*-related diseases by noting a patient with myopathy with cytoplasmic aggregates and respiratory insufficiency who was diagnosed by NGS, is an example of such occasions [50]. In the following sections we mention some published work conducted by other investigators as examples of how NGS has been applied in NMDs testing.

6 Targeted Panel Approaches

6.1 Comprehensive NMDs Test: Resolving the Heterogeneities

Vasli et al. reported an efficient mutation screening strategy for heterogeneous NMD genes using DNA samples of patients with or without known mutations (Table 1) [4]. In the NMD panel, an average of 125 variants affecting splice sites or predicted to change the amino acid sequence, were described. Importantly, clinical, histological and molecular data were necessary for matching the genetic data with the phenotype. All known and novel mutations were identified in previously characterized patients and in patients without a molecular diagnosis. By comparing the number of reads in these regions with control DNA samples, a large deletion (exons 18–44) in *DMD* was detected in a patient with Duchenne muscular dystrophy (DMD). Mutations in *SETX* were found in two patients with ataxia. Similarly, samples from patients with heterogeneous NMDs without molecular characterization were sequenced. Potential disease-causing mutations in *RYR1*, *TTN* and *COL6A3* were identified in several patients. These genes were consistent with the clinical information. This demonstrates that this strategy can identify mutations in a wide range of genetically heterogeneous diseases.

In a more recent study by Chae et al. 43 patients with early onset neuromuscular disorders from unknown genetic origin were examined by performing NGS on 579

Table 1 Enrichment and next generation sequencing technologies applied to muscular dystrophy panels

Muscular dystrophy type	# of genes	# of exons	Target interval (kb)	Enrichment platform	NGS platform	References
Inherited NMD	267	4604	1600	Solution-based capture[a]	Illumina's Genome Analyzer IIx	[4]
DMD	1	79	30	Solid phase capture[b]	HiSeq 2000	[78]
CMD	12	321	65	RainDance[c], Solution-based capture[a]	SOLiD 3	[3]
DMD, CMD, LGMD	26	747	1069	Solution-based capture[a]	Illumina's Genome Analyzer IIx	[79]
Early onset neuromuscular disorders	579	10,706	388	Haloplex	GAIIx	[35]

[a]*Solution-based capture* Agilent's hybridization-based SureSelect
[b]*Solid phase capture* NimbleGen's hybridization-based microarray capture
[c]*RainDance* Highly multiplex PCR amplification, *Inherited NMD* inherited neuromuscular disorders, *DMD* Duchenne muscular dystrophy, *CMD* congenital muscular dystrophies, *LGMD* limb girdle muscular dystrophies

nuclear genes associated with myopathy [35]. In 21 of the 43 patients, the authors identified the definite genetic causes (48.8%). Additionally, likely pathogenic variants were identified in seven cases and variants of uncertain significance (VUCS) were suspected in four cases. In total, 19 novel and 15 known pathogenic variants in 17 genes were identified in 32 patients. A targeted NGS approach can offer cost effective, safe and fairly rapid turnaround time, which can improve quality of care for patients with early onset myopathies and muscular dystrophies. Tian et al. examined 35 unrelated NMD families (38 patients) with clinical and/or muscle pathologic diagnoses but without a genetic cause. Importantly, mutations were found in 29 families (83%); causative mutations were identified in 21 families (60%) and likely cause was found in eight families (23%) [80].

6.2 NGS Approach for DMD/ BMD: For a Single Versatile Assay

DMD and BMD are the most common forms of childhood muscular dystrophy [1]. They are both X-linked inherited neuromuscular disorders caused by mutations in the dystrophin gene (*DMD*; locus Xp21.2). The mutation spectrum of *DMD* is unique in that 65% of causative mutations are intragenic deletions, with intragenic duplications and point mutations (along with other sequence variants) accounting for 6–10% and 30–35%, respectively. The strategy for molecular diagnostic testing for *DMD* involves initial screening for deletions/duplications using microarray-based comparative genomic hybridization (array-CGH) followed by full sequence analysis of *DMD* for sequence variants [35, 81–85]. Recently, Lim et al. reported the application of NGS as a singleton assay for the diagnosis of DMD/BMD addressing the diagnostic issues related to the large size of the *DMD* gene and its complex mutational spectrum (Table 1) [79]. Individuals with deficient dystrophin expression were analyzed by a single NGS assay and were found to have either a large deletion/duplication or a point mutation, thereby establishing the diagnosis. This was a successful demonstration of the clinical utility of NGS assay as a single platform test for DMD/BMD. Similarly, substituting Sanger sequencing of all the 79 exons of *DMD* by NGS assay, Xie et al. detected a nonsense mutation p.R3381* located in exon 70 [78].

7 NMDs-Related Gene Variants Detected by WES: Diagnostic Tool

In certain circumstances clinical exome sequencing may be indicated, for example, if the patient's symptoms or family history suggests a genetic etiology but does not correspond to a specific genetic disorder. In another scenario, the patient may have symptoms of a well-defined genetic disorder that is caused by multiple genes (genetic heterogeneity) for which a multi-gene panel is not clinically available. In a third situation, a patient has a likely genetic disorder but clinical genetic testing did not yield a genetic diagnosis. Reports of the use of whole-exome sequencing in clinical practice are limited [86]. Yang et al. presented data on the first 250 probands for whom referring physicians ordered whole-exome sequencing [87]. Approximately 80% were children with neurologic phenotypes. They identified 86 mutated alleles that were highly likely to be causative in 62 of the 250 patients, achieving a 25% molecular diagnostic rate. A total of four probands received two non-overlapping molecular diagnoses, which potentially challenged the clinical diagnosis. For example, patient four had pathogenic mutations in *POMT2* (autosomal recessive, muscular dystrophy dystroglycanopathy) and *SCN2A* (autosomal dominant, seizure disorder). Similarly, Iglesias et al. report whole-exome sequencing in 115 patients to evaluate its clinical usefulness in clinical care [86]. They identified four new candidate human disease genes and possibly expanded the disease phenotypes associated with five different genes. Establishing a diagnosis led to discontinuation of additional planned testing in all patients, screening for additional manifestations in eight, altered management in 14, novel therapy in two, identification of other familial mutation carriers in five, and reproductive planning in six.

8 NMD-Related Gene Variants Detected by WES: Discovery Tool

8.1 Facioscapulohumeral Muscular Dystrophy (FSHD)

FSHD is an adult muscular dystrophy, which is characterized by progressive muscle weakness (Table 2). FSHD1 is clinically indistinguishable from FSHD2. FSHD is divided into types 1 (FSHD1) and 2 (FSHD2) based on genetic mutations. FSHD1, which is clinically indistinguishable from FSHD2, is associated with contraction of the D4Z4 macrosatellite repeat in the subtelomeric region of chromosome 4q35. FSHD2 shows digenic inheritance, requiring the inheritance of two independent genetic variations: a mutation in the *SMCHD1* gene on chromosome 18p that results in D4Z4 chromatin relaxation and an FSHD-permissive DUX4 allele on chromosome 4. *SMCHD1* mutations segregate independently from the FSHD-permissive DUX4 allele. By using trio exome and linkage analysis, Mitsuhashi et al. found a novel p.L275del mutation in *SMCHD1* altering a highly conserved amino acid in the ATPase domain [99]. Linkage analysis was also performed in this family with the assumption of autosomal dominant inheritance. The evidence for *SMCHD1* mutations in FSHD2 and the clinical presentations permitted Mitsuhashi et al. to conclude that the *SMCHD1* mutation is the likely cause of the disease in this family.

Table 2 Exome capture and NGS technologies applied to muscular dystrophies

NM type	Gene	Inheritance	Mutations	Study type	Enrichment	NGS platform	References
CMD	*B3GALNT2*	AR	c.740G>A (p.Gly247Glu) c.875G>C (p.Arg292Pro) c.51_73dup (p.Ser25Cysfs*38) c.308del (p.Val103Glyfs*10) c.755T>G (p.Val252Gly) c.802G>A (p.Val268Met) c.1423C>T (p.Gln475*)	102 individuals with dystroglycanopathy	SureSelect Exon 50 Mb	Illumina HiSeq	[88]
CMD	*DYSF, FKTN* *ISPD*	AR AR AD	c.2779del (p.Ala927Leufs*21) c.[915G>C]; [920G>A] (p.[Trp305Cys];[Arg307Gln]) c.458T>C (p.Ile153Thr)	Three independent trios	Illumina Exome Enrichment	Illumina HiSeq2000	[89]
CMD	*COL6A1*	AD	c.1056+1G>A	Four generation family	SeqCap EZ Human exome library,v2.0 (NimbleGen)	Illumina GAIIx	[90]
LGMD	*TRAPPC11*	AR	c.2938G>A (p.Gly980Arg) C.1287+5G>A (p.Ala372_Ser429del)	Homozygosity mapping & WES	SureSelect Human All Exon 50 MB Agilent	Illumina GAll	[91]
LGMD	*DES, FLNC*	AR	DES: c.3+3A>G FLNC: c.8130G>A(p.Trp2710*)	Two unrelated families	SureSelect Exon 50 Mb	Illumina HiSeq2000	[92]
LGMD	*DNAJB6*	AD	c.277T>C (p.Phe93Leu) c.287C>G (p.Pro96Arg)	Linkage analysis & WES	Illumina TruSeq	Illumina HiSeq2000	[93]

(continued)

Table 2 (continued)

NM type	Gene	Inheritance	Mutations	Study type	Enrichment	NGS platform	References
LGMD1F	*TNPO3*	AD	c.G2453A (p.Arg818Gln)	Four individuals	Agilent SureSelect Exon50Mb NimbleGen SeqCap EX v2.0	SOLiD system v4, Illumina HiSeq2000	[94]
CMD LGMD	*GMPPB*	AR	c.1000G>A, c.64C>T, c.95C>T c.860G>A, c.988G>A	Eight unrelated individuals	Illumina TruSeq Exome & SureSelect v.2	Illumina HiScanSQ 100x2	[95]
LGMD	*POMK*	AR	c.325C>T, (p.Q109*)	Two siblings with consanguineous parents	SureSelect V4	Illumina HiSeq2000	[96]
LGMD	*DNAJB6*	AD	c.265T>A (p.F89I)	Four generation family	SureSelect	Illumina MiSeq	[97]
EDMD	*LMNA*	AD	c.674G>A (p. Arg225Gln)	Generational exome analysis	SeqCap EZ exome & TruSeq	Illumina HiSeq2000	[98]
FSHD2	*SMCHD1*	AD	c. c.823_825del (p. Lys275del)	Trio analysis	SureSelect Human All Exon V4 Agilent	Illumina HiSeq 100bpX2	[99]
FSHD2	*SMCHD1*	Digenic	c.1058A>G (p.Tyr353Cys) c.1436G>C (p.Arg479Pro) c.1474T>C (p.Cys492Arg) c.2068C>T (p.Pro690Ser) c.2603G>A (p.Ser868Asn) c.4661T>C (p.Phe1554Ser)	14 patients	Short gun library	Illumina GAll	[57]

NM type	Gene	Inheritance	Mutations	Study type	Enrichment	NGS platform	References
WWS	*GTDC2, FKTN, POMT2*	AR	GTDC2: c.590G>A (p.Trp197*) FKTN: c.642dup (p. Asp215*) POMT2: c.1433A>G (p.His478Arg)	19 consanguinous pedigrees	SureSelect Exon37Mb	Illumina HiSeq	[100]
MD SPG	*CAPN3*	AR	c.550delA (p.Thr184Argfs*36)	125 individuals	SureSelect & TruSeq	Illumina GAll	[101]
MD	*CHKB*	AD	c.1031+3G>C	Trio		Ion proton	[102]
WWS	*B3GNT1*	AR	c.821_822insTT (p.Glu274Aspfs*94)	One consanguinous family	NA	NA	[103]
WDM	*TIA1*	AD	c.1150G>A (p.Glu384Lys)	43 patients with haplotype analysis and WES	SureSelect Human All Exon 50mb Agilent	SOLiD5500xI	[104]

LGMD limb girdle muscular dystrophy, *WWS* Walker-Warburg syndrome, *WDM* welander distal myopathy, *EDMD* emery-dreifuss muscular dystrophy, *FSHD2* facioscapulohumeral muscular dystrophy, *CMD* congenital muscular dystrophy, *MD* muscular disease, *SPG* spastic paraplegia, *AD* autosomal dominant, *AR* autosomal recessive

8.2 *Limb Girdle Muscular Dystrophy (LGMD)*

LGMDs compose of a large group of genetic myopathies, which are characterized by skeletal muscle weakness and atrophy restricted to the limbs, with relative sparing of the bulbar muscles. LGMDs are highly heterogeneous diseases (>50 genes) with two types LGMD1 (autosomal dominant inheritance) and LGMD2 (autosomal recessive inheritance). However, reaching a clinical diagnosis is always challenging because of the clinical variability of disease onset, progression, and distribution of the weakness as well as muscle atrophy among individuals and genetic subtypes in different LGMD patients. Bögershausen et al. performed WES on a consanguineous Syrian family with LGMD and five individuals of Huttcrite descent presenting with myopathy, infantile hyperkinetic movements, ataxia, and intellectual disability [91]. They identified a homozygous missense variant p.G980A in *TRAPPC11*, which encodes a component of the multiprotein TRAPP complex involved in membrane trafficking. Additionally, they investigated two affected siblings and their parents in a Hutterite family. The segregation study confirmed by Sanger sequencing, identified and verified a homozygous c.1287+5G>A splice-site mutation, which resulted in a 58 amino acid in-frame deletion (p.Ala372_Ser429del) in the foie gras domain of TRAPPC11. Furthermore, immunostaining of fibroblasts from patients showed disrupted Golgi morphology and changed protein transport along the secretory pathway. This study indicated that abnormal membrane trafficking could be the possible molecular mechanism of LGMD resulted from *TRAPPC11* mutants and that WES is an efficient technology to detect the multiple gene variants involved in LGMD at one time.

Harms et al. combined WES with genome-wide linkage analysis to identify the causative gene in an autosomal dominant LGMD with skeletal muscle vacuoles in a Caucasian family of 3 patients [93]. They found that p.F93L substitution in a highly conserved residue within the G/F domain of *DNAJB6* was confirmed by Sanger sequencing and segregated with disease in the broader pedigree. Furthermore, the linkage analysis excluded all variants except the p.F93L mutation in the G/F domain of the *DNAJB6* gene. In addition, abnormal accumulation of DNAJB6 in patient muscles, suggested a damage caused by the mutant. Their study has provided genetic and pathologic evidence that mutations in *DNAJB6* are a novel cause of this dominantly-inherited myopathy.

McDonald et al. performed WES for three affected individuals in family LGMD2359 and two affected individuals in family LGMD2692, respectively [92]. In addition, one married-in spouse from each family and one unrelated individual who did not show any signs of muscle disease were submitted for WES as controls. A splice variant, c.3+3A>G, in desmin (*DES*) segregated with disease in the family. In the second family, a second variant in filamin C (*FLNC*) causing protein truncation p.W2710X was identified.

Torella et al. performed WES on 19 LGMD1F patients and eight healthy relatives. LGMD1F is a subgroup disease of LGMD1 presenting with an early onset of disease without massive protein aggregates of myofibrillar myopathy [94]. Linkage analysis followed by WES analysis was performed on four individuals. A missense change p.R818Q in *TNPO3* gene was identified. Immunoblotting analyses of skeletal muscle biopsies using anti-TNPO3 antibody showed higher molecular weight of TNPO3 indicating muscle damage of patients. This study demonstrated that muted *TNPO3* gene is associated with LGMD1F. Similarly, Couthouis et al. identified the genetic basis of limb-girdle muscular dystrophy type 1 in an American family of Northern European descent using exome sequencing [97]. They reported a p.F89I mutation in *DNAJB6*. Exome sequencing provided an unbiased and effective method for identifying the genetic etiology of limb-girdle muscular dystrophy type 1 in a previously genetically uncharacterized family.

8.3 Congenital Muscular Dystrophy

CMD with hypoglycosylation of α-dystroglycan (α-DG) are a heterogeneous group of disorders often associated with brain and eye defects in addition to muscular dystrophy. α-DG is a peripheral-membrane protein that is part of the dystrophin-associated glycoprotein complex, providing a link between proteins located in the extracellular matrix and proteins located in the cytoplasm immediately beneath the plasma membrane (Fig. 1). There are 14 genes with different mutations that are associated with CMD with α-DG and cause dystroglycanopathy. Using WES combined with a Sanger sequencing confirmation, Carss et al. screened eight unrelated individuals for mutations in 14 genes of the glycosylation of α-DG [95]. They identified two compound heterozygous variants in *GMPPB* which are p.A74X and p.D334N, which are consistent with an autosomal recessive inheritance pattern. Furthermore, the muscle fibers were sparse and disordered frequently spanning two somites when *GMPPB* was knocked out in zebrafish embryos, indicating damage or incomplete development of the myosepta. The *GMPPB* functional study suggested that these variants are pathogenic and related to dystroglycanopathy.

WWS is a genetically heterogenous disease which is characterized by CMD, lissencephaly, hydrocephalus, cerebellar malformation and eye abnormalities. Mutations in seven genes are known to cause Walker-Warburg syndrome (WWS). However, mutations in these genes are only responsible for 50–60% of cases; therefore, new genes are expected to be associated with the rest of cases. Manzini et al. performed WES on seven probands with WWS [100]. Authors identified one homozygous frameshift insertion p.D215fs in *FKTN* gene in two families and a homozygous missense change p.H478A in *POMT2* gene. In addition to two homozygous variants in *GTDC2* genes, a nonsense change p.A445X and a missense variant p. A158H were found, which were predicted to be deleterious. To test how *GTDC2* loss of function would affect brain and muscle development, they used zebrafish as an animal model. They found that *GTDC2* knockdown not only severely affected survival, but also recapitulated WWS phenotypes in the zebrafish embryo. In another study, Shaheen et al. screened first cousin family with autosomal recessive WWS [103]. Exome sequencing followed by autozygome filtration revealed a homozygous insertion change c.821_822insTT causing frameshift p.E274Dfs in

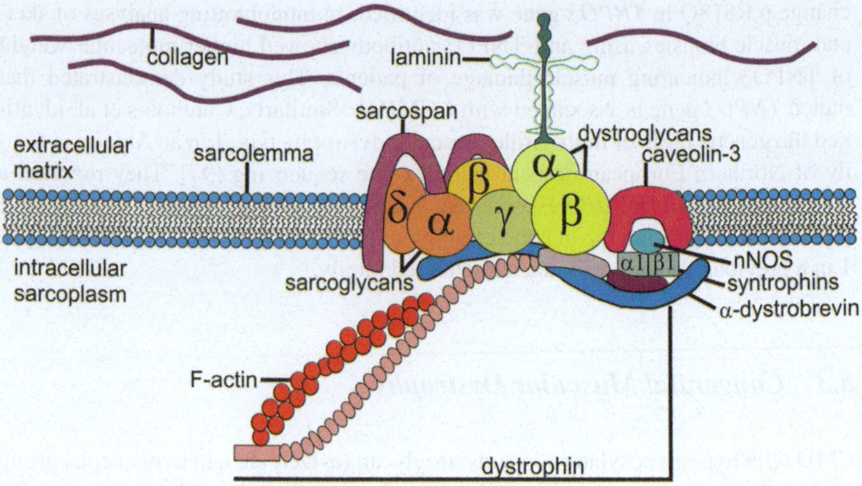

Fig. 1 Schematic diagram of the muscle cell structural proteins. Dystrophin, which is localized at the sarcolemma, has a rod domain consisting of spectrin repeats. The C terminus is preceded by a cysteine-rich domain. Dystrophin binds to the dystrophin-associated protein complex (*DAPC*) through its C terminus. The DAPC consists of sarcoplasmic proteins (alpha-dystrobrevin, syntrophins and neuronal nitric oxide synthase (nNOS), transmembrane proteins (beta-dystroglycan, the sarcoglycans, caveolin-3 and sarcospan) and extracellular proteins (alpha-dystroglycan and laminin). Many members of the DAPC are also associated with muscular dystrophy, either owing to mutations in the genes that encode them (for example, alpha-, beta-, gamma- and delta-sarcoglycan, laminin or caveolin-3), or through mutant binding partners (for example, nNOS, syntrophin, alpha-dystrobrevin, alpha-dystroglycan, beta-dystroglycan or sarcospan). The N terminus of dystrophin associates with the cytoskeleton by interacting with filamentous (F)-actin. The DAPC provides a strong mechanical link between the intracellular cytoskeleton and the extracellular matrix (Adapted from Davies and Nowak) [105]

B3GNT1 gene. The patient died at age six of life and no neuroimaging study was performed.

Stevens et al. found two new mutations in a different gene, *B3GALNT2,* in patients with dystroglycanopathy [88]. *B3GALNT2* encodes a protein β-1,3-N-acetylgalactosaminyltransferase two which transfers N-acetyl galactosamine (GalNAc) in a β-1,3 linkage to N-acetyl glucosamine (GlcNAc). Two cohorts of individuals with dystroglycanopathy were analyzed by WES. Affected individuals P1 and P2 were identified from one cohort and P3 to P7 were from another cohort. WES was carried out by three different laboratories in three different countries using different methods, resulting in a very variable mean coverage. In this study, they found mutations in *B3GALNT2* in four patients with CMD with α-DG. More recently, an autozygosity mapping and exome sequencing study identified homozygous truncating mutations in *POMK* which lead to CMD with secondary merosin deficiency, hypomyelination and intellectual disability, in a consanguineous family from Lebanon with two affected children [96]. Furthermore, they detected co-localization of *POMK* with desmin at the costameres in healthy muscle, and a substantial loss of desmin from the patient muscle. Similarly, Ceyha-Birsoy et al. performed whole exome sequencing in five CMD patients from three families who presented with highly similar clinical features, including early-onset rapidly progressive weakness without brain or eye abnormalities [89]. Pathogenic mutations in three different genes, *DYSF*, *FKTN*, and *ISPD* were identified in each family.

8.4 Spastic Paraplegia (SPG) and Muscle Disease

The hereditary SPG is another clinically and genetically heterogeneous disorder, with a pathophysiologic hallmark of length-dependent distal axonal degeneration of the corticospinal tracts. To date, 54 SPG genes have been identified. Because of the heterogeneity and complexity, precise diagnosis of HSPG has been challenged by using current techniques including Sanger sequencing. To evaluate WES as a generalizable clinical diagnostic tool as well as for gene discovery, Dias et al. used three exome-capture kits on 125 individuals for assessing the targeting efficiency and sequencing coverage of 88 genes associated with muscle disease (MD) and HSPG [101]. Of note, all three exome capture kits fully targeted the CCDS-annotated exons of 14 SPG genes. However, only one frameshift mutation p.T184fs in *CAPN3* was identified demonstrating a causative mutant in *LGMD2A* in the Croatian population. This study indicated that WES is a rapid and efficient for screening mutations in these diseases.

8.5 Emery-Dreifuss Muscular Dystrophy (EDMD)

EDMD is characterized by three clinical manifestations: early- contractures of elbow flexors, ankles and neck extensors; progressive wasting and weakness of the humero-peroneal/scapulo-peroneal muscles from the early stages; and cardiac disease with conduction defects and arrhythmias. There are three genes in which mutations are known to cause EDMD with three different inheritance patterns, which are *EMD* (X-linked pattern), *FHL1* (X-linked pattern); and *LMNA* (autosomal dominant or autosomal recessive pattern). Jimenes-Escrig et al. carried out WES on one proband in a three generation with autosomal dominant EDMD [98]. By cross-referencing candidate genes for muscular dystrophy, they identified a homozygote mutation p.R225Q in exon 4 of *LMNA*. The *LMNA* gene encodes two lamins, A and C, by differential maturation of the 3' end of the mRNA. However, the p.R225Q variant has not been reported so far and no further functional study available to support the pathogenic mechanism.

8.6 Welander Distal Myopathy (WDM)

WDM is one of the distal myopathies, which has an autosomal dominant inheritance pattern and late onset of disease. All of patients of Swedish and Finnish origin share a haplotype on chromosome 2p13, suggesting a founder effect for a single mutation. However, this haplotype spans >60 genes and no mutation has been found in any of them. Klar et al. used haplotype analysis of 21 microsatellite markers on chromosome 2p13 followed by exome sequencing to identify WDM-associated gene(s) [104]. They identified one novel heterozygous missense p.E384L in gene *TIA1* within the core haplotype which is segregates with WDM in familial cases. In addition to filtering against in-house exomes, the *TIA1* variant was excluded for presence in 200 Swedish control chromosomes by Sanger sequencing. TIA1 contains four functional domains including three RNA recognition motifs and a Q-rich C-terminal auxiliary domain, regulates translation of mRNAs involved in apoptotic pathways. The mutation is located in exon 13 of the TIA1 cytotoxic granule-associated RNA binding protein. Immunohistochemistry of WDM muscle biopsies revealed strong staining of TIA1, TDP43, G3BP, and P62 that co-localize to areas adjacent to rimmed vacuoles, suggesting a critical role of TIA1 and RNA metabolism in skeletal muscle homeostasis.

8.7 *Muscular Dystrophy*

Oliveira et al. reported WES analysis on a patient (trio) with childhood-onset progressive MD, also presenting mental retardation and dilated cardiomyopathy using the ion proton sequencing system [102]. Inspection of sequence alignments ultimately identified the causal variant (*CHKB*: c.1031+3G>C). This splice site mutation was confirmed using Sanger sequencing and its effect was further evaluated with gene expression analysis. On reassessment of the muscle biopsy, typical abnormal mitochondrial oxidative changes were observed. This study demonstrates exome's utility in solving cases from highly heterogeneous groups of diseases.

9 Conclusions

The NMD NGS panels offer cost-effective and more rapid molecular diagnostic testing than the conventional sequential Sanger sequencing of associated genes. However, in specific aforementioned situations clinical exome sequencing may be indicated. A faster molecular diagnosis, by panel or exome sequencing, of NMDs will have major impacts on patients as it offers timely diagnosis, management and genetic counseling, and will allow access to therapy or inclusion into therapeutic trials. Finally, this technology has its own limitations and challenges. However, molecular geneticists are aware of them and have devised clever ways to overcome them. They are continually adopting, validating and implementing the newest version of existing enrichment and sequencing technologies or acquiring new ones. In addition, they employ complementary platforms that address the drawbacks of existing next generation sequencing technologies, such as the use of Sanger sequencing to rescue low coverage areas. Without doubt NGS has been become the mainstay for the diagnosis of NMDs and its popularity will only continue to grow.

References

1. Emery, A.E.: Population frequencies of inherited neuromuscular diseases – a world survey. Neuromuscul. Disord. NMD. **1**, 19–29 (1991)
2. North, K.: What's new in congenital myopathies? Neuromuscul. Disord. NMD. **18**, 433–442 (2008). doi:10.1016/j.nmd.2008.04.002
3. Valencia, C.A., Ankala, A., Rhodenizer, D., et al.: Comprehensive mutation analysis for congenital muscular dystrophy: a clinical PCR-based enrichment and next-generation sequencing panel. PLoS One. **8**, e53083 (2013). doi:10.1371/journal.pone.0053083
4. Vasli, N., Böhm, J., Le Gras, S., et al.: Next generation sequencing for molecular diagnosis of neuromuscular diseases. Acta Neuropathol. (Berl). **124**, 273–283 (2012). doi:10.1007/s00401-012-0982-8
5. Flanigan, K.M.: The muscular dystrophies. Semin. Neurol. **32**, 255–263 (2012). doi:10.1055/s-0032-1329199
6. Mercuri, E., Muntoni, F.: Muscular dystrophies. Lancet. **381**, 845–860 (2013). doi:10.1016/S0140-6736(12)61897-2
7. Rocha, C.T., Hoffman, E.P.: Limb-girdle and congenital muscular dystrophies: current diagnostics, management, and emerging technologies. Curr. Neurol. Neurosci. Rep. **10**, 267–276 (2010). doi:10.1007/s11910-010-0119-1
8. Mercuri, E., Muntoni, F.: The ever-expanding spectrum of congenital muscular dystrophies. Ann. Neurol. **72**, 9–17 (2012). doi:10.1002/ana.23548
9. Saporta, A.S.D., Sottile, S.L., Miller, L.J., et al.: Charcot-Marie-Tooth disease subtypes and genetic testing strategies. Ann. Neurol. **69**, 22–33 (2011). doi:10.1002/ana.22166
10. North, K.N., Wang, C.H., Clarke, N., et al.: Approach to the diagnosis of congenital myopathies. Neuromuscul. Disord. NMD. **24**, 97–116 (2014). doi:10.1016/j.nmd.2013.11.003
11. Nance, J.R., Dowling, J.J., Gibbs, E.M., Bönnemann, C.G.: Congenital myopathies: an update. Curr. Neurol. Neurosci. Rep. **12**, 165–174 (2012). doi:10.1007/s11910-012-0255-x
12. Richard, P., Charron, P., Carrier, L., et al.: Hypertrophic cardiomyopathy: distribution of disease genes, spectrum of mutations, and implications for a molecular diagnosis strategy. Circulation. **107**, 2227–2232 (2003). doi:10.1161/01.CIR.0000066323.15244.54
13. Villard, E., Duboscq-Bidot, L., Charron, P., et al.: Mutation screening in dilated cardiomyopathy: prominent role of the beta myosin heavy chain gene. Eur. Heart J. **26**, 794–803 (2005). doi:10.1093/eurheartj/ehi193
14. Meredith, C., Herrmann, R., Parry, C., et al.: Mutations in the slow skeletal muscle fiber myosin heavy chain gene (MYH7) cause laing early-onset distal myopathy (MPD1). Am. J. Hum. Genet. **75**, 703–708 (2004). doi:10.1086/424760
15. Tajsharghi, H., Thornell, L.-E., Lindberg, C., et al.: Myosin storage myopathy associated with a heterozygous missense mutation in MYH7. Ann. Neurol. **54**, 494–500 (2003). doi:10.1002/ana.10693
16. Tajsharghi, H., Oldfors, A., Macleod, D.P., Swash, M.: Homozygous mutation in MYH7 in myosin storage myopathy and cardiomyopathy. Neurology. **68**, 962 (2007). doi:10.1212/01.wnl.0000257131.13438.2c
17. Muelas, N., Hackman, P., Luque, H., et al.: MYH7 gene tail mutation causing myopathic profiles beyond Laing distal myopathy. Neurology. **75**, 732–741 (2010). doi:10.1212/WNL.0b013e3181eee4d5
18. Pegoraro, E., Gavassini, B.F., Borsato, C., et al.: MYH7 gene mutation in myosin storage myopathy and scapulo-peroneal myopathy. Neuromuscul. Disord. NMD. **17**, 321–329 (2007). doi:10.1016/j.nmd.2007.01.010
19. Ortolano, S., Tarrío, R., Blanco-Arias, P., et al.: A novel MYH7 mutation links congenital fiber type disproportion and myosin storage myopathy. Neuromuscul. Disord. NMD. **21**, 254–262 (2011). doi:10.1016/j.nmd.2010.12.011
20. Homayoun, H., Khavandgar, S., Hoover, J.M., et al.: Novel mutation in MYH7 gene associated with distal myopathy and cardiomyopathy. Neuromuscul. Disord. NMD. **21**, 219–222 (2011). doi:10.1016/j.nmd.2010.12.005

21. Gazzerro, E., Bonetto, A., Minetti, C.: Caveolinopathies: translational implications of caveolin-3 in skeletal and cardiac muscle disorders. Handb. Clin. Neurol. **101**, 135–142 (2011). doi:10.1016/B978-0-08-045031-5.00010-4
22. Fee, D.B., So, Y.T., Barraza, C., et al.: Phenotypic variability associated with Arg26Gln mutation in caveolin3. Muscle Nerve. **30**, 375–378 (2004). doi:10.1002/mus.20092
23. Rosen, D.R., Siddique, T., Patterson, D., et al.: Mutations in Cu/Zn superoxide dismutase gene are associated with familial amyotrophic lateral sclerosis. Nature. **362**, 59–62 (1993). doi:10.1038/362059a0
24. Chen, Y.-Z., Bennett, C.L., Huynh, H.M., et al.: DNA/RNA helicase gene mutations in a form of juvenile amyotrophic lateral sclerosis (ALS4). Am. J. Hum. Genet. **74**, 1128–1135 (2004). doi:10.1086/421054
25. Hadano, S., Hand, C.K., Osuga, H., et al.: A gene encoding a putative GTPase regulator is mutated in familial amyotrophic lateral sclerosis 2. Nat. Genet. **29**, 166–173 (2001). doi:10.1038/ng1001-166
26. Yang, Y., Hentati, A., Deng, H.X., et al.: The gene encoding alsin, a protein with three guanine-nucleotide exchange factor domains, is mutated in a form of recessive amyotrophic lateral sclerosis. Nat. Genet. **29**, 160–165 (2001). doi:10.1038/ng1001-160
27. Gitcho, M.A., Bigio, E.H., Mishra, M., et al.: TARDBP 3′-UTR variant in autopsy-confirmed frontotemporal lobar degeneration with TDP-43 proteinopathy. Acta Neuropathol. (Berl). **118**, 633–645 (2009). doi:10.1007/s00401-009-0571-7
28. Kwiatkowski, T.J., Bosco, D.A., Leclerc, A.L., et al.: Mutations in the FUS/TLS gene on chromosome 16 cause familial amyotrophic lateral sclerosis. Science. **323**, 1205–1208 (2009). doi:10.1126/science.1166066
29. Greenway, M.J., Alexander, M.D., Ennis, S., et al.: A novel candidate region for ALS on chromosome 14q11.2. Neurology. **63**, 1936–1938 (2004)
30. Hayward, C., Colville, S., Swingler, R.J., Brock, D.J.: Molecular genetic analysis of the APEX nuclease gene in amyotrophic lateral sclerosis. Neurology. **52**, 1899–1901 (1999)
31. DeJesus-Hernandez, M., Mackenzie, I.R., Boeve, B.F., et al.: Expanded GGGGCC hexanucleotide repeat in noncoding region of C9ORF7 auses chromosome 9p-linked FTD and ALS. Neuron. **72**, 245–256 (2011). doi:10.1016/j.neuron.2011.09.011
32. Renton, A.E., Majounie, E., Waite, A., et al.: A hexanucleotide repeat expansion in C9ORF72 is the cause of chromosome 9p21-linked ALS-FTD. Neuron. **72**, 257–268 (2011). doi:10.1016/j.neuron.2011.09.010
33. Lo, H.P., Cooper, S.T., Evesson, F.J., et al.: Limb-girdle muscular dystrophy: diagnostic evaluation, frequency and clues to pathogenesis. Neuromuscul. Disord. NMD. **18**, 34–44 (2008). doi:10.1016/j.nmd.2007.08.009
34. Peat, R.A., Smith, J.M., Compton, A.G., et al.: Diagnosis and etiology of congenital muscular dystrophy. Neurology. **71**, 312–321 (2008). doi:10.1212/01.wnl.0000284605.27654.5a
35. Chae, J.H., Vasta, V., Cho, A., et al.: Utility of next generation sequencing in genetic diagnosis of early onset neuromuscular disorders. J. Med. Genet. **52**, 208–216 (2015). doi:10.1136/jmedgenet-2014-102819
36. Jungbluth, H., Sewry, C.A., Counsell, S., et al.: Magnetic resonance imaging of muscle in nemaline myopathy. Neuromuscul. Disord. NMD. **14**, 779–784 (2004). doi:10.1016/j.nmd.2004.08.005
37. Mercuri, E., Jungbluth, H., Muntoni, F.: Muscle imaging in clinical practice: diagnostic value of muscle magnetic resonance imaging in inherited neuromuscular disorders. Curr. Opin. Neurol. **18**, 526–537 (2005)
38. Fischer, D., Kley, R.A., Strach, K., et al.: Distinct muscle imaging patterns in myofibrillar myopathies. Neurology. **71**, 758–765 (2008). doi:10.1212/01.wnl.0000324927.28817.9b
39. Klein, A., Jungbluth, H., Clement, E., et al.: Muscle magnetic resonance imaging in congenital myopathies due to ryanodine receptor type 1 gene mutations. Arch. Neurol. **68**, 1171–1179 (2011). doi:10.1001/archneurol.2011.188
40. Ghaoui, R., Clarke, N., Hollingworth, P., Needham, M.: Muscle disorders: the latest investigations. Intern. Med. J. **43**, 970–978 (2013). doi:10.1111/imj.12234

41. Bornemann, A., Petersen, M.B., Schmalbruch, H.: Fatal congenital myopathy with actin filament deposits. Acta Neuropathol. (Berl). **92**, 104–108 (1996)
42. Laing, N.G., Wallgren-Pettersson, C.: 161st ENMC international workshop on nemaline myopathy and related disorders, Newcastle upon Tyne, 2008. Neuromuscul. Disord. NMD. **19**, 300–305 (2009). doi:10.1016/j.nmd.2009.02.002
43. Nowak, K.J., Wattanasirichaigoon, D., Goebel, H.H., et al.: Mutations in the skeletal muscle alpha-actin gene in patients with actin myopathy and nemaline myopathy. Nat. Genet. **23**, 208–212 (1999). doi:10.1038/13837
44. Goebel, H.H., Piirsoo, A., Warlo, I., et al.: Infantile intranuclear rod myopathy. J. Child Neurol. **12**, 22–30 (1997)
45. Sewry, C.A., Holton, J.L., Dick, D.J., et al.: Zebra body myopathy is caused by a mutation in the skeletal muscle actin gene (ACTA1). Neuromuscul. Disord. NMD. (2015). doi:10.1016/j.nmd.2015.02.003
46. Laing, N.G., Clarke, N.F., Dye, D.E., et al.: Actin mutations are one cause of congenital fibre type disproportion. Ann. Neurol. **56**, 689–694 (2004). doi:10.1002/ana.20260
47. Hung, R.M., Yoon, G., Hawkins, C.E., et al.: Cap myopathy caused by a mutation of the skeletal alpha-actin gene ACTA1. Neuromuscul. Disord. NMD. **20**, 238–240 (2010). doi:10.1016/j.nmd.2010.01.011
48. Zvaritch, E., Kraeva, N., Bombardier, E., et al.: Ca2+ dysregulation in Ryr1(I4895T/wt) mice causes congenital myopathy with progressive formation of minicores, cores, and nemaline rods. Proc. Natl. Acad. Sci. U. S. A. **106**, 21813–21818 (2009). doi:10.1073/pnas.0912126106
49. Tennyson, C.N., Klamut, H.J., Worton, R.G.: The human dystrophin gene requires 16 hours to be transcribed and is cotranscriptionally spliced. Nat. Genet. **9**, 184–190 (1995). doi:10.1038/ng0295-184
50. Hackman, P., Vihola, A., Haravuori, H., et al.: Tibial muscular dystrophy is a titinopathy caused by mutations in TTN, the gene encoding the giant skeletal-muscle protein titin. Am. J. Hum. Genet. **71**, 492–500 (2002). doi:10.1086/342380
51. Donner, K., Sandbacka, M., Lehtokari, V.-L., et al.: Complete genomic structure of the human nebulin gene and identification of alternatively spliced transcripts. Eur. J. Hum. Genet. EJHG. **12**, 744–751 (2004). doi:10.1038/sj.ejhg.5201242
52. Phillips, M.S., Fujii, J., Khanna, V.K., et al.: The structural organization of the human skeletal muscle ryanodine receptor (RYR1) gene. Genomics. **34**, 24–41 (1996). doi:10.1006/geno.1996.0238
53. Abbs, S., Bobrow, M.: Analysis of quantitative PCR for the diagnosis of deletion and duplication carriers in the dystrophin gene. J. Med. Genet. **29**, 191–196 (1992)
54. Muntoni, F., Torelli, S., Ferlini, A.: Dystrophin and mutations: one gene, several proteins, multiple phenotypes. Lancet Neurol. **2**, 731–740 (2003)
55. Schouten, J.P., McElgunn, C.J., Waaijer, R., et al.: Relative quantification of 40 nucleic acid sequences by multiplex ligation-dependent probe amplification. Nucleic Acids Res. **30**, e57 (2002)
56. van der Steege, G., Grootscholten, P.M., van der Vlies, P., et al.: PCR-based DNA test to confirm clinical diagnosis of autosomal recessive spinal muscular atrophy. Lancet. **345**, 985–986 (1995)
57. Lemmers, R.J.F.L., Wohlgemuth, M., Frants, R.R., et al.: Contractions of D4Z4 on 4qB subtelomeres do not cause facioscapulohumeral muscular dystrophy. Am. J. Hum. Genet. **75**, 1124–1130 (2004). doi:10.1086/426035
58. Warner, J.P., Barron, L.H., Goudie, D., et al.: A general method for the detection of large CAG repeat expansions by fluorescent PCR. J. Med. Genet. **33**, 1022–1026 (1996)
59. Radvansky, J., Ficek, A., Kadasi, L.: Repeat-primed polymerase chain reaction in myotonic dystrophy type 2 testing. Genet. Test. Mol. Biomarkers. **15**, 133–136 (2011). doi:10.1089/gtmb.2010.0127
60. Catalli, C., Morgante, A., Iraci, R., et al.: Validation of sensitivity and specificity of tetraplet-primed PCR (TP-PCR) in the molecular diagnosis of myotonic dystrophy type 2 (DM2). J. Mol. Diagn. JMD. **12**, 601–606 (2010). doi:10.2353/jmoldx.2010.090239

61. Mardis, E.R.: The impact of next-generation sequencing technology on genetics. Trends Genet. TIG. **24**, 133–141 (2008). doi:10.1016/j.tig.2007.12.007
62. Shendure, J., Ji, H.: Next-generation DNA sequencing. Nat. Biotechnol. **26**, 1135–1145 (2008). doi:10.1038/nbt1486
63. Voelkerding, K.V., Dames, S.A., Durtschi, J.D.: Next-generation sequencing: from basic research to diagnostics. Clin. Chem. **55**, 641–658 (2009). doi:10.1373/clinchem.2008.112789
64. Albert, T.J., Molla, M.N., Muzny, D.M., et al.: Direct selection of human genomic loci by microarray hybridization. Nat. Methods. **4**, 903–905 (2007). doi:10.1038/nmeth1111
65. Hodges, E., Xuan, Z., Balija, V., et al.: Genome-wide in situ exon capture for selective resequencing. Nat. Genet. **39**, 1522–1527 (2007). doi:10.1038/ng.2007.42
66. Okou, D.T., Steinberg, K.M., Middle, C., et al.: Microarray-based genomic selection for high-throughput resequencing. Nat. Methods. **4**, 907–909 (2007). doi:10.1038/nmeth1109
67. Porreca, G.J., Zhang, K., Li, J.B., et al.: Multiplex amplification of large sets of human exons. Nat. Methods. **4**, 931–936 (2007). doi:10.1038/nmeth1110
68. Ding, L., Getz, G., Wheeler, D.A., et al.: Somatic mutations affect key pathways in lung adenocarcinoma. Nature. **455**, 1069–1075 (2008). doi:10.1038/nature07423
69. Kim, P.M., Lam, H.Y.K., Urban, A.E., et al.: Analysis of copy number variants and segmental duplications in the human genome: evidence for a change in the process of formation in recent evolutionary history. Genome Res. **18**, 1865–1874 (2008). doi:10.1101/gr.081422.108
70. Tewhey, R., Warner, J.B., Nakano, M., et al.: Microdroplet-based PCR enrichment for large-scale targeted sequencing. Nat. Biotechnol. **27**, 1025–1031 (2009). doi:10.1038/nbt.1583
71. Eid, J., Fehr, A., Gray, J., et al.: Real-time DNA sequencing from single polymerase molecules. Science. **323**, 133–138 (2009). doi:10.1126/science.1162986
72. Borgström, E., Lundin, S., Lundeberg, J.: Large scale library generation for high throughput sequencing. PLoS One. **6**, e19119 (2011). doi:10.1371/journal.pone.0019119
73. Margulies, M., Egholm, M., Altman, W.E., et al.: Genome sequencing in microfabricated high-density picolitre reactors. Nature. **437**, 376–380 (2005). doi:10.1038/nature03959
74. Ruparel, H., Bi, L., Li, Z., et al.: Design and synthesis of a 3′-O-allyl photocleavable fluorescent nucleotide as a reversible terminator for DNA sequencing by synthesis. Proc. Natl. Acad. Sci. U. S. A. **102**, 5932–5937 (2005). doi:10.1073/pnas.0501962102
75. Rothberg, J.M., Hinz, W., Rearick, T.M., et al.: An integrated semiconductor device enabling non-optical genome sequencing. Nature. **475**, 348–352 (2011). doi:10.1038/nature10242
76. Clark, M.J., Chen, R., Lam, H.Y.K., et al.: Performance comparison of exome DNA sequencing technologies. Nat. Biotechnol. **29**, 908–914 (2011). doi:10.1038/nbt.1975
77. Lin, X., Tang, W., Ahmad, S., et al.: Applications of targeted gene capture and next-generation sequencing technologies in studies of human deafness and other genetic disabilities. Hear. Res. **288**, 67–76 (2012). doi:10.1016/j.heares.2012.01.004
78. Xie, S., Lan, Z., Qu, N., et al.: Detection of truncated dystrophin lacking the C-terminal domain in a Chinese pedigree by next-generation sequencing. Gene. **499**, 139–142 (2012). doi:10.1016/j.gene.2012.03.029
79. Lim, B.C., Lee, S., Shin, J.-Y., et al.: Genetic diagnosis of Duchenne and Becker muscular dystrophy using next-generation sequencing technology: comprehensive mutational search in a single platform. J. Med. Genet. **48**, 731–736 (2011). doi:10.1136/jmedgenet-2011-100133
80. Tian, X., Liang, W.-C., Feng, Y., et al.: Expanding genotype/phenotype of neuromuscular diseases by comprehensive target capture/NGS. Neurol. Genet. **1**, e14 (2015). doi:10.1212/NXG.0000000000000015
81. Bennett, R.R., den Dunnen, J., O'Brien, K.F., et al.: Detection of mutations in the dystrophin gene via automated DHPLC screening and direct sequencing. BMC Genet. **2**, 17 (2001)
82. Flanigan, K.M., von Niederhausern, A., Dunn, D.M., et al.: Rapid direct sequence analysis of the dystrophin gene. Am. J. Hum. Genet. **72**, 931–939 (2003)
83. Lalic, T., Vossen, R.H.A.M., Coffa, J., et al.: Deletion and duplication screening in the DMD gene using MLPA. Eur. J. Hum. Genet. EJHG. **13**, 1231–1234 (2005). doi:10.1038/sj.ejhg.5201465
84. Mendell, J.R., Buzin, C.H., Feng, J., et al.: Diagnosis of Duchenne dystrophy by enhanced detection of small mutations. Neurology. **57**, 645–650 (2001)

85. Prior, T.W., Bridgeman, S.J.: Experience and strategy for the molecular testing of Duchenne muscular dystrophy. J. Mol. Diagn. JMD. **7**, 317–326 (2005). doi:10.1016/S1525-1578(10)60560-0
86. Iglesias, A., Anyane-Yeboa, K., Wynn, J., et al.: The usefulness of whole-exome sequencing in routine clinical practice. Genet. Med. Off. J. Am. Coll. Med. Genet. **16**, 922–931 (2014). doi:10.1038/gim.2014.58
87. Yang, Y., Muzny, D.M., Reid, J.G., et al.: Clinical whole-exome sequencing for the diagnosis of mendelian disorders. N. Engl. J. Med. **369**, 1502–1511 (2013). doi:10.1056/NEJMoa1306555
88. Stevens, E., Carss, K.J., Cirak, S., et al.: Mutations in B3GALNT2 cause congenital muscular dystrophy and hypoglycosylation of α-dystroglycan. Am. J. Hum. Genet. **92**, 354–365 (2013). doi:10.1016/j.ajhg.2013.01.016
89. Ceyhan-Birsoy, O., Talim, B., Swanson, L.C., et al.: Whole exome sequencing reveals DYSF, FKTN, and ISPD mutations in congenital muscular dystrophy without brain or eye involvement. J. Neuromuscul. Dis. **2**, 87–92 (2015). doi:10.3233/JND-140038
90. Park, H.J., Choi, Y.C., Kim, S.M., et al.: Molecular genetic diagnosis of a Bethlem myopathy family with an autosomal-dominant COL6A1 mutation, as evidenced by exome sequencing. J. Clin. Neurol. Seoul Korea. **11**, 183–187 (2015). doi:10.3988/jcn.2015.11.2.183
91. Bögershausen, N., Shahrzad, N., Chong, J.X., et al.: Recessive TRAPPC11 mutations cause a disease spectrum of limb girdle muscular dystrophy and myopathy with movement disorder and intellectual disability. Am. J. Hum. Genet. **93**, 181–190 (2013). doi:10.1016/j.ajhg.2013.05.028
92. McDonald, K.K., Stajich, J., Blach, C., et al.: Exome analysis of two limb-girdle muscular dystrophy families: mutations identified and challenges encountered. PLoS One. **7**, e48864 (2012). doi:10.1371/journal.pone.0048864
93. Harms, M.B., Sommerville, R.B., Allred, P., et al.: Exome sequencing reveals DNAJB6 mutations in dominantly-inherited myopathy. Ann. Neurol. **71**, 407–416 (2012). doi:10.1002/ana.22683
94. Torella, A., Fanin, M., Mutarelli, M., et al.: Next-generation sequencing identifies transportin 3 as the causative gene for LGMD1F. PLoS One. **8**, e63536 (2013). doi:10.1371/journal.pone.0063536
95. Carss, K.J., Stevens, E., Foley, A.R., et al.: Mutations in GDP-mannose pyrophosphorylase B cause congenital and limb-girdle muscular dystrophies associated with hypoglycosylation of α-dystroglycan. Am. J. Hum. Genet. **93**, 29–41 (2013). doi:10.1016/j.ajhg.2013.05.009
96. von Renesse, A., Petkova, M.V., Lützkendorf, S., et al.: POMK mutation in a family with congenital muscular dystrophy with merosin deficiency, hypomyelination, mild hearing deficit and intellectual disability. J. Med. Genet. **51**, 275–282 (2014). doi:10.1136/jmedgenet-2013-102236
97. Couthouis, J., Raphael, A.R., Siskind, C., et al.: Exome sequencing identifies a DNAJB6 mutation in a family with dominantly-inherited limb-girdle muscular dystrophy. Neuromuscul. Disord. NMD. **24**, 431–435 (2014). doi:10.1016/j.nmd.2014.01.014
98. Jimenez-Escrig, A., Gobernado, I., Garcia-Villanueva, M., Sanchez-Herranz, A.: Autosomal recessive Emery-Dreifuss muscular dystrophy caused by a novel mutation (R225Q) in the Lamin A/C gene identified by exome sequencing. Muscle Nerve. **45**, 605–610 (2012). doi:10.1002/mus.22324
99. Mitsuhashi, S., Boyden, S.E., Estrella, E.A., et al.: Exome sequencing identifies a novel SMCHD1 mutation in facioscapulohumeral muscular dystrophy 2. Neuromuscul. Disord. NMD. **23**, 975–980 (2013). doi:10.1016/j.nmd.2013.08.009
100. Manzini, M.C., Tambunan, D.E., Hill, R.S., et al.: Exome sequencing and functional validation in zebrafish identify GTDC2 mutations as a cause of Walker-Warburg syndrome. Am. J. Hum. Genet. **91**, 541–547 (2012). doi:10.1016/j.ajhg.2012.07.009
101. Dias, C., Sincan, M., Cherukuri, P.F., et al.: An analysis of exome sequencing for diagnostic testing of the genes associated with muscle disease and spastic paraplegia. Hum. Mutat. **33**, 614–626 (2012). doi:10.1002/humu.22032

102. Oliveira, J., Negrão, L., Fineza, I., et al.: New splicing mutation in the choline kinase beta (CHKB) gene causing a muscular dystrophy detected by whole-exome sequencing. J. Hum. Genet. (2015). doi:10.1038/jhg.2015.20
103. Shaheen, R., Faqeih, E., Ansari, S., Alkuraya, F.S.: A truncating mutation in B3GNT1 causes severe Walker-Warburg syndrome. Neurogenetics. **14**, 243–245 (2013). doi:10.1007/s10048-013-0367-8
104. Klar, J., Sobol, M., Melberg, A., et al.: Welander distal myopathy caused by an ancient founder mutation in TIA1 associated with perturbed splicing. Hum. Mutat. **34**, 572–577 (2013). doi:10.1002/humu.22282
105. Davies, K.E., Nowak, K.J.: Molecular mechanisms of muscular dystrophies: old and new players. Nat. Rev. Mol. Cell Biol. **7**, 762–773 (2006). doi:10.1038/nrm2024

Next-Generation Sequencing for the Diagnosis of Monogenic Disorders of Insulin Secretion

Amy E. Knight Johnson, Siri Atma W. Greeley, and Daniela del Gaudio

Abstract Studies of monogenic disorders of β-cell function have led to a greater understanding of the β-cell physiology and have improved the diagnosis and treatment of patients with these rare conditions. These disorders include single-gene defects associated with increased insulin secretion, causing hypoglycemia, and decreased insulin secretion, resulting in diabetes.

Mutations in at least 40 genes have been identified through studies using genetic linkage, candidate gene sequencing and most recently, exome sequencing. Concurrent with the expansion in the number of known genes has been an extension of the known phenotypic spectrum of various syndromic etiologies as well as overlap between genes associated with each condition. Despite these challenges, making a molecular diagnosis of these disorders can be greatly beneficial, as it enables more appropriate treatment, better prediction of disease prognosis and progression, and counseling and screening of family members.

In the recent years, the exponential growth in the field of high-throughput capture and sequencing technology has made it possible and cost-effective to sequence many genes simultaneously, making it an efficient diagnostic tool for clinically and genetically heterogeneous disorders including monogenic disorders of insulin secretion.

This chapter provides an overview of the genetic etiology of these conditions and outlines the major advances arising from the identification of the genetic and molecular mechanisms underlying these disorders. In addition, we discuss the clinical utility of next generation sequencing either for the purpose of a clinical diagnostic test or as a pre-screen in the search for novel disease-causing genes.

Keywords Monogenic diabetes • Neonatal diabetes • Maturity-onset diabetes of the young (MODY) • Congenital hyperinsulinism • Next-generation sequencing

A.E.K. Johnson • D. del Gaudio, Ph.D. (✉)
Department of Human Genetics, The University of Chicago,
5841 S. Maryland Ave. MC0077, Chicago, IL, 60637 USA
e-mail: ddelgaudio@bsd.uchicago.edu

S.A.W. Greeley
Departments of Medicine and Pediatrics, Section of Adult and Pediatric Endocrinology, Diabetes, & Metabolism, The University of Chicago, Chicago, IL, USA

L.-J.C. Wong (ed.), *Next Generation Sequencing Based Clinical Molecular Diagnosis of Human Genetic Disorders*, DOI 10.1007/978-3-319-56418-0_11

1 Insulin Secretion

Insulin is an important hormone required for normal metabolism. In healthy subjects, insulin release is exquisitely exact to meet the metabolic demand. Specifically, β-cells sense changes in plasma glucose concentration and respond by releasing corresponding amounts of insulin [1] (Fig. 1). Following the ingestion of a meal, glucose enters the β-cells via glucose transporters (GLUT-2) and is then phosphorylated by glucokinase, to generate glucose-6-phosphate: the first rate-limiting step in glucose metabolism. Glucokinase functions as the glucose sensor in the β-cell by controlling the rate of entry of glucose into the glycolytic pathway and its subsequent metabolism. Following phosphorylation, glucose is metabolized, eventually resulting in an increase of the intracellular concentration ration of ATP/ADP, subsequently leading to the closure of a specialized potassium channel. This channel, known as the ATP-sensitive potassium channel or K_{ATP} channel, is the principal gatekeeper of electrical activity on the β-cell. The K_{ATP} channel is a hetero-octameric complex composed of four pore-forming Kir6.2 subunits encoded by the *KCNJ11* gene and four regulatory sulphonylurea receptor (SUR) 1 subunits encoded by the *ABCC8* gene [3]. The closure of the K_{ATP} channels induces a depolarization of the plasma cell membrane, leading to Ca^{++} entry via voltage-gated calcium channels. The rise in intracellular Ca^{++} concentration then triggers the exocytosis of insulin. Circulating insulin stimulates uptake in insulin-sensitive tissues (mostly liver, adipose tissue and muscle) lowering the blood glucose concentration.

2 Monogenic Diabetes

Monogenic diabetes mellitus includes a heterogeneous group of diabetes types where a mutation or mutations in a single gene result in diabetes [4]. As of today, mutations in over 40 genes (Table 1) causing monogenic forms of diabetes have been identified though studies using candidate gene sequencing, linkage analysis and more recently exome sequencing. Inheritance may be autosomal dominant, autosomal recessive, X-linked, maternally inherited or mutations may be *de novo*. It is estimated that 1–2% of all patients with diabetes have monogenic diabetes [63]. The main phenotypes suggestive of an underlying monogenic cause include transient or permanent neonatal diabetes mellitus (TNDM or PNDM), maturity-onset diabetes of the young (MODY) and rare diabetes-associated syndromes. Several etiological mechanisms of dysfunction are involved including impairment of pancreatic β-cell development and/or gene expression, failure of glucose sensing, disruption of insulin synthesis, disorders of ion channels and increased endoplasmic reticulum stress leading to destruction of the β-cell [63–65]. The vast majority of patients with monogenic diabetes go unrecognized [66] and continue to be misdiagnosed as type 1 or type 2 diabetes [67–69]. Defining the genetic subtype of monogenic diabetes has considerable implications for patient care as

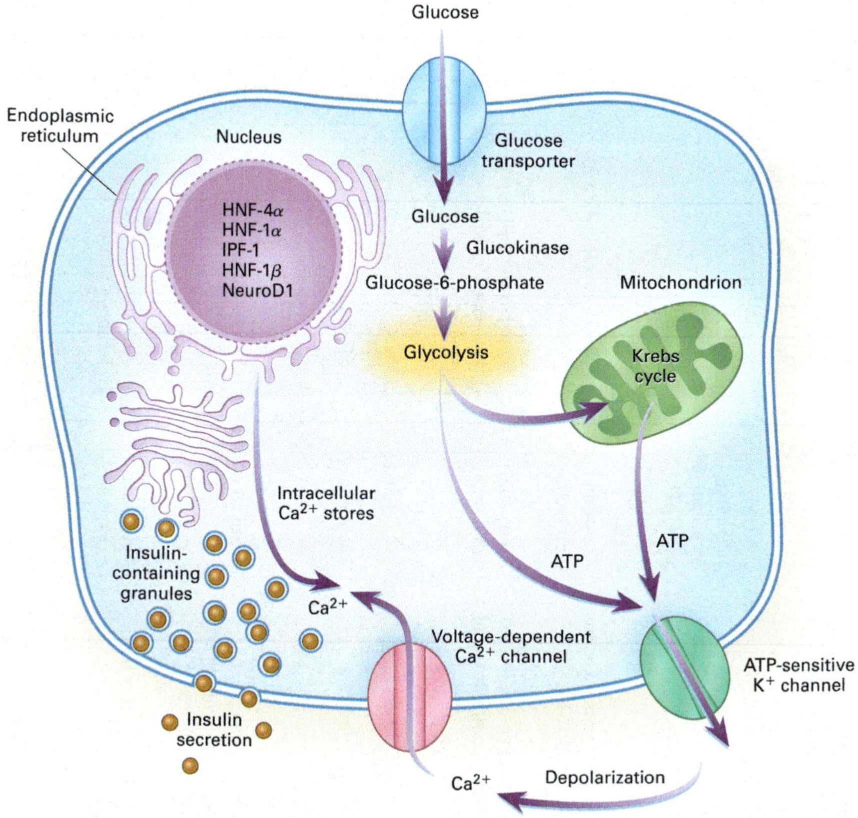

Fig. 1 Model of a Pancreatic Beta Cell and the Proteins Implicated in Maturity-Onset Diabetes of the Young (MODY). Glucose enters the β-cells via glucose transporters (GLUT-2) and is then phosphorylated by glucokinase, to generate glucose-6-phosphate. Following phosphorylation, glucose is metabolized, thus resulting in an increase of the intracellular concentration ration of ATP/ADP, subsequently leading to the closure the K_{ATP} channels. The closure of the K_{ATP} channels induces a depolarization of the plasma cell membrane, leading to Ca^{++} entry via voltage-gated calcium channels. The rise in intracellular Ca^{++} concentration then triggers the exocytosis of insulin into the circulation. A mutation in one of the alleles of the gene encoding glucokinase leads to a reduction in beta-cell glucokinase activity, resulting in decreased glucose phosphorylation in the beta cell and glucose-stimulated insulin release at any blood glucose concentration. The MODY-associated transcription factors — hepatocyte nuclear factor (HNF) 4a (associated with MODY1), HNF-1a (MODY3), insulin promoter factor 1 (IPF-1 [MODY4]), HNF-1b (MODY5), and neurogenic differentiation factor 1 (NeuroD1), or beta-cell E-box transactivator 2 (BETA2 [MODY6]) — function in the nucleus of the beta cell and regulate the transcription of the insulin gene (either directly, as in the case of HNF-1a, HNF-1b, IPF-1, and NeuroD1 or BETA2, or indirectly, through effects on the expression of other transcription factors, as in the case of HNF-4a); they also regulate the transcription of genes encoding enzymes involved in the transport and metabolism of glucose as well as other proteins required for normal beta-cell function (Reprinted, by permission, from Fajans et al. [2] © 2001, Massachusetts Medical Society)

Table 1 Monogenic causes of diabetes and congenital hyperinsulinism

Gene	Protein / function	Phenotypes / syndromes	Inheritance	Onset of diabetes/CHI	References
ABCC8	Sulfonylurea receptor 1 (SUR1) subunit of ATP-sensitive potassium channel	PNDM (less often) or TNDM (more often); DEND; CHI	Spontaneous (PNDM) or autosomal dominant (PNDM and CHI); autosomal recessive (CHI)	Most cases <6 months/ infancy or childhood	[5–8]
CEL	Carboxyl-Ester Lipase / involved in duodenal hydrolysis of cholesterol esters	Diabetes and pancreatic exocrine dysfunction (MODY8)	Autosomal dominant	Most cases less than 25 years	[9]
DNAJC3	DNAJ/HSP40 homolog, subfamily 3, member 3/BiP (immunoglobulin heavy-chain binding protein) co-chaperone that attenuates late phases of ER stress	Ataxia, combined cerebellar and peripheral, with hearing loss and diabetes mellitus	Autosomal recessive	One family with onset between 15 and 18 years	[10]
EIF2AK3	Eukaryotic translation initiation factor 2-alpha kinase 3/kinase involved in regulation of translation	Wolcott-Rallison syndrome (WRS)	Autosomal recessive	Most cases within weeks	[11, 12]
FOXP3	Forkhead box protein P3 / transcription factor involved in development and function of T-regulatory cells (T-regs)	Immunodysregulation polyendocrinopathy, enteropathy, X-linked (IPEX) syndrome	X-linked recessive	Days – infancy	[13, 14]
LRBA	Lypopolysaccharide-responsive, beige-like anchor protein / implicated in regulating endosomal trafficking	Immune deficiency-8 (CVID8) with autoimmunity	Autosomal recessive	Autoimmune DM in infancy (1 case)	[15]

GATA4	GATA-binding protein 4/ transcription factor essential for pancreatic organogenesis	PNDM or TNDM	Autosomal dominant	Infancy to childhood	[16, 17]
GATA6	GATA-binding protein 6/ transcription factor essential for pancreatic organogenesis	PNDM or TNDM with cardiac, biliary or gut malformations and/ or other endocrine abnormalities or MODY	Often spontaneous or autosomal dominant	Infancy to adulthood	[18, 19]
GCK	Glucokinase / glycolytic enzyme	PNDM; Mild fasting hyperglycemia (MODY2); CHI	Autosomal recessive (PNDM); or autosomal dominant (MODY2 and CHI)	PNDM days of life; MODY2 present from birth but not usually detected until later/ CHI median age of onset 1 year	[2021]
GLIS3	Glioma-associated oncogene-similar family zinc finger 3/ Krüppel-like transcription factor	PNDM with congenital hypothyroidism (NDH)	Autosomal recessive	Days	[22, 23]
GLUD1	Glutamate dehydrogenase 1/ involved in nitrogen metabolism	Hyperinsulinism-hyperammonemia syndrome	Autosomal dominant	Infancy to adulthood	[24, 25]
HADH	3-Hydroxyacyl-CoA dehydrogenase / catalyzes dehydrogenation in β-oxidation of fatty acids	CHI	Autosomal recessive	Neonatal hypoglycaemia	[26]
HNF1A	Hepatocyte nuclear factor 1-alpha (HNF-1α) / transcription factor	MODY3; diazoxide-responsive CHI	Spontaneous or autosomal dominant	Most cases less than 25 years; some cases late onset/ hypoglycaemia in infancy	[2728, 29]

(continued)

Table 1 (continued)

Gene	Protein / function	Phenotypes / syndromes	Inheritance	Onset of diabetes/CHI	References
HNF4A	Hepatocyte nuclear factor 4-alpha (HNF-4α) / transcription factor	MODY1; diazoxide-responsive CHI	Spontaneous or autosomal dominant	Most cases less than 25 years; some cases late onset/ hypoglycaemia in infancy	[30, 31]
HNF1B	Hepatocyte nuclear factor 1-beta (HNF-1β) / transcription factor	Renal cysts and diabetes (MODY5); rare cases with TNDM/PNDM	Spontaneous or autosomal dominant	Most cases less than 25 years; some cases late onset	[32]
IER3IP1	Immediate early response 3 interacting protein 1/involved in endoplasmic reticulum stress response	PNDM with microcephaly	Autosomal recessive	Infancy	[33, 34]
IL2RA	Interleukin-2 receptor alpha (IL2R or CD25) / involved in maintenance of immune system	Immunodysregulation polyendocrinopathy, enteropathy (IPEX-like syndrome)	Autosomal recessive	One case with neonatal diabetes	[35]
INS	Insulin / hormone	PNDM (more often) or TNDM (rarely)	Spontaneous (80%), autosomal dominant or recessive (rarely)	<6 months; less often later	[36, 37]
KCNJ11	Inward rectifier K(+) channel (Kir6.2) subunit of ATP-sensitive potassium channel	PNDM (more often) or TNDM (less often); DEND; CHI	Spontaneous (80%) or autosomal dominant (PNDM and CHI); autosomal recessive (CHI)	<6 months; rarely later/ infancy or childhood	[7, 8]
MNX1	Motor neuron and pancreas homeobox-1/transcription factor involved in pancreatic development	PNDM with sacral agenesis, neurologic defects	Autosomal recessive	Neonatal	[38]

NEUROD1	(NeuroD1 or BETA2) / bHLH transcription factor	PNDM with cerebellar (but not pancreatic) hypoplasia; MODY6	Autosomal recessive (PNDM) or dominant (MODY)	Neonatal; MODY cases less than 25 years	[39, 40]
NEUROG3	(NeuroG3 or NGN3) / bHLH transcription factor	PNDM with severe congenital diarrhea	Autosomal recessive	Days-infancy	[41]
NKX2-2	NK2 homeobox 2/transcription factor involved in pancreatic development	PNDM with agenesis of corpus callosum	Autosomal recessive	Neonatal	[38]
PAX6	Paired box gene 6/paired box and homeo domain box containing transcription factor	PNDM with severe microcephaly and eye defects	Autosomal recessive	Days	[42]
PCBD1	Pterin-4-alpha carbinolamine dehydratase/ acts a dimerization cofactor of transcriptionfactors HNF1α and HNF1β	Early-onset diabetes	Autosomal recessive	Variable	[43]
PDX1	Pancreas/duodenum homeobox protein 1 (PDX1 or IPF1) / transcription factor	PNDM with pancreatic agenesis/ hypoplasia	Autosomal recessive	Days-infancy	[44–46]
PLAGL1 HYMAI (6q24)	Over-expression of paternally expressed genes *PLAGL1* (zinc finger protein or ZAC tumor suppressor) and *HYMAI* (non-protein coding) within the imprinted region of chromosome 6q24 / unknown function	TNDM	UPD6 (40%; de-novo, non-recurrent), paternal duplication (40%, may be inherited) or maternal methylation defect (20%; autosomal recessive, e.g. *ZFP57*)	Within days; remission within months;relapse during adolescence	[47–49]

(continued)

Table 1 (continued)

Gene	Protein / function	Phenotypes / syndromes	Inheritance	Onset of diabetes/CHI	References
PTF1A	Pancreas transcription factor 1, subunit alpha / bHLH transcription factor	PNDM with cerebellar and pancreatic agenesis	Autosomal recessive	Days	[50]
RFX6	DNA-binding protein / winged-helix transcription factor	PNDM with intestinal atresia, gall bladder hypoplasia	Autosomal recessive	Days	[51, 52]
SLC19A2	Thiamine transporter 1/ transports thiamine across the plasma membrane	Thiamine-responsive megaloblastic anemia (TRMA) syndrome	Autosomal recessive	Infancy; Others: DM later	[53, 54]
SLC16A1	Monocarboxylate transporter (MCT1)/ mediates lactate transport	Exercise-induced CHI	Autosomal dominant mutations in regulatory regions of the gene	Onset is variable and is triggered by exercise	[55]
SLC2A2	GLUT2 / facilitates glucose transporter	Fanconi Bickel syndrome (FBS)	Autosomal recessive	Infancy-childhood	[56]
STAT1	Signal transducer and activator of transcription 1/signal transduction and activation of transcription	IPEX-like syndrome	Autosomal dominant	Infancy	[57]
STAT3	Signal transducer and activator of transcription 3/transcription factor	IPEX-like syndrome	Autosomal dominant	Neonatal	[58]

UCP2	Uncoupling protein 2/ mitochondrial protein that acts as a negative regulator of insulin secretion	CHI	Autosomal dominant	Infancy CHI that resolves by age 7	[59]
WFS1	Wolframin / membrane glycoprotein	Wolfram syndrome (DIDMOAD)	Autosomal recessive	Childhood to adolescence	[60]
ZFP57	Zinc finger protein 57/ transcription factor with a role in maintenance of imprinted DNA methylation	TNDM	Autosomal recessive	Similar to 6q24	[61, 62]

TNDM transient neonatal diabetes, *PNDM* permanent neonatal diabetes, *CHI* congenital hyperinsulinism, *DEND* developmental delay, epilepsy, neonatal diabetes, *WRS* Wolcott-Rallison syndrome, *IPEX* immunodysregulation polyendocrinopathy, enteropathy, x-linked, *MODY* maturity-onset diabetes of the young, *DIDMOAD* diabetes insipidus, diabetes mellitus, optic atrophy and deafness *TRMA* Thiamine-responsive megaloblastic anemia syndrome, *FBS* Fanconi-Bickel syndrome, *NDH* Neonatal diabetes with congenital hypothyroidism, *bHLH* basic Helix-Loop-Helix obtaining a genetic diagnosis provides accurate information regarding inheritance, prognosis, can explain clinical features and may guide patient treatment. As monogenic diabetes is a genetically heterogeneous group of disorders, the ability to use next-generation sequencing (NGS) technology allows for the identification of mutations in any of the known genes in a single test for clinical diagnostic purposes, or for the analysis of the whole exome in selected cases for novel disease gene discovery [10, 18, 70–76].

2.1 *Neonatal Diabetes Mellitus*

Neonatal diabetes mellitus (NDM) occurs in approximately of 1:100,000–300,000 live births [63]. Traditionally, it has been defined as persistent hyperglycemia, with onset within the first months of life requiring insulin management. Many patients with NDM are born small for gestational age, which reflects a prenatal deficiency of insulin secretion as insulin exerts potent growth-promoting effects during intrauterine development [77]. Clinically, NDM can be divided into three subgroups: (i) permanent NDM (PNDM); (ii) transient NDM (TNDM); and (iii) NDM existing as part of a syndrome. Approximately 50% of NDM is permanent and 50% is transient [78]. There are no distinguishing clinical features that can predict whether a neonate with diabetes, in the absence of other dysmorphic features, will manifest PNDM or TNDM [47, 79]. Recently, advances have been made in the understanding of the molecular mechanisms of pancreatic development that are relevant to PNDM and TNDM [80].

2.1.1 Permanent Neonatal Diabetes Mellitus (PNDM)

Nearly half of all cases of PNDM are due to heterozygous activating mutations in *KCNJ11* and *ABCC8* [81, 82], the two subunits of the ATP-sensitive potassium channel, Kir6.2 and SUR1 respectively. These mutations result in increase in the open probability of the channel by making it less sensitive to ATP [83] and overall result in gain of function of K_{ATP} channels so that they are persistently open, leading to β-cell hyperpolarization even in the presence of elevated plasma glucose levels. Hyperpolarization prevents the secretion of insulin, thus resulting in the diabetic phenotype. About 80% of *KCNJ11* mutations and 50% of *ABCC8* mutations arise *de novo* [84] and almost all reported mutations are missense.

The same K_{ATP} channels that are present on insulin-secreting β-cells are also found in brain as well as other electrically active tissues including brain and muscle. As a consequence, some patients with *KCNJ11* mutations exhibit developmental delay, epilepsy, muscle weakness in addition to neonatal diabetes (DEND syndrome) [85]. Patients with a milder form, termed intermediate DEND (iDEND), do not have epilepsy. The p.Val59Met mutation in Kir6.2 is the most common cause of iDEND [81] but several other mutations, including those at the Arg201 residue, seem to be associated with learning disabilities. In contrast to *KCNJ11*-related PNDM, the DEND/iDEND syndrome occurs more rarely with *ABCC8* mutations.

Heterozygous, dominantly inherited mutations in the insulin gene (*INS*) are the second most common cause of PNDM, with diagnosis of diabetes sometimes occurring after 6 months of age [36, 37]. The mutations usually results in a misfolded proinsulin molecule that is trapped and accumulated in the endoplasmatic reticulum (ER), leading to ER stress and β-cell apoptosis [86]. *De novo INS* mutations occur in 80% of cases. These patients require insulin therapy and typically their birth weight is reduced as a consequence of decreased *in utero* insulin secretion. Recessive mutations that impair *INS* gene expression also cause PNDM [87]. Biallelic mutations do not cause progressive β-cell destruction but result in lack of insulin biosynthesis before and after birth, which explains the earlier presentation of diabetes and lower birth weight [88]. Recessive mutations in the gene encoding the glycolytic enzyme glucokinase (*GCK*) cause complete glucokinase deficiency leading to PNDM [21, 89, 90]. Autosomal dominant *GCK* mutations lead to stable, non-progressive mild fasting hyperglycemia that rarely requires treatment (MODY2) [20]. Although mutations in *GCK* are a rare cause of neonatal diabetes, this molecular etiology should be considered in cases with isolated diabetes, especially if consanguinity is suspected and/or family members have a *GCK*-MODY phenotype.

2.1.2 Transient Neonatal Diabetes Mellitus (TNDM)

TNDM is characterized by severe intrauterine growth retardation (IUGR) and typically presents within the first several days to week of life. TNDM remits, on average, by 12 weeks-of-age; however approximately 50% of individuals will relapse, typically in adolescence or young adulthood [91]. Approximately 70% of TNDM cases are caused by genetic and epigenetic abnormalities of genes in the TNDM locus on chromosome 6q24 [92], including paternal uniparental disomy (patUPD6), paternal duplications and methylation defects. These abnormalities result in the overexpression of at least two imprinted genes: *PLAGL1* (pleomorphic adenoma gene-like 1) also known as *ZAC* (zinc finger protein associated with apoptosis and cell-cycle arrest), the major TNDM candidate gene [93] and *HYMAI*, an untranslated transcript of unknown function [94]. *PLAGL1* is a transcriptional regulator of the type 1 receptor for pituitary adenylate cyclase-activating polypeptide, which is important in insulin secretion regulation [93]. Studies to understand the role of *PLAGL1* in diabetes have been limited, and at present it is not well understood how *PLAGL1* overexpression is related to TNDM pathogenesis. *HYMAI* encodes an untranslated mRNA ubiquitously expressed from the same imprinted promoter and in the same orientation as *PLAGL1*, but its role in TNDM pathogenesis is poorly understood [94]. Both *PLAGL1* and *HYMAI* are expressed in a parent-of-origin specific manner with only the paternally inherited allele expressed and the maternal allele silenced in most human tissues. This mono-allelic expression is regulated by a shared promoter, which coincides with a differentially methylated CpG island (TNDM DMR) [95]. Loss of the maternal methylation pattern at the TNDM DMR results in biallelic expression of these genes in TNDM patients [96]. Recently, Mackay et al. [97] noted that loss of methylation at 6q24 could be part of a more complex loss of methylation pattern at multiple imprinted loci (termed hypomethylation at imprinted loci (HIL)). Furthermore a pattern emerged in that the same imprinted loci were involved in most cases and all involved loss of methylation at DMRs, which were normally methylated on the maternal allele. Using genome-wide SNP testing in 6 consanguineous pedigrees, 7/13 (54%) of these complex TND HIL cases were found to be homozygous for loss of function mutations of the zinc finger transcription factor gene, *ZFP57*, on chromosome 6p22.3 [61].

In addition, mutations in *KCNJ11* and *ABCC8,* account for approximately 25% of TNDM and recessive mutations in the *INS* gene have been found to be a cause of both PNDM and TNDM [88]. However, 6q24-related diabetes seems to be most consistently characterized by a long remission phase, whereas other genetic causes seem to involve inconsistent episodes during which treatment may not be required.

2.1.3 Other Causes of PNDM

There are several syndromes that include NDM as a feature. Age of onset of diabetes, disease severity and the extra-pancreatic features observed can vary greatly depending on the gene involved. In consanguineous families, the most common form of syndromic PNDM is Wolcott-Rallison syndrome, a rare recessive disorder caused by mutations in the *EIF2AK3* gene encoding the eukaryotic translation initiation factor 2-alpha kinase 3 (*EIF2AK3*). The function of *EIF2AK3* is to inactivate the eukaryotic translation factor 2, which leads to the reduction of translational initiation and suppression of global protein synthesis. Affected patients manifest extremely early onset diabetes, epiphyseal dysplasia, renal impairment, acute hepatic failure, and developmental delay [11, 12]. Wolfram syndrome (sometimes known by the acronym DIDMOAD) is an autosomal recessive syndrome in which the association of diabetes with progressive optic atrophy under 16 years of age is diagnostic [98]. Patients require insulin treatment from the time of diagnosis. Mutations in the *WFS1* gene encoding wolframin have been identified in 90% of patients with a clinical diagnosis of Wolfram syndrome [60]. Rarely, the phenotype of PNDM may be due to pancreatic hypoplasia or aplasia; diabetes develops in these cases as a result of either a lack of, or a significantly reduced mass of pancreatic β-cells. This can result from homozygous or compound heterozygous mutations in *PDX1* (also called *IPF1*) a key factor in pancreatic development and function [99]. Pancreatic agenesis and PNDM may also result from homozygous mutations in the pancreatic transcription factor 1- α (*PTF1A*) gene. Additional features include cerebellar aplasia or hypoplasia [50]. Furthermore, recessive mutations in the *NEUROD1* gene, another transcription factor critical in β-cell development, cause PNDM with neurological abnormalities without pancreatic exocrine dysfunction [39, 40]. Another form of early onset PNDM including features of hypoplastic pancreas, intestinal atresia and gall bladder hypoplasia is due to biallelic mutations in transcription factor *RFX6* [51, 52]. A few cases with neonatal diabetes with renal anomalies have been reported to be caused by heterozygous mutations in the transcription factor *HNF1B*, although such mutations have long been described as a cause of later-onset diabetes with renal and/or genitourinary abnormalities (renal cysts and diabetes syndrome or MODY5) [100].

PNDM can also occur in association with other autoimmune conditions as part of a monogenic polyautoimmune disorder. PNDM is a common feature of immunodysregulation, polyendocrinopathy, enteropathy, X-linked (IPEX) syndrome, which is caused by *FOXP3* mutations [13, 14], and autoimmune polyendocrinopathy syndrome 1 (APS1), which is caused by recessive mutations in *AIRE* [101]. In these polyendocrinopathy syndromes, autoimmune disease has a very young age of onset, with autoimmune diabetes usually diagnosed before 3 months of age in IPEX syndrome, which is in contrast to polygenic Type 1 diabetes that is very rare before 6 months of age [13]. The investigation of individuals with multiple early-onset autoimmune features has identified new monogenic causes of autoimmunity including biallelic mutations in the interleukin-2 receptor alpha chain gene *IL2RA* (also known as *CD25*) [35] and dominant inactivating mutations in the transcription factor *STAT3* [58]. Additional causes of syndromic PNDM are listed in Table 1.

Maternal transmission of mutated or deleted mitochondrial DNA (mtDNA) can result in maternally inherited diabetes [102], which typically does not develop until adulthood. In addition to diabetes, these patients can also have sensorineural deafness and short stature. Mitochondria-associated diabetes is therefore often denoted as maternally inherited diabetes and deafness (MIDD) syndrome [103]. The degree of heteroplasmy for the mutation in the mitochondrial DNA varies between affected individuals, leading to different clinical expression of the disease.

While some of these rare diabetes-related disorders are relatively easy to diagnose due to the presence of comorbidities, others are frequently under recognized due to incomplete clinical expression of the specific symptoms at the time of diabetes onset. For patients with suspected monogenic diabetes in addition to other seemingly unrelated symptoms or disorders, a monogenic syndrome that could explain the diabetes and the patient's other additional symptoms should be considered [104].

2.2 *Maturity Onset Diabetes of the Young (MODY)*

The term "maturity-onset diabetes of the young" (MODY) refers to a heterogeneous group of disorders associated with non-ketotic diabetes, that are caused by mutations in genes that are involved in pancreatic β-cell function [105]. Mutations in these genes lead to impaired glucose-stimulated insulin secretion by pancreatic β-cells, thus resulting in diabetes. MODY is inherited in an autosomal dominant manner, and is traditionally suspected in families with multiple generations affected by diabetes, with age of disease onset younger than 25 years [2, 105, 106]. Unlike patients with type 2 diabetes mellitus (T2DM), patients with MODY are typically not obese, which can help clinically distinguish cases of MODY from childhood-onset T2DM [107]. Younger patients with MODY are frequently asymptomatic, and may go undiagnosed until adulthood [2, 108]. Residual insulin secretion may be maintained for several years after the time of diagnosis [109, 110]. MODY is considered rare in young individuals with diabetes, with estimates of prevalence within the pediatric diabetic population ranging from 2.4% [111] to less than 1% [107, 112]. However, MODY is likely underdiagnosed in this population, with a study of a pediatric diabetes cohort finding that out of 47 patients found to carry a mutation in a MODY gene, only 3 (6%) had previously been clinically diagnosed with MODY, with most being originally diagnosed with either type 1 or type 2 diabetes [106]. Although a family history of diabetes consistent with autosomal dominant inheritance is considered for the clinical diagnosis of MODY, not all patients with genetically confirmed MODY have an affected parent, revealing the limitations of using family history as a criterion for identifying affected individuals [106]. To date at least 13 genes have been associated with MODY, and the clinical presentation including age of onset, pattern of hypoglycemia, typical response to treatment and associated extra-pancreatic manifestations can vary based on the underlying molecular defect in the family [105].

The term MODY was first used when type 1 diabetes was referred to as "juvenile onset diabetes", and type 2 diabetes was referred to as "maturity onset diabetes". Now that understanding of the underlying etiology of this disorder has advanced, some have proposed replacing the MODY terminology with more specific classifications that reflect increased knowledge [63]. However, transition to new terminology is difficult, given the pervasive use of this nomenclature in the literature, and for consistency with the current body of literature on these disorders, the term MODY continues to be used.

2.2.1 *HNF1A* (MODY3) and *HNF4A* (MODY1)

Mutations in the *HNF1A* gene [27] are the most common cause of MODY (MODY3). Mutations in this gene have been reported in multiple populations including Caucasian, Asian and African populations [2], and account for an estimated 55% of cases of MODY overall [106]. The *HNF1A* gene encodes for transcription factor HNF-1α, which is expressed in multiple different tissues including the pancreas, kidney and liver and can form a homodimer, or can form a heterodimer with related protein HNF-1β [27]. HNF-1α is involved in pancreatic β-cell function and growth through the regulation of expression of genes involved in glucose transport and metabolism, including the insulin gene [2, 28, 29]. Patients with a mutation in *HNF1A* have progressive reduction in β-cell function overtime, leading to progressive hyperglycemia due to impaired secretion of insulin in response to high glucose levels [113]. In families with a mutation in *HNF1A,* the penetrance is high with >90% of mutation carriers being diagnosed with diabetes [114]. The majority (60%) of *HNF1A* mutation carriers develop diabetes before age 25 years, however overall the age at diagnosis ranges widely from early childhood to late adulthood [115].

The *HNF4A* gene is also associated with MODY (MODY1) [30], and accounts for an estimated 10–16% of MODY cases overall [105, 106]. The *HNF4A* gene encodes a transcription factor HNF-4α, which like HNF-1α is expressed in multiple organs including the pancreas, liver and kidneys [116]. HNF4α has been shown to regulate HNF-1α expression, and the clinical characteristics of MODY patients with a mutation in either *HNF1A* or *HNF4A* are very similar, due to the related underlying mechanism of disease [105, 116, 117]. As a point of differentiation between the two, *HNF1A*-associated MODY has been reported in association with glycosuria due to decreased renal absorption of glucose [118], whereas mutations in *HNF4A* may impact expression of genes expressed in the liver and lead to reduced triglyceride levels [119].

2.2.2 *HNF1B* (MODY5)

Mutations in the *HNF1B* gene are associated with MODY5. In addition to early-onset diabetes, MODY5 is also characterized by renal cysts [120]. The *HNF1B* gene encodes for HNF-1β, a transcription factor that is structurally related to HNF-1α and is involved in the tissue-specific regulation of gene expression in a number of organs including the pancreas and kidneys [121]. Renal disease is typically recognized in adolescence or adulthood, and in addition to cysts, other common renal defects include decreased kidney size and mild pelvic dilatation, and patients can have decreased creatinine clearance and mild to moderate renal failure [122]. Genital anomalies have also been reported in individuals with MODY5, including vaginal aplasia, bicornate uterus, epidydimal cysts, and atresia of the vas deferens [122, 123].

2.2.3 *GCK* (MODY2)

Heterozygous inactivating mutations in the *GCK* gene are associated with MODY2 [124]. *GCK* encodes glucokinase, which is expressed in the pancreas and liver, and converts glucose to glucose-6-phosphate, the first step in glucose metabolism [2]. The phosphorylation of glucose by glucokinase acts as a glucose sensor for pancreatic β-cells, and activity of glucokinase in response of increased glucose levels leads to insulin secretion [105]. Heterozygous mutations in *GCK* lead to partial deficiency of glucokinase, leading to a higher set point for insulin secretion in response to increased blood sugar, and causing persistent mild non-progressive fasting hyperglycemia [105]. In patients with MODY2, hyperglycemia is present from birth but is typically subclinical, and may be identified incidentally during routine screening, such as screening for gestational diabetes during pregnancy [124, 125]. Individuals with MODY2 may have a family history of individuals diagnosed with "borderline" diabetes or gestational diabetes, or may have no known family history of diabetes or hyperglycemia, given that carriers of a *GCK* mutation are typically asymptomatic. As with other forms of MODY, individuals with *GCK* mutations can also be misdiagnosed with have type 1 diabetes [126]. Although classified as a type of MODY, less than 50% of carriers of a *GCK* mutation have overt diabetes, and many of those individuals have additional risk factors for diabetes such as obesity or increased age [2]. Complications associated with diabetes such as microvascular disease are rare in patients with *GCK*-associated mild fasting hyperglycemia [2, 124]. Like women with gestational diabetes [127], women with *GCK*-associated hyperglycemia may be at risk for complications such as fetal macrosomia, however the risk of complications is impacted by fetal genotype. In pregnant women with *GCK*-associated hyperglycemia birth weight is typically normal in fetuses who also inherited the *GCK* mutation, whereas fetuses who do not inherit the *GCK* mutation are at increased risk for macrosomia [128, 129].

2.2.4 Rare Forms of MODY

In addition to *HNF1A, HNF4A* and *GCK*, a number of other genes have also been reported in association with MODY. Like *HNF1A* and *HNF4A*, some of these genes encode for transcription factors expressed in pancreatic β-cells. The *PDX1* gene (previously known as *IPF1*) encodes a transcription factor that is involved in regulating expression of the insulin and glucokinase genes [2]. Homozygous or compound heterozygous mutations in this gene have been reported in a small number of patients with pancreatic agenesis [44, 45, 99, 130], and within these families, some heterozygous *PDX1* mutation carriers with a mild form of early-onset diabetes, designated MODY4, have been identified. In two large pedigrees that included 10 heterozygous carriers for a *PDX1* mutation with MODY4, the age of diabetes diagnosis ranged from 2 to 67 years [46]. Obesity and hyperinsulinemia were prominent features in these two families, in both mutation carriers and non-carriers who were affected with type 2 diabetes.

Other rare causes of MODY include mutations in transcription factor genes *KLF11* (MODY7) [131], *NEUROD1* (MODY6) [40] and *PAX4* (MODY9) [132]. MODY has also been associated with mutations in *CEL* (MODY8) [9], which encodes a bile salt-stimulated lipase, and *BLK* (MODY11) [133], which encodes a tyrosine kinase. Mutations in *ABCC8, KCNJ11,* and *INS* were originally described in association with a permanent neonatal diabetes phenotype, however more recently these genes have also been reported in association with MODY in a small number of cases [5, 75, 134, 135], indicating that these genes can exhibit variable phenotypic expression depending on the mutation(s) present. Mutations in these rare MODY genes have been reported in only a small number of families to date, and clinical genetic testing for MODY frequently focuses on testing for the most common MODY genes, meaning that not all MODY patients are evaluated for mutations in these rare genes. Thus, although these genes are predicted to each account for only a small proportion of cases of MODY, their overall contribution to the disorder has not been fully elucidated. Depending on the criteria used for screening patients, some studies of patients with a clearly defined MODY phenotype have estimated that at least 10–20% will not have a mutation in any of the known MODY genes [136].

3 Congenital Hyperinsulinism (CHI)

Congenital hyperinsulinism (CHI) is a genetically heterogeneous disorder characterized by unregulated insulin secretion from pancreatic β-cells, leading to inadequate suppression of insulin secretion in response to low blood sugar levels and hyperinsulinemic hypoglycemia [137, 138]. CHI is the most frequent cause of severe persistent hypoglycemia in infancy, and has a incidence of 1 in 50,000 in the general population [138]. In some populations the incidence may be higher, such as the Ashkenazi population, where the incidence is estimate at 1 in 7800 [139]. The incidence is also increased in consanguineous families [138]. Overall, molecular testing for the known genes associated with CHI identifies causative mutations in approximately half of all cases, the underlying molecular etiology remains unknown in the remaining cases [140].

CHI most commonly presents in the neonatal period [141]. Symptoms of hyperinsulinemic hypoglycemia are non-specific, and can include irritability, hypotonia, poor feeding, seizures and coma [141]. Untreated hypoglycemia can lead to permanent brain injury. In individuals with childhood onset CHI, hypoglycemia can lead to symptoms including fainting, tachycardia, sweating and seizures [141]. Histologically, CHI can be caused by diffuse disease, involving abnormally functioning β-cells throughout the pancreas, or focal disease where unregulated insulin secretion involves a limited region of abnormally functioning β-cells within the pancreas, surrounded by normally functioning pancreatic tissue [141].

3.1 K_{ATP}-Associated Congenital Hyperinsulinism

In contrast to activating mutations in *ABCC8* and *KCNJ11*, which are associated with diabetes [6, 81, 142], inactivating mutations are associated with CHI [7, 8]. In normally functioning β-cells, the K_{ATP} channel closes in response to increased glucose levels, leading to membrane depolarization and secretion of insulin. Inactivating mutations in these genes can lead to non-functional or dysfunctional K_{ATP} channels that fail to open when glucose levels are low, leading to continued inappropriate insulin secretion [143].

Recessive mutations in *ABCC8* and *KCNJ11* are the most common cause of CHI, and account for the majority of cases of severe CHI that is unresponsive to medical management (diazoxide therapy) [140]. These severely affected patients have diffuse disease that most commonly presents on the first day of life, and are typically large for gestational age [140]. Presentation later in infancy can also occur [140]. Diazoxide-unresponsive disease can also be observed in patients with focal disease; in these cases the patients carry a paternally inherited mutation in either *ABCC8* or *KCNJ11*, and somatic loss of the maternal allele of the gene occurs in the affected pancreatic tissue [144]. Dominantly inherited mutations in *ABCC8* and *KCNJ11* can also be associated with CHI. The severity of autosomal dominant K_{ATP}-related CHI is variable, however compared to patients with recessive disease, patients with dominant disease typically have much milder disease that is responsive to medical management [141, 145, 146]. However, severe cases of medically unresponsive CHI have also been observed in patients with a dominant *ABCC8* or *KCNJ11* mutation [140].

3.2 *GLUD1*-Associated Hyperinsulinism/Hyperammonemia Syndrome

The *GLUD1* gene encodes for glutamate dehydrogenase (GDH), and heterozygous activating mutations in this gene are associated with hyperinsulinism/hyperammonemia syndrome (HI/HA), the second most common form of CHI [24]. GDH catalyzes the conversion of glutamate to α-ketoglutarate and ammonia, and is highly expressed in the liver, pancreas, liver and brain [24, 25]. Activating mutations in *GLUD1* reduce inhibition of the enzyme by GTP, leading to excess α-ketoglutarate entering the Krebs cycle, a glucose-independent increase in the intracellular ATP/ADP ratio, and unregulated insulin secretion [24]. Hypoglycemia in patients with HI/HA is typically mild, and responsive to diet manipulation and medical therapy [147]. Seizures are a common presenting symptom, and an estimated 43% of patients develop epilepsy with generalized seizures [147]. GDH is activated by leucine and patients with HI/HA have leucine-sensitivity, which manifests as postprandial hypoglycemia following a high protein meal [147].

3.3 *HNF4A*-Associated Congenital Hyperinsulinism

Heterozygous mutations in *HNF4A* have been identified in 5% of patients with diazoxide-responsive hyperinsulinemic hypoglycemia [148]. Affected infants are commonly macrosomic, and hypoglycemia has onset in the neonatal period and can be transient, or may require longer-term medical therapy [148, 149]. CHI-associated *HNF4A* mutations may occur *de novo* or be inherited; for cases with an inherited mutation the patient may have a family history of *HNF4A*-associated MODY1 [149]. Pearson et al. (2007) identified transient neonatal hypoglycemia in 15% of *HNF4A* mutation carriers with a family history of MODY1 [31], suggesting that this finding may be an under-recognized feature of MODY1.

3.4 *GCK*-Associated Congenital Hyperinsulinism

Unlike inactivating mutations in *GCK* that lead to a diabetes phenotype, heterozygous activating mutations in *GCK* have been associated with CHI due to a lowered threshold for glucose stimulated insulin secretion. Most *GCK*-associated CHI results in a mild hypoglycemia that is responsive to diazoxide, and *GCK* accounts for an estimated 7% of medically responsive CHI cases [150]. *GCK*-associated CHI varies is severity, and mutations in this gene may also account for an estimated 2% of diazoxide unresponsive patients [151].

3.5 Rare Forms of Congenital Hyperinsulinism

A number of genes have been identified as rare causes of CHI, including *HNF1A* [152], *HADH* [26], *UCP2* [59], *INSR* [153], and *SLC16A1* [55], or overexpression of genes at chromosome 6q24 that initially causes transient neonatal diabetes but in some cases can be followed soon after diabetes remission by protracted hyperinsulinemic hypoglycemia requiring treatment [154]. Recessive mutations in *HADH* are associated with mild, medically responsive hypoglycemia, in addition to elevated urine 3-hydroxyglutaric acid and serum 3-hydroxybutyryl-carnitine. Heterozygous mutations in *HNF1A* have been observed in a small number of cases of medically responsive CHI [151, 152]. Heterozygous mutations in *UCP2* have been reported in two unrelated children with CHI who had previously tested negative for other known causes of the disease [155]. Overall, the *HADH, HNF1A* and *UCP2* genes may each account for 1–2% of all cases of diazoxide responsive CHI [151]. A mutation in the insulin receptor gene *INSR* has been associated with autosomal dominant hyperinsulinemic hypoglycemia in one large pedigree. Affected individuals had episodes of hypoglycemia with onset between 3 and 30 years, which resulted in loss of consciousness and convulsions in the most severely affected family members [153]. Mutations in *SLC16A1*, which encodes monocarboxylate transporter-1, have been identified in three families with exercise-induced hyperinsulinism [55].

4 Benefits of Genetic Testing for Monogenic Disorders of Insulin Secretion

4.1 Treatment

4.1.1 Treatment of Neonates and Infants with Diabetes

Regardless of what the cause of diabetes might be, it is essential that infants with diabetes be treated initially with insulin in order to normalize glucose levels. The cardinal signs of diabetes (polyuria and polydipsia) can be inappropriately reassuring when such babies are brought to medical attention by parents who often have vague concerns about their well being. Diabetes is therefore very difficult to diagnose clinically and consequently neonates and infants will often not have any diagnostic laboratory testing until late in the disease process, by which time they often have diabetic ketoacidosis and other complications [156].

Insulin therapy is similar in infants compared to older children, except that they will often be more insulin sensitive and require considerably less insulin, even on a per-weight basis (as low as 0.2–0.3 units/kg/day). Intravenous insulin drip rates might be as low as 0.02–0.05 units/kg/h, whereas subcutaneous insulin administration will often require dilution of insulin, although rates of delivery by continuous subcutaneous insulin infusion (CSII) are often low enough (as low as 0.025 units/h, and can on some pumps be programmed as 0.0 units/h during periods of time overnight if needed) to suffice without dilution [157]. If CSII therapy is not possible, a regimen of multiple daily injections (MDI) of insulin utilizing long-acting insulin (such as glargine or detemir in doses as low as 0.5 unit once per day) in conjunction with short-acting insulin to cover carbohydrate intake and correct hyperglycemia remains the standard of care. The fast-acting insulin will often need to be diluted (10% or U-10 is usually most convenient and least confusing). The appropriate diluent for either aspart or lispro can be obtained directly from the manufacturer and the dilution can be prepared either by a pharmacy or by competent trained care providers at home. Clinicians may be tempted to cover frequent feedings with intermediate-acting insulins such as NPH; however, this will often make it difficult for the baby to tolerate periods of fasting without hypoglycemia and preclude transition to a normal sleep schedule, as the infant gets older. While babies are feeding frequently, insulin doses can often be given with every other feed so as to minimize the number of injections per day. The overall goal is generally to minimize extreme highs (above 400 mg/dL) or extreme lows (below 40 mg/dL) rather than to achieve perfect control. Of note, HbA1c values are falsely low due to the presence of fetal hemoglobin but become more reliable after 6 months of age [158].

Genetic testing is mandatory in any baby diagnosed with diabetes under 6 months of age, as a monogenic diagnosis will be found in more than 80% of cases and will be essential in guiding long-term management [159]. Because infants diagnosed between 6 and 12 months of age may have as much as a 5% chance of a monogenic diagnosis [160] it is also likely to be cost effective to test such patients [161]. Of note, even autoantibody positivity may not preclude such testing, since the number of known monogenic causes of autoimmune diabetes is expanding [13, 58].

Regardless of initial management considerations, genetic testing is mandatory and should be done as soon as possible, given the potential for major ramifications on treatment decisions and long-term outcome. Because a newly diagnosed infant has a nearly 50% chance of having a KATP mutation, it may be reasonable to consider an empiric trial of oral sulfonylureas (SU) while awaiting results of genetic testing because of the great potential benefit of treatment if it is successful [162]. Reasonable glucose control with insulin therapy must be maintained during the SU trial with the guidance of published protocols suggesting rapid increase to an SU dose sufficient to affect blood glucose levels [163]. If insulin continues to be required after 5–7 days, then the SU should be discontinued until genetic testing results are available, because SU treatment of non-KATP gene causes could in theory be detrimental. Pancreatic ultrasound may be considered to evaluate for pancreatic hypoplasia, and such patients or those with other clinical complications such as diarrhea should not be started on SU as they are more likely to have non-KATP gene causes.

4.1.2 Sulfonylurea Treatment of K_{ATP}-Associated Diabetes

Patients with mutations in either of the genes (*KCNJ11* or *ABCC8*) encoding the two subunits of the KATP channel have about 95% chance of being able to switch from insulin injections to oral SU therapy, even when attempted decades after diabetes diagnosis [163, 164]. Using published protocols [163], the general approach is to cautiously decrease insulin doses while rapidly increasing the SU – usually glyburide – started at 0.2 mg/kg/day in two divided doses and increased by 0.2 mg/kg/day each day until insulin can be discontinued. The transition can be done most expeditiously in the inpatient setting (particularly for the youngest patients), whereas it can also be done at home if parents (or adult patient) are capable of frequent blood glucose monitoring and can remain in close contact with their medical provider. Although insulin should continue to be given to maintain reasonable glycemic control, it is often better to decrease the insulin somewhat aggressively to avoid hypoglycemia from the combined effects of SU and insulin. Once the insulin is discontinued there is very little risk of hypoglycemia, whereas mild hyperglycemia will usually improve over the first several weeks it may take for the full effect of the SU to be realized. The tablets can be easily suspended in aqueous solution either by parents who can crush tablets at home, or by a pharmacy; although the stability of suspensions has not been well studied, it appears that they can be used for 2–4 weeks without noticeable difficulty [165]. The high dose of SU often required (usually around 1 mg/kg/day but can be up to 2–2.5 mg/kg/day) can be up to ten times higher than used for type 2 diabetes (usually about 0.2 mg/kg/day). Interestingly, review of data from our Registry has shown that the dose required is directly correlated to the age at which they are switched from insulin to SU [164]. The youngest babies will often require much lower doses and still achieve excellent glycemic control, whereas patients over 13 years of age will usually need maximal doses and often require additional non-insulin medications for optimal glycemic control. Of note, even the rare patients who continue to require small doses of insulin in addition to SU still have much better glycemic control on SU than they did when on insulin therapy alone. Very importantly, there is also good evidence that SU treatment can help with the spectrum of neurodevelopmental disability exhibited by a large fraction of patients with KATP channel mutations [166–172]. For this reason, many experts suggest maintaining a dose of up to 1 mg/kg/day or greater even when a lower dose would suffice for glycemic control (so long as hypoglycemia is avoided), as it appears that maintenance of glyburide levels across the blood brain barrier may be a limiting factor [173]. Glyburide (glibenclamide outside the US) has been the mainstay of treatment, in part because it appears to have more beneficial effect on neurodevelopmental problems [169]. Of note, early evidence suggests that those started on treatment at a very young age may have better eventual neurodevelopmental outcome, presumably because the intervention is occurring before brain development is complete [171171].

4.1.3 Treatment Considerations for HNF1A/4A MODY

Heterozygous mutations or deletions in the pancreatic transcription factors HNF1A and HNF4A lead to significant impairment of beta cell function, although the precise mechanism for diabetes and factors affecting the age of onset remain incompletely understood [174–176]. Because these cases are characterized by defects in insulin secretion, it makes sense that insulin secretagogues have long been the treatment of choice for HNF1A/4A MODY, in contrast to first line agents for type 2 diabetes (metformin) that improve insulin sensitivity [177, 178]. In fact, these patients classically respond to very low doses of sulfonylureas (in contrast to KATP mutation patients who require very high doses) and not uncommonly have a history of hypoglycemia in response to such agents [179]. HNF1A/4A MODY patients will thus usually have good glycemic control in response to minimal treatment with a variety of drugs, including insulin in those who were presumed to have type 1 diabetes. Once a genetic diagnosis is established in such insulin-treated cases they can be transitioned to SU therapy, but will be at risk of hypoglycemia during concurrent use of insulin and SU. Therefore, it is very important to monitor glucose levels closely during the transition, to start with a very low dose of SU, and to decrease insulin by at least 50% or even stop it entirely before the first dose of SU is given (depending on the level of glycemic control and current insulin dose). The disease course in HNF1A/4A MODY is progressive and may limit the responsiveness to SU in patients who have had diabetes for many years; however, one study has shown that patients who were transitioned from insulin to SU after a variable number of years had an average decrease in their HbA1c of 0.8% [180]. The potential improvements in glycemic control on cheaper therapy with presumed quality of life benefits suggest that it is cost-effective to screen selected cases who otherwise might have been considered to have type 2 diabetes [181]. Recent studies have suggested that there may also be a role for incretin-based therapies; however, it is not yet clear that they have any advantage over SU agents [182].

4.1.4 GCK MODY Usually Does Not Require Any Treatment

The real importance of making a diagnosis of GCK-MODY is that it generally does not require any treatment at all [63]. In fact, some experts prefer the term *GCK-related hyperglycemia* to emphasize that the mild level of non-progressive fasting hyperglycemia is not equivalent to having diabetes and generally appears to have no long-term detrimental consequences [183]. A recent careful examination of a large cohort of older patients with GCK-MODY revealed only very mild retinopathy in a small fraction of patients [184]. Of note, the degree of retinopathy was not even to the level that required any intervention. Other studies have shown that even when medical treatment of the hyperglycemia is initiated, it can be very difficult to achieve normal blood sugar levels and attempts to do so can lead patients to experience hypoglycemia [185]. The one clinical scenario in which treatment can reasonably be considered is pregnancy, where the potential for overgrowth in a genetically unaffected fetus in an affected mother justifies close monitoring prenatally, with a consideration of insulin treatment to avoid macrosomia; however, it is not clear how effectively this approach either normalizes glycemia or avoids birth complications [129]. Studies of patients with diabetes in which some were found to have GCK-MODY revealed that 20% or more had been unnecessarily treated with medical therapy, including insulin injections, both in Europe [185] and especially in the US [106186]. The dramatic potential difference in treatment suggests that screening for *GCK* defects may be cost-effective in appropriately selected patients who may have been misdiagnosed as having type 1 or type 2 diabetes [181].

4.1.5 Treatment Considerations in Congenital Hyperinsulinism

Neonates who exhibit significant hypoglycemia despite adequate feedings require initial treatment with intravenous glucose infusion, continued monitoring and thorough evaluation, regardless of the underlying cause [187]. Untreated hypoglycemia leads to significant neurological impairment that likely can be avoided with appropriate monitoring and treatment [188]. Neonatal hypoglycemia may be due to a wide variety of causes that may be due to the lack of counter-regulatory response to maintain normal glycemia, (such as pan-hypopituitarism), or more often from hyperinsulinism that is not necessarily due to specific gene mutations, but rather may result from causes such as maternal diabetes or adaptation to perinatal stress [189]. Those with genetic forms of congenital HI will often require very high glucose infusion rates (GIR; up to 20 mg/kg/h or higher). A diagnosis of hyperinsulinism is not always made by measurement of inappropriately elevated insulin levels in the setting of hypoglycemia but rather may be inferred from inappropriately low beta-hydroxybutyrate and/or free fatty acid levels in setting of hypoglycemia, often confirmed by a significant rise in glucose level following glucagon administration [190]. Once hyperinsulinism is confirmed, initial medical therapy is diazoxide, which keeps the KATP channels open and reduces insulin secretion. Diazoxide can be started every 8 h with a dose as low as 5 mg/kg/day or as high as 15 mg/kg/day depending on the severity of presentation, but is not usually increased above 20 mg/kg/day [191]. If hypoglycemia is severe and/or it persists after at least 5 days of diazoxide, urgent genetic testing should be ordered as it suggests a KATP channel mutation or other rarer genetic cause. Glucagon infusion of 1 mg/day may be added to the glucose infusion as needed while the patient is transferred to a center capable of specialized workup for focal or diffuse hyperinsulinism, including F-DOPA PET imaging [192]. As described below, genetic testing will clarify whether a focal lesion amenable to targeted resection and complete cure is possible or whether near-total pancreatectomy with high risk of diabetes may be required. If partial response is seen with diazoxide or if pancreatectomy is hoped to be avoided, other agents have also been used with some success in certain scenarios [193], including long-acting somatostatin analogues (such as octreotide or lanreotide) [194], the mammalian target of rapamycin (mTOR) inhibitor sirolimus [195] or the calcium channel blocker nifedipine [196].

4.2 *Predicting Disease Progression and Associated Features*

Identifying the underlying molecular basis of disease in an individual with a monogenic disorder of insulin secretion can help predict disease severity and likelihood of disease progression. Even within a specific gene, the type of mutation and its overall impact on gene function can affect a patient's clinical features and prognosis. In addition, some genes may be associated with a single disorder, such as diabetes or congenital hyperinsulinism, whereas others may be multi-systemic disorders associated with additional features. The use of genetic testing to confirm a diagnosis can therefore be beneficial by predicting disease course and severity, and identifying additional associated features of the disorder that a patient may be at risk of develop in the future [197]. For patients already exhibiting disease involving multiple organs or tissues, genetic testing can provide a diagnosis that ties a patient's varied features together under one genetic disorder [197].

4.2.1 Predicting Disease Course of CHI Associated with *GLUD1*

GLUD1-associated hyperinsulinism (HI/HA) is associated with persistent mild-to-moderate hyperammonemia, in addition to hyperinsulinemic hypoglycemia. Ammonia levels in affected patients are typically stable, and do not fluctuate in relation to fasting, protein intake, or glucose levels [198]. Patients with *GLUD1*-associated hyperammonemia do not have typical signs of hyperammonemia such as lethargy and disorientation [198]. Patients with HI/HA have an increased frequency of generalized seizures, and unlike seizures that are observed in other types of CHI, these can occur in the absence of hypoglycemia [198]. Seizures appear to be most common in patients with mutations in exons 6 and 7 of the *GLUD1* gene, indicating a genotype-phenotype correlation [147]. Some patients also have developmental delays or behavioral issues [147, 198, 199]. Episodes of hyperinsulinemic hypoglycemia can be triggered by protein intake in patients with HI/HA, as leucine is an activator of the GDH enzyme encoded by *GLUD1* [24, 198]. The use of genetic testing to confirm a diagnosis of *GLUD1*-associated HI/HA in a patient with hyperinsulinemic hypoglycemia can therefore be useful in predicting risk of developing other associated features such as epilepsy, and identification of unique triggers for hypoglycemia not present in other forms of CHI. Molecular identification of a *GLUD1* mutation can also provide a unifying diagnosis for patients with a history of unexplained hypoglycemia, hyperammonemia, and epilepsy.

4.2.2 Distinguishing Diffuse and Focal Hyperinsulinism

Biallelic inactivating mutations in *ABCC8* or *KCNJ11* can lead to diffuse CHI with disease involvement throughout the pancreas, whereas focal disease is associated with a single paternally inherited mutation in one of these genes [140, 144]. In the focal cases, a somatic event leads to the loss of the maternal allele in a limited region of the pancreas [144]. Fluorodopa positron emission tomography (F-DOPA-PET) scanning can be used to help distinguish between focal and diffuse disease [200], however this testing requires access to [76] F-DOPA which is not readily available in all medical centers. Molecular genetic testing can be used as an alternative to imaging studies in patients with severe diazoxide unresponsive CHI the identification of two mutations in *ABCC8* or *KCNJ11* is predictive of diffuse disease, whereas the presence of one paternally inherited variant from an unaffected father is predictive of focal disease. Prediction of histology of disease in patients with CHI can lead to changes in management and affects overall clinical course and outcomes. Preoperative identification of a focal lesion can lead to partial pancreatectomy and removal of the lesion, which is curative in these patients [200]. By contrast, patients with diffuse disease who are unresponsive to medical management require extensive pancreatic resection leading to risk of developing postoperative insulin-dependent diabetes mellitus [200].

4.2.3 Identifying Syndromic Forms of Monogenic Diabetes

Monogenic diabetes may be observed as part of multisystem disease in some patients. An example of this is Wolcott-Rallison syndrome (WRS), a recessive disorder caused by mutations in *EIF2AK3* and associated with PNDM, skeletal dysplasia and hepatic dysfunction [12]. In patients with WRS, diabetes typically presents in the first 6 months of life, however skeletal abnormalities may not present until 1–2 years of age [201]. Recurrent episodes of acute hepatic dysfunction can occur at any age and may be triggered by concurrent diseases such as airway infections [201]. WRS is a rare condition, however in consanguineous pedigrees it represents the most common genetic cause of PNDM [12]. Similarly, extra-pancreatic manifestations of several syndromic causes may follow long after the diabetes diagnosis and thus the use of genetic testing to confirm a diagnosis in an infant with seemingly isolated PNDM can provide important prognostic information for the family and clinician [159]. The availability of NGS panels is thus changing the paradigm from individual Sanger gene testing based on symptoms to utilization of comprehensive genetic testing to reveal a monogenic diagnosis that can help to predict other syndromic features before they are clinically apparent.

4.3 Genetic Counseling, Family Studies and Recurrence Risks

Confirming a molecular diagnosis in patients with monogenic disorders of insulin secretion including diabetes and CHI can lead to the identification of other affected or at-risk family members. As MODY is a dominant disorder, children who are diagnosed with the disorder typically have an affected parent, who may have been misdiagnosed as having either T1DM or T2DM, and may be receiving insulin therapy [104]. Families with mild fasting hyperglycemia due to a *GCK* mutation (MODY 2) typically have asymptomatic non-progressive disease, however affected family members may have been identified as having hyperglycemia during routine screening, such as screening for gestational diabetes during pregnancy [124]. In infants with *HNF4A*-associated CHI, the identification of the causative mutation and testing of family members may lead to the identification of family members who have MODY caused by the same *HNF4A* mutation [149]. Milder form of CHI such as *GLUD1*-associated HI/HA may go unrecognized in some cases, and thus confirming the diagnosis in an affected child may lead to the identification of an affected parent [198]. The molecular confirmation of a disorder of insulin secretion in one individual allows other family members to undergo testing for the same familial mutation. Affected family members identified in this way then have the opportunity to have their condition managed appropriately, with a tailored treatment approach based on an understanding of the underlying molecular defect [197].

Identifying the underlying molecular basis of disease in a patient with a disorder of insulin secretion can also provide families with information about recurrence risk. For patients confirmed to have an autosomal recessive disorder such as *GCK*-associated PNDM or diffuse *ABCC8*-associated CHI, the recurrence risk for their siblings is 25%. In contrast, other disorders such as *KCNJ11*-associated PNDM and *HNF4A*-associated CHI are frequently caused by *de novo* mutations [81, 148], and for confirmed *de novo* cases the associated recurrence risk is <1%. For focal CHI associated with *ABCC8* or *KCNJ11,* the recurrence risk is based on the likelihood of the child inheriting the paternally inherited mutation, plus the likelihood of a somatic event causing loss of the maternal allele in the pancreas, which has been estimated to be a 1 in 540 risk [139]. For dominant disorders such as MODY and HI/HA, the recurrence risk for children of a mutation carrier is 50%. Some genes can exhibit both dominant and recessive phenotypes; for example heterozygous mutations in *GCK* can be associated with mild fasting hyperglycemia (MODY2), whereas biallelic mutations can be associated with PNDM [2]. Although biallelic *GCK* mutations are rare, it may be important to counsel *GCK* mutation carriers on the risk of having a child with PNDM if both parents are carriers, particularly in the setting of consanguinity where the likelihood of this occurring is higher. For patients with diabetes associated with mitochondrial mutations, recurrence risk for children of affected females may depend on the degree of heteroplasmy. Children of an affected male are not at risk of inheriting the mutation, as the mitochondrial DNA is inherited only from females.

5 Clinical Utility of Genetic Testing for Monogenic Disorders of Insulin Secretion Using Next-Generation Sequencing

Due to the clinical variability and the genetic heterogeneity of monogenic disorders of insulin secretion, the selection of appropriate gene(s) to test is challenging when based solely on phenotype. Pathogenic variants within several genes can present with similar clinical features, while other features that are more gene specific may not yet have manifested at the time of presentation of hyperglycemia or hypoglycemia. When genetic testing is performed according to phenotype, routine analysis is often restricted to the most common subtypes of MODY, NDM and CHI it is primarily based on Sanger sequencing of individual genes in a sequential manner. If the first round of gene sequencing is negative, the cost of sequencing the other genes may preclude further testing. Furthermore, the choice of genes to be tested using this approach depends on the availability of reliable and comprehensive phenotypic information although such features may not have yet manifested at the time of diagnosis.

Next-generation sequencing (NGS) technology has been rapidly adapted to clinical testing and has redefined the practice of molecular diagnosis of human diseases. For clinically and genetically heterogeneous diseases like monogenic disorders of insulin secretion, NGS offers the advantage of testing all known genes including rare causes simultaneously, thus increasing the rate of molecular confirmation of the diagnosis in affected individuals. Recently, targeted NGS sequencing assays for known causes of monogenic diabetes have been developed to facilitate the diagnosis of these rare disorders [70, 73]. Ellard et al. [70] developed a capture-based targeted NGS assay to identify mutations in 29 genes known to cause monogenic diabetes. This assay was used to analyze a cohort of 82 patients in whom previous testing for MODY or neonatal diabetes had failed to confirm a genetic diagnosis. Previously unidentified mutations were found in nine patients with neonatal diabetes (18%) and in five patients with MODY (15%). Most of these patients had mutations in genes that had not been previously tested because extra-pancreatic features characteristic of the genetic subtype were either not present or not noted at the time of diagnosis. Notably, in three neonatal diabetes cases, mutations were found in genes where neonatal diabetes is usually observed as part of a syndrome (*EIF2AK3* and *SLC19A2*). In these cases, the testing was performed soon after the diagnosis of diabetes, before other features had manifested. Within the MODY category, a mitochondrial m.3243A>G mutation associated with MIDD was found in two patients in whom *HNF1α* testing had been the original test requested, and a *HNF1β* mutation was identified in a patient previously tested for *HNF1α* and *HNF4α*, highlighting the limitation of selecting single genes to test based on the patient's clinical phenotype. In addition, five partial/whole gene deletions/duplications were detected using the NGS data by cross-sample normalization and comparison, demonstrating the feasibility of using NGS data to also perform exonic copy number analysis. A similar targeted panel for the detection of mutations in 36 genes known to cause monogenic forms of diabetes has been developed by the authors [73]. Their assay identified pathogenic sequence changes in 19/76 (25%) patients in whom testing had been limited to a subset of genes or had not yet been performed. Also in this study, in a number of patients the causal mutation was not expected based on the available phenotypic details. The authors identified a mutation in *EIF2AK3* in a neonatal patient who did not yet exhibit epiphyseal dysplasia and other characteristic features associated with Wolcott Rallison syndrome at the time of diabetes diagnosis. Similarly, another infant without a history of consanguinity was found to have a homozygous *GCK* mutation causing PNDM at birth. These examples clearly illustrate the limitation of the candidate gene approach in that the diagnosis of some forms of monogenic diabetes is not always clear-cut and may be complicated by incomplete or absent clinical and/or family history information.

Recently, Flanagan et al. [202] employed NGS to analyze the entire genomic regions of *ABCC8* and *HADH* to investigate individuals with hyperinsulinemic hypoglycemia in whom there was genetic evidence to suggest deep intronic mutations and identified two deep intronic mutations, c.1333-1013A>G in *ABCC8* and c.636+471G>T *HADH*. Both mutations were predicted to create a cryptic splice donor site and an out-of-frame pseudoexon. Sequence analysis of mRNA from affected individuals' fibroblasts or lymphoblastoid cells confirmed mutant transcripts with pseudoexon inclusion and premature termination codons. Testing of 56 additional individuals showed that these are founder mutations in the Irish and Turkish populations, accounting for 14% of focal hyperinsulinism cases and 32% of subjects with *HADH* mutations in their cohort. This study suggests that analyzing noncoding regions of known genes by targeted NGS might prove useful before starting to search for mutations in unknown genes.

The feasibility of whole exome sequencing to study patients with both MODY and NDM has also been investigated. The first published study in monogenic diabetes performed exome sequencing in a patient with permanent neonatal diabetes, for whom mutations in *KCNJ11, ABCC8* and *INS* and 6q24 abnormalities had previously been excluded [74]. Likely deleterious variants were detected in 407 genes and included a novel heterozygous *ABCC8* missense mutation, p.Gln485His, that was confirmed by Sanger sequencing and determined to be in the *de novo* state in the patient. Re-analysis of the earlier Sanger sequencing results revealed that the mutation was present but had not been reported. The same group later sequenced the exomes of four relatives from a large French family with MODY and identified a causative *KCNJ11* missense mutation, p.Glu227Lys, which has previously been reported to cause transient neonatal diabetes [92, 135]. Prior to this study *KCNJ11* had not previously been described in patients with a MODY phenotype; this again illustrates the phenotypic heterogeneity that can be observed with monogenic diabetes genes. Johansson et al. performed exome sequencing in nine patients with a suspected diagnosis of MODY [75]. They analyzed variants identified in a predefined set of 111 genes implicated in glucose metabolism and identified three pathogenic variants (in *ABCC8*, *HNF4α* and *PPARG* respectively), leading to a genetic diagnosis in three patients. On follow-up, the sulphonylurea-treated adult-onset diabetes of the proband with the novel p.Ala1366Thr *ABCC8* mutation was consistent with previous reports [5, 203]. The patient with the truncating *PPARG* mutation was noted to be insulin resistant and the same mutation had been previously reported in patients with a similar phenotype [204, 205]. The patient with the novel p.Arg89Gln mutation in *HNF4α* had previously undergone Sanger sequencing and re-analysis of the data demonstrated that the mutation was detected but it had been overlooked.

To date a number of studies have described the discovery of a novel monogenic diabetes gene by exome sequencing. The first study by Lango Allen et al. [18]

employed a trio analysis strategy to investigate the cause of pancreatic agenesis in two unrelated probands. After filtering synonymous variants, variants present in the dbSNP or 1000 Genomes Project databases and variants identified in either parent, they identified two potentially pathogenic *de novo* mutations in the transcription factor gene *GATA6* in each subject. Sanger sequencing confirmed that the mutations were present in the two affected subjects but not in their unaffected parents. Further analysis by Sanger sequencing of the coding exons and intron boundaries of *GATA6* identified mutations in 13 additional patients, providing evidences of this being the most common cause of pancreatic agenesis. More recently, Synofzik et al. [10] employed exome sequencing to investigate three siblings from a 'likely consanguineous' Turkish family, with juvenile-onset insulin-dependent diabetes mellitus and central and peripheral nervous system abnormalities and identified a homozygous truncating mutation in the *DNAJC3* gene (p.Arg194*). *DNAJC3* encodes a protein that acts as a co-chaperone of BiP (immunoglobulin heavy-chain binding protein), a major endoplasmic reticulum-localized member of the HSP70 family of molecular chaperones that promote normal protein folding [206]. In mice, loss of *DNAJC3* leads to hyperglycemia and glucosuria associated with increasing apoptosis of pancreatic β cells and reduced insulin levels [207]. Screening of additional patients with a similar phenotype revealed only one further loss-of-function allele in *DNAJC3* and no further associations in subjects with only a subset of the features of the main phenotype. A similar approach has been taken by Prudente et al. [76] to investigate two large families with a high prevalence of diabetes not due to mutations in known genes involved in MODY. This study identified two loss-of-function mutations (p. Leu552* and p.Asp94Asn) in the gene for the Adaptor Protein, Phosphotyrosine Interaction, PH domain, and leucine zipper containing 1 (*APPL1*). APPL1 binds to AKT2, a key molecule in the insulin signaling pathway, thereby enhancing insulin-induced AKT2 activation and downstream signaling leading to insulin action and secretion. Functional studies demonstrated that both mutations cause APPL1 loss of function. Given the central role of AKT in insulin signaling, these results support a detrimental role of both mutations on insulin action and, potentially, insulin secretion.

A variety of molecular diagnostic approaches are now available because NGS technology has entered the clinical diagnostic arena, including single-gene tests, gene panel tests and exome/genome sequencing. The use of whole exome sequencing (WES) and its implementation as a diagnostic tool is exciting, especially with the rapidly decreasing costs, and provides a significant cost benefit for patients who remain undiagnosed after a few traditional approaches [208]. The "hypothesis free" approach does not focus on a set of genes, making it a better diagnostic tool for certain clinical scenarios. However the clinical utilization of WES needs to be done through a careful and thoughtful process. WES may not provide adequate coverage of the genes of interest, and can result in a large number of sequence variants of unknown clinical significance. There is also the potential for the identification of incidental findings unrelated to the patient's phenotype. The use of more focused approaches, such as targeted gene panels, or perhaps a systems-based approach to prioritize analysis of a subset of genes (i.e. genes involved in glucose metabolism for monogenic diabetes/congenital hyperinsulinism), provides a 'bridge' between single-gene/panel gene sequencing and WES for heterogeneous conditions such as disorders of insulin secretion. This approach allows the benefits of next generation sequencing technology to be leveraged to improve first pass diagnostic yield, while limiting data sets to genes of known diagnostic value.

References

1. Schmitz, O., Rungby, J., Edge, L., Juhl, C.B.: On high-frequency insulin oscillations. Ageing Res. Rev. **7**, 301–305 (2008)
2. Fajans, S.S., Bell, G.I., Polonsky, K.S.: Molecular mechanisms and clinical pathophysiology of maturity-onset diabetes of the young. N. Engl. J. Med. **345**, 971–980 (2001)
3. Inagaki, N., Gonoi, T., Clement 4th, J.P., Namba, N., Inazawa, J., Gonzalez, G., Aguilar-Bryan, L., Seino, S., Bryan, J.: Reconstitution of IKATP: an inward rectifier subunit plus the sulfonylurea receptor. Science. **270**, 1166–1170 (1995)
4. Greeley, S.A., Naylor, R.N., Philipson, L.H., Bell, G.I.: Neonatal diabetes: an expanding list of genes allows for improved diagnosis and treatment. Curr. Diab. Rep. **11**, 519–532 (2011)
5. Bowman, P., Flanagan, S.E., Edghill, E.L., Damhuis, A., Shepherd, M.H., Paisey, R., Hattersley, A.T., Ellard, S.: Heterozygous ABCC8 mutations are a cause of MODY. Diabetologia. **55**, 123–127 (2012)
6. Babenko, A.P., Polak, M., Cave, H., Busiah, K., Czernichow, P., Scharfmann, R., Bryan, J., Aguilar-Bryan, L., Vaxillaire, M., Froguel, P.: Activating mutations in the ABCC8 gene in neonatal diabetes mellitus. N. Engl. J. Med. **355**, 456–466 (2006)
7. Thomas, P., Ye, Y., Lightner, E.: Mutation of the pancreatic islet inward rectifier Kir6.2 also leads to familial persistent hyperinsulinemic hypoglycemia of infancy. Hum. Mol. Genet. **5**, 1809–1812 (1996)
8. Thomas, P.M., Cote, G.J., Wohllk, N., Haddad, B., Mathew, P.M., Rabl, W., Aguilar-Bryan, L., Gagel, R.F., Bryan, J.: Mutations in the sulfonylurea receptor gene in familial persistent hyperinsulinemic hypoglycemia of infancy. Science. **268**, 426–429 (1995)
9. Raeder, H., Johansson, S., Holm, P.I., Haldorsen, I.S., Mas, E., Sbarra, V., Nermoen, I., Eide, S.A., Grevle, L., Bjørkhaug, L., et al.: Mutations in the CEL VNTR cause a syndrome of diabetes and pancreatic exocrine dysfunction. Nat. Genet. **38**, 54–62 (2006)
10. Synofzik, M., Haack, T.B., Kopajtich, R., Gorza, M., Rapaport, D., Greiner, M., Schonfeld, C., Freiberg, C., Schorr, S., Holl, R.W., et al.: Absence of BiP co-chaperone DNAJC3 causes diabetes mellitus and multisystemic neurodegeneration. Am. J. Hum. Genet. **95**, 689–697 (2014)
11. Delepine, M., Nicolino, M., Barrett, T., Golamaully, M., Lathrop, G.M., Julier, C.: EIF2AK3, encoding translation initiation factor 2-alpha kinase 3, is mutated in patients with Wolcott-Rallison syndrome. Nat. Genet. **25**, 406–409 (2000)
12. Rubio-Cabezas, O., Patch, A.M., Minton, J.A., Flanagan, S.E., Edghill, E.L., Hussain, K., Balafrej, A., Deeb, A., Buchanan, C.R., Jefferson, I.G., et al.: Wolcott-Rallison syndrome is the most common genetic cause of permanent neonatal diabetes in consanguineous families. J. Clin. Endocrinol. Metab. **94**, 4162–4170 (2009)
13. Rubio-Cabezas, O., Minton, J.A., Caswell, R., Shield, J.P., Deiss, D., Sumnik, Z., Cayssials, A., Herr, M., Loew, A., Lewis, V., et al.: Clinical heterogeneity in patients with FOXP3 mutations presenting with permanent neonatal diabetes. Diabetes Care. **32**, 111–116 (2009)
14. Wildin, R.S., Ramsdell, F., Peake, J., Faravelli, F., Casanova, J.L., Buist, N., Levy-Lahad, E., Mazzella, M., Goulet, O., Perroni, L., et al.: X-linked neonatal diabetes mellitus, enteropathy and endocrinopathy syndrome is the human equivalent of mouse scurfy. Nat. Genet. **27**, 18–20 (2001)

15. Charbonnier, L.M., Janssen, E., Chou, J., Ohsumi, T.K., Keles, S., Hsu, J.T., Massaad, M.J., Garcia-Lloret, M., Hanna-Wakim, R., Dbaibo, G., et al.: Regulatory T-cell deficiency and immune dysregulation, polyendocrinopathy, enteropathy, X-linked-like disorder caused by loss-of-function mutations in LRBA. J. Allergy Clin. Immunol. **135**, 217–227 (2015)
16. D'Amato, E., Giacopelli, F., Giannattasio, A., D'Annunzio, G., Bocciardi, R., Musso, M., Lorini, R., Ravazzolo, R.: Genetic investigation in an Italian child with an unusual association of atrial septal defect, attributable to a new familial GATA4 gene mutation, and neonatal diabetes due to pancreatic agenesis. Diabet. Med. **27**, 1195–1200 (2010)
17. Shaw-Smith, C., De Franco, E., Lango Allen, H., Batlle, M., Flanagan, S.E., Borowiec, M., Taplin, C.E., van Alfen-van der Velden, J., Cruz-Rojo, J., Perez de Nanclares, G., et al.: GATA4 mutations are a cause of neonatal and childhood-onset diabetes. Diabetes. **63**, 2888–2894 (2014)
18. Lango Allen, H., Flanagan, S.E., Shaw-Smith, C., De Franco, E., Akerman, I., Caswell, R., International Pancreatic Agenesis Consortium, Ferrer, J., Hattersley, A.T., Ellard, S.: GATA6 haploinsufficiency causes pancreatic agenesis in humans. Nat. Genet. **44**, 20–22 (2012)
19. De Franco, E., Shaw-Smith, C., Flanagan, S.E., Shepherd, M.H., International NDM Consortium, Hattersley, A.T., Ellard, S.: GATA6 mutations cause a broad phenotypic spectrum of diabetes from pancreatic agenesis to adult-onset diabetes without exocrine insufficiency. Diabetes. **62**, 993–997 (2013)
20. Njolstad, P.R., Sovik, O., Cuesta-Munoz, A., Bjorkhaug, L., Massa, O., Barbetti, F., Undlien, D.E., Shiota, C., Magnuson, M.A., Molven, A., et al.: Neonatal diabetes mellitus due to complete glucokinase deficiency. N. Engl. J. Med. **344**, 1588–1592 (2001)
21. Gloyn, A.L.: Glucokinase (GCK) mutations in hyper- and hypoglycemia: maturity-onset diabetes of the young, permanent neonatal diabetes, and hyperinsulinemia of infancy. Hum. Mutat. **22**, 353–362 (2003)
22. Dimitri, P., Warner, J.T., Minton, J.A., Patch, A.M., Ellard, S., Hattersley, A.T., Barr, S., Hawkes, D., Wales, J.K., Gregory, J.W.: Novel GLIS3 mutations demonstrate an extended multisystem phenotype. Eur. J. Endocrinol. **164**, 437–443 (2011)
23. Senee, V., Chelala, C., Duchatelet, S., Feng, D., Blanc, H., Cossec, J.C., Charon, C., Nicolino, M., Boileau, P., Cavener, D.R., et al.: Mutations in GLIS3 are responsible for a rare syndrome with neonatal diabetes mellitus and congenital hypothyroidism. Nat. Genet. **38**, 682–687 (2006)
24. Stanley, C.A., Lieu, Y.K., Hsu, B.Y., Burlina, A.B., Greenberg, C.R., Hopwood, N.J., Perlman, K., Rich, B.H., Zammarchi, E., Poncz, M.: Hyperinsulinism and hyperammonemia in infants with regulatory mutations of the glutamate dehydrogenase gene. N. Engl. J. Med. **338**, 1352–1357 (1998)
25. Hudson, R.C., Daniel, R.M.: L-glutamate dehydrogenases: distribution, properties and mechanism. Comp. Biochem. Physiol. B. **106**, 767–792 (1993)
26. Clayton, P.T., Eaton, S., Aynsley-Green, A., Edginton, M., Hussain, K., Krywawych, S., Datta, V., Malingre, H.E., Berger, R., van den Berg, I.E.: Hyperinsulinism in short-chain L-3-hydroxyacyl-CoA dehydrogenase deficiency reveals the importance of beta-oxidation in insulin secretion. J. Clin. Invest. **108**, 457–465 (2001)
27. Yamagata, K., Oda, N., Kaisaki, P.J., Menzel, S., Furuta, H., Vaxillaire, M., Southam, L., Cox, R.D., Lathrop, G.M., Boriraj, V.V., et al.: Mutations in the hepatocyte nuclear factor-1alpha gene in maturity-onset diabetes of the young (MODY3). Nature. **384**, 455–458 (1996)
28. Ellard, S., Colclough, K.: Mutations in the genes encoding the transcription factors hepatocyte nuclear factor 1 alpha (HNF1A) and 4 alpha (HNF4A) in maturity-onset diabetes of the young. Hum. Mutat. **27**, 854–869 (2006)
29. Yamagata, K.: Roles of HNF1α and HNF4α in pancreatic β-cells: lessons from a monogenic form of diabetes (MODY). Vitam. Horm. **95**, 407–423 (2014)
30. Yamagata, K., Furuta, H., Oda, N., Kaisaki, P.J., Menzel, S., Cox, N.J., Fajans, S.S., Signorini, S., Stoffel, M., Bell, G.I.: Mutations in the hepatocyte nuclear factor-4alpha gene in maturity-onset diabetes of the young (MODY1). Nature. **384**, 458–460 (1996)

31. Pearson, E.R., Boj, S.F., Steele, A.M., Barrett, T., Stals, K., Shield, J.P., Ellard, S., Ferrer, J., Hattersley, A.T.: Macrosomia and hyperinsulinaemic hypoglycaemia in patients with heterozygous mutations in the HNF4A gene. PLoS Med. **4**, e118 (2007)
32. Edghill, E.L., Bingham, C., Slingerland, A.S., Minton, J.A., Noordam, C., Ellard, S., Hattersley, A.T.: Hepatocyte nuclear factor-1 beta mutations cause neonatal diabetes and intrauterine growth retardation: support for a critical role of HNF-1beta in human pancreatic development. Diabet. Med. **23**, 1301–1306 (2006)
33. Abdel-Salam, G.M., Schaffer, A.E., Zaki, M.S., Dixon-Salazar, T., Mostafa, I.S., Afifi, H.H., Gleeson, J.G.: A homozygous IER3IP1 mutation causes microcephaly with simplified gyral pattern, epilepsy, and permanent neonatal diabetes syndrome (MEDS). Am. J. Med. Genet. A. **158A**, 2788–2796 (2012)
34. Poulton, C.J., Schot, R., Kia, S.K., Jones, M., Verheijen, F.W., Venselaar, H., de Wit, M.C., de Graaff, E., Bertoli-Avella, A.M., Mancini, G.M.: Microcephaly with simplified gyration, epilepsy, and infantile diabetes linked to inappropriate apoptosis of neural progenitors. Am. J. Hum. Genet. **89**, 265–276 (2011)
35. Caudy, A.A., Reddy, S.T., Chatila, T., Atkinson, J.P., Verbsky, J.W.: CD25 deficiency causes an immune dysregulation, polyendocrinopathy, enteropathy, X-linked-like syndrome, and defective IL-10 expression from CD4 lymphocytes. J. Allergy Clin. Immunol. **119**, 482–487 (2007)
36. Stoy, J., Edghill, E.L., Flanagan, S.E., Ye, H., Paz, V.P., Pluzhnikov, A., Below, J.E., Hayes, M.G., Cox, N.J., Lipkind, G.M., et al.: Insulin gene mutations as a cause of permanent neonatal diabetes. Proc. Natl. Acad. Sci. U. S. A. **104**, 15040–15044 (2007)
37. Edghill, E.L., Flanagan, S.E., Patch, A.M., Boustred, C., Parrish, A., Shields, B., Shepherd, M.H., Hussain, K., Kapoor, R.R., Malecki, M., et al.: Insulin mutation screening in 1,044 patients with diabetes: mutations in the INS gene are a common cause of neonatal diabetes but a rare cause of diabetes diagnosed in childhood or adulthood. Diabetes. **57**, 1034–1042 (2008)
38. Flanagan, S.E., De Franco, E., Lango Allen, H., Zerah, M., Abdul-Rasoul, M.M., Edge, J.A., Stewart, H., Alamiri, E., Hussain, K., Wallis, S., et al.: Analysis of transcription factors key for mouse pancreatic development establishes NKX2-2 and MNX1 mutations as causes of neonatal diabetes in man. Cell Metab. **19**, 146–154 (2014)
39. Rubio-Cabezas, O., Minton, J.A., Kantor, I., Williams, D., Ellard, S., Hattersley, A.T.: Homozygous mutations in NEUROD1 are responsible for a novel syndrome of permanent neonatal diabetes and neurological abnormalities. Diabetes. **59**, 2326–2331 (2010)
40. Malecki, M.T., Jhala, U.S., Antonellis, A., Fields, L., Doria, A., Orban, T., Saad, M., Warram, J.H., Montminy, M., Krolewski, A.S.: Mutations in NEUROD1 are associated with the development of type 2 diabetes mellitus. Nat. Genet. **23**, 323–328 (1999)
41. Rubio-Cabezas, O., Jensen, J.N., Hodgson, M.I., Codner, E., Ellard, S., Serup, P., Hattersley, A.T.: Permanent neonatal diabetes and enteric Anendocrinosis associated with biallelic mutations in NEUROG3. Diabetes. **60**, 1349–1353 (2011)
42. Solomon, B.D., Pineda-Alvarez, D.E., Balog, J.Z., Hadley, D., Gropman, A.L., Nandagopal, R., Han, J.C., Hahn, J.S., Blain, D., Brooks, B., et al.: Compound heterozygosity for mutations in PAX6 in a patient with complex brain anomaly, neonatal diabetes mellitus, and microophthalmia. Am. J. Med. Genet. A. **149A**, 2543–2546 (2009)
43. Simaite, D., Kofent, J., Gong, M., Ruschendorf, F., Jia, S., Arn, P., Bentler, K., Ellaway, C., Kuhnen, P., Hoffmann, G.F., et al.: Recessive mutations in PCBD1 cause a new type of early-onset diabetes. Diabetes. **63**, 3557–3564 (2014)
44. Thomas, I.H., Saini, N.K., Adhikari, A., Lee, J.M., Kasa-Vubu, J.Z., Vazquez, D.M., Menon, R.K., Chen, M., Fajans, S.S.: Neonatal diabetes mellitus with pancreatic agenesis in an infant with homozygous IPF-1 Pro63fsX60 mutation. Pediatr. Diabetes. **10**, 492–496 (2009)
45. Nicolino, M., Claiborn, K.C., Senee, V., Boland, A., Stoffers, D.A., Julier, C.: A novel hypomorphic PDX1 mutation responsible for permanent neonatal diabetes with subclinical exocrine deficiency. Diabetes. **59**, 733–740 (2010)

46. Fajans, S.S., Bell, G.I., Paz, V.P., Below, J.E., Cox, N.J., Martin, C., Thomas, I.H., Chen, M.: Obesity and hyperinsulinemia in a family with pancreatic agenesis and MODY caused by the IPF1 mutation Pro63fsX60. Transl. Res. **156**, 7–14 (2010)
47. Gardner, R.J., Mackay, D.J., Mungall, A.J., Polychronakos, C., Siebert, R., Shield, J.P., Temple, I.K., Robinson, D.O.: An imprinted locus associated with transient neonatal diabetes mellitus. Hum. Mol. Genet. **9**, 589–596 (2000)
48. Temple, I.K., James, R.S., Crolla, J.A., Sitch, F.L., Jacobs, P.A., Howell, W.M., Betts, P., Baum, J.D., Shield, J.P.: An imprinted gene(s) for diabetes? Nat. Genet. **9**, 110–112 (1995)
49. Temple, I.K., Shield, J.P.: 6q24 transient neonatal diabetes. Rev. Endocr. Metab. Disord. **11**, 199–204 (2010)
50. Sellick, G.S., Barker, K.T., Stolte-Dijkstra, I., Fleischmann, C., Coleman, R.J., Garrett, C., Gloyn, A.L., Edghill, E.L., Hattersley, A.T., Wellauer, P.K., et al.: Mutations in PTF1A cause pancreatic and cerebellar agenesis. Nat. Genet. **36**, 1301–1305 (2004)
51. Mitchell, J., Punthakee, Z., Lo, B., Bernard, C., Chong, K., Newman, C., Cartier, L., Desilets, V., Cutz, E., Hansen, I.L., et al.: Neonatal diabetes, with hypoplastic pancreas, intestinal atresia and gall bladder hypoplasia: search for the aetiology of a new autosomal recessive syndrome. Diabetologia. **47**, 2160–2167 (2004)
52. Spiegel, R., Dobbie, A., Hartman, C., de Vries, L., Ellard, S., Shalev, S.A.: Clinical characterization of a newly described neonatal diabetes syndrome caused by RFX6 mutations. Am. J. Med. Genet. A. **155A**, 2821–2825 (2011)
53. Bay, A., Keskin, M., Hizli, S., Uygun, H., Dai, A., Gumruk, F.: Thiamine-responsive megaloblastic anemia syndrome. Int. J. Hematol. **92**, 524–526 (2010)
54. Shaw-Smith, C., Flanagan, S.E., Patch, A.M., Grulich-Henn, J., Habeb, A.M., Hussain, K., Pomahacova, R., Matyka, K., Abdullah, M., Hattersley, A.T., et al.: Recessive SLC19A2 mutations are a cause of neonatal diabetes mellitus in thiamine-responsive megaloblastic anaemia. Pediatr. Diabetes. **13**, 314–321 (2012)
55. Otonkoski, T., Jiao, H., Kaminen-Ahola, N., Tapia-Paez, I., Ullah, M.S., Parton, L.E., Schuit, F., Quintens, R., Sipila, I., Mayatepek, E., et al.: Physical exercise-induced hypoglycemia caused by failed silencing of monocarboxylate transporter 1 in pancreatic beta cells. Am. J. Hum. Genet. **81**, 467–474 (2007)
56. Sansbury, F.H., Flanagan, S.E., Houghton, J.A., Shuixian Shen, F.L., Al-Senani, A.M., Habeb, A.M., Abdullah, M., Kariminejad, A., Ellard, S., Hattersley, A.T.: SLC2A2 mutations can cause neonatal diabetes, suggesting GLUT2 may have a role in human insulin secretion. Diabetologia. **55**, 2381–2385 (2012)
57. Uzel, G., Sampaio, E.P., Lawrence, M.G., Hsu, A.P., Hackett, M., Dorsey, M.J., Noel, R.J., Verbsky, J.W., Freeman, A.F., Janssen, E., et al.: Dominant gain-of-function STAT1 mutations in FOXP3 wild-type immune dysregulation-polyendocrinopathy-enteropathy-X-linked--like syndrome. J. Allergy Clin. Immunol. **131**, 1611–1623 (2013)
58. Flanagan, S.E., Haapaniemi, E., Russell, M.A., Caswell, R., Lango Allen, H., De Franco, E., McDonald, T.J., Rajala, H., Ramelius, A., Barton, J., et al.: Activating germline mutations in STAT3 cause early-onset multi-organ autoimmune disease. Nat. Genet. **46**, 812–814 (2014)
59. Gonzalez-Barroso, M.M., Giurgea, I., Bouillaud, F., Anedda, A., Bellanne-Chantelot, C., Hubert, L., de Keyzer, Y., de Lonlay, P., Ricquier, D.: Mutations in UCP2 in congenital hyperinsulinism reveal a role for regulation of insulin secretion. PLoS One. **3**, e3850 (2008)
60. Strom, T.M., Hortnagel, K., Hofmann, S., Gekeler, F., Scharfe, C., Rabl, W., Gerbitz, K.D., Meitinger, T.: Diabetes insipidus, diabetes mellitus, optic atrophy and deafness (DIDMOAD) caused by mutations in a novel gene (wolframin) coding for a predicted transmembrane protein. Hum. Mol. Genet. **7**, 2021–2028 (1998)
61. Mackay, D.J., Callaway, J.L., Marks, S.M., White, H.E., Acerini, C.L., Boonen, S.E., Dayanikli, P., Firth, H.V., Goodship, J.A., Haemers, A.P., et al.: Hypomethylation of multiple imprinted loci in individuals with transient neonatal diabetes is associated with mutations in ZFP57. Nat. Genet. **40**, 949–951 (2008)

62. Mackay, D.J., Temple, I.K.: Transient neonatal diabetes mellitus type 1. Am. J. Med. Genet. C Semin. Med. Genet. **154C**, 335–342 (2010)
63. Murphy, R., Ellard, S., Hattersley, A.T.: Clinical implications of a molecular genetic classification of monogenic beta-cell diabetes. Nat. Clin. Pract. Endocrinol. Metab. **4**, 200–213 (2008)
64. Philipson, L.H., Murphy, R., Ellard, S., Hattersley, A.T., Stoy, J., Greeley, S.A., Bell, G.I., Polonsky, K.S.: Genetic testing in diabetes mellitus: a clinical guide to monogenic diabetes. In: Weiss, R.E., Refetoff, S. (eds.) Genetic Diagnosis of Endocrine Disorders. Elsevier, London (2010)
65. Porter, J.R., Barrett, T.G.: Monogenic syndromes of abnormal glucose homeostasis: clinical review and relevance to the understanding of the pathology of insulin resistance and beta cell failure. J. Med. Genet. **42**, 893–902 (2005)
66. Shields, B.M., Hicks, S., Shepherd, M.H., Colclough, K., Hattersley, A.T., Ellard, S.: Maturity-onset diabetes of the young (MODY): how many cases are we missing? Diabetologia. **53**, 2504–2508 (2010)
67. Slingerland, A.S., Hattersley, A.T.: Mutations in the Kir6.2 subunit of the KATP channel and permanent neonatal diabetes: new insights and new treatment. Ann. Med. **37**, 186–195 (2005)
68. Lambert, A.P., Ellard, S., Allen, L.I., Gallen, I.W., Gillespie, K.M., Bingley, P.J., Hattersley, A.T.: Identifying hepatic nuclear factor 1alpha mutations in children and young adults with a clinical diagnosis of type 1 diabetes. Diabetes Care. **26**, 333–337 (2003)
69. Møller, A.M., Dalgaard, L.T., Pociot, F., Nerup, J., Hansen, T., Pedersen, O.: Mutations in the hepatocyte nuclear factor-1alpha gene in Caucasian families originally classified as having type I diabetes. Diabetologia. **41**, 1528–1531 (1998)
70. Ellard, S., Lango Allen, H., De Franco, E., Flanagan, S.E., Hysenaj, G., Colclough, K., Houghton, J.A., Shepherd, M., Hattersley, A.T., Weedon, M.N., et al.: Improved genetic testing for monogenic diabetes using targeted next-generation sequencing. Diabetologia. **56**, 1958–1963 (2013)
71. Bonnefond, A., Philippe, J., Durand, E., Muller, J., Saeed, S., Arslan, M., Martínez, R., De Graeve, F., Dhennin, V., Rabearivelo, I., et al.: Highly sensitive diagnosis of 43 monogenic forms of diabetes or obesity through one-step PCR-based enrichment in combination with next-generation sequencing. Diabetes Care. **37**, 460–467 (2014)
72. Gao, R., Liu, Y., Gjesing, A.P., Hollensted, M., Wan, X., He, S., Pedersen, O., Yi, X., Wang, J., Hansen, T.: Evaluation of a target region capture sequencing platform using monogenic diabetes as a study-model. BMC Genet. **15**, 13 (2014)
73. Alkorta-Aranburu, G., Carmody, D., Cheng, Y.W., Nelakuditi, V., Ma, L., Dickens, J.T., Das, S., Greeley, S.A., del Gaudio, D.: Phenotypic heterogeneity in monogenic diabetes: the clinical and diagnostic utility of a gene panel-based next-generation sequencing approach. Mol. Genet. Metab. **113**, 315–320 (2014)
74. Bonnefond, A., Durand, E., Sand, O., De Graeve, F., Gallina, S., Busiah, K., Lobbens, S., Simon, A., Bellanne-Chantelot, C., Letourneau, L., et al.: Molecular diagnosis of neonatal diabetes mellitus using next-generation sequencing of the whole exome. PLoS One. **5**, e13630 (2010)
75. Johansson, S., Irgens, H., Chudasama, K.K., Molnes, J., Aerts, J., Roque, F.S., Jonassen, I., Levy, S., Lima, K., Knappskog, P.M., et al.: Exome sequencing and genetic testing for MODY. PLoS One. **7**, e38050 (2012)
76. Prudente, S., Jungtrakoon, P., Marucci, A., Ludovico, O., Buranasupkajorn, P., Mazza, T., Hastings, T., Milano, T., Morini, E., Mercuri, L., et al.: Loss-of-function mutations in APPL1 in familial diabetes mellitus. Am. J. Hum. Genet. **97**, 177–185 (2015)
77. Gicquel, C., Le Bouc, Y.: Hormonal regulation of fetal growth. Horm. Res. **65**(Suppl 3), 28–33 (2006)
78. Edghill, E.L., Hattersley, A.T.: Genetic disorders of the pancreatic beta cell and diabetes (permanent neonatal diabets and maturity-onset diabetes of the young). In: Seino, S., Bell, G. (eds.) Pancreatic Beta Cell in Health and Disease, pp. 389–420. Springer, Japan (2008)

79. Metz, C., Cave, H., Bertrand, A.M., Deffert, C., Gueguen-Giroux, B., Czernichow, P., Polak, M., NDM French Study Group. Neonatal Diabetes Mellitus: Neonatal diabetes mellitus: chromosomal analysis in transient and permanent cases. J. Pediatr. **141**, 483–489 (2002)
80. Ashcroft, F.M., Rorsman, P.: Diabetes mellitus and the β cell: the last ten years. Cell. **148**, 1160–1171 (2012)
81. Gloyn, A.L., Pearson, E.R., Antcliff, J.F., Proks, P., Bruining, G.J., Slingerland, A.S., Howard, N., Srinivasan, S., Silva, J.M., Molnes, J., et al.: Activating mutations in the gene encoding the ATP-sensitive potassium-channel subunit Kir6.2 and permanent neonatal diabetes. N. Engl. J. Med. **350**, 1838–1849 (2004)
82. Edghill, E.L., Gloyn, A.L., Gillespie, K.M., Lambert, A.P., Raymond, N.T., Swift, P.G., Ellard, S., Gale, E.A., Hattersley, A.T.: Activating mutations in the KCNJ11 gene encoding the ATP-sensitive K+ channel subunit Kir6.2 are rare in clinically defined type 1 diabetes diagnosed before 2 years. Diabetes. **53**, 2998–3001 (2004)
83. Ashcroft, F.M.: ATP-sensitive potassium channelopathies: focus on insulin secretion. J. Clin. Invest. **115**, 2047–2058 (2005)
84. Remedi, M.S., Koster, J.C.: K(ATP) channelopathies in the pancreas. Pflugers Arch. **460**, 307–320 (2010)
85. De Leon, D.D., Stanley, C.A.: Permanent neonatal diabetes mellitus. In: Pagon, R.A., Adam, M.P., Ardinger, H.H., Wallace, S.E., Amemiya, A., Bean, L.J.H., Bird, T.D., Dolan, C.R., Fong, C.T., Smith, R.J.H., et al. (eds.) GeneReviews(R). University of Washington, Seattle (1993)
86. Eizirik, D.L., Cardozo, A.K., Cnop, M.: The role for endoplasmic reticulum stress in diabetes mellitus. Endocr. Rev. **29**, 42–61 (2008)
87. Stoy, J., Steiner, D.F., Park, S.Y., Ye, H., Philipson, L.H., Bell, G.I.: Clinical and molecular genetics of neonatal diabetes due to mutations in the insulin gene. Rev. Endocr. Metab. Disord. **11**, 205–215 (2010)
88. Garin, I., Edghill, E.L., Akerman, I., Rubio-Cabezas, O., Rica, I., Locke, J.M., Maestro, M.A., Alshaikh, A., Bundak, R., del Castillo, G., et al.: Recessive mutations in the INS gene result in neonatal diabetes through reduced insulin biosynthesis. Proc. Natl. Acad. Sci. U. S. A. **107**, 3105–3110 (2010)
89. Njølstad, P.R., Søvik, O., Cuesta-Muñoz, A., Bjørkhaug, L., Massa, O., Barbetti, F., Undlien, D.E., Shiota, C., Magnuson, M.A., Molven, A., et al.: Neonatal diabetes mellitus due to complete glucokinase deficiency. N. Engl. J. Med. **344**, 1588–1592 (2001)
90. Turkkahraman, D., Bircan, I., Tribble, N.D., Akçurin, S., Ellard, S., Gloyn, A.L.: Permanent neonatal diabetes mellitus caused by a novel homozygous (T168A) glucokinase (GCK) mutation: initial response to oral sulphonylurea therapy. J. Pediatr. **153**, 122–126 (2008)
91. Temple, I.K., Gardner, R.J., Mackay, D.J., Barber, J.C., Robinson, D.O., Shield, J.P.: Transient neonatal diabetes: widening the understanding of the etiopathogenesis of diabetes. Diabetes. **49**, 1359–1366 (2000)
92. Flanagan, S.E., Patch, A.M., Mackay, D.J., Edghill, E.L., Gloyn, A.L., Robinson, D., Shield, J.P., Temple, K., Ellard, S., Hattersley, A.T.: Mutations in ATP-sensitive K+ channel genes cause transient neonatal diabetes and permanent diabetes in childhood or adulthood. Diabetes. **56**, 1930–1937 (2007)
93. Kamiya, M., Judson, H., Okazaki, Y., Kusakabe, M., Muramatsu, M., Takada, S., Takagi, N., Arima, T., Wake, N., Kamimura, K., et al.: The cell cycle control gene ZAC/PLAGL1 is imprinted–a strong candidate gene for transient neonatal diabetes. Hum. Mol. Genet. **9**, 453–460 (2000)
94. Arima, T., Drewell, R.A., Oshimura, M., Wake, N., Surani, M.A.: A novel imprinted gene, HYMAI, is located within an imprinted domain on human chromosome 6 containing ZAC. Genomics. **67**, 248–255 (2000)
95. Varrault, A., Bilanges, B., Mackay, D.J., Basyuk, E., Ahr, B., Fernandez, C., Robinson, D.O., Bockaert, J., Journot, L.: Characterization of the methylation-sensitive promoter of the

imprinted ZAC gene supports its role in transient neonatal diabetes mellitus. J. Biol. Chem. **276**, 18653–18656 (2001)
96. Mackay, D.J., Coupe, A.M., Shield, J.P., Storr, J.N., Temple, I.K., Robinson, D.O.: Relaxation of imprinted expression of ZAC and HYMAI in a patient with transient neonatal diabetes mellitus. Hum. Genet. **110**, 139–144 (2002)
97. Mackay, D.J., Boonen, S.E., Clayton-Smith, J., Goodship, J., Hahnemann, J.M., Kant, S.G., Njolstad, P.R., Robin, N.H., Robinson, D.O., Siebert, R., et al.: A maternal hypomethylation syndrome presenting as transient neonatal diabetes mellitus. Hum. Genet. **120**, 262–269 (2006)
98. Barrett, T.G., Bundey, S.E., Macleod, A.F.: Neurodegeneration and diabetes: UK nationwide study of Wolfram (DIDMOAD) syndrome. Lancet. **346**, 1458–1463 (1995)
99. Stoffers, D.A., Zinkin, N.T., Stanojevic, V., Clarke, W.L., Habener, J.F.: Pancreatic agenesis attributable to a single nucleotide deletion in the human IPF1 gene coding sequence. Nat. Genet. **15**, 106–110 (1997)
100. Yorifuji, T., Kurokawa, K., Mamada, M., Imai, T., Kawai, M., Nishi, Y., Shishido, S., Hasegawa, Y., Nakahata, T.: Neonatal diabetes mellitus and neonatal polycystic, dysplastic kidneys: phenotypically discordant recurrence of a mutation in the hepatocyte nuclear factor-1beta gene due to germline mosaicism. J. Clin. Endocrinol. Metab. **89**, 2905–2908 (2004)
101. Finnish-German, A.C.: An autoimmune disease, APECED, caused by mutations in a novel gene featuring two PHD-type zinc-finger domains. Nat. Genet. **17**, 399–403 (1997)
102. van den Ouweland, J.M., Lemkes, H.H., Ruitenbeek, W., Sandkuijl, L.A., de Vijlder, M.F., Struyvenberg, P.A., van de Kamp, J.J., Maassen, J.A.: Mutation in mitochondrial tRNA(Leu) (UUR) gene in a large pedigree with maternally transmitted type II diabetes mellitus and deafness. Nat. Genet. **1**, 368–371 (1992)
103. Laloi-Michelin, M., Meas, T., Ambonville, C., Bellanné-Chantelot, C., Beaufils, S., Massin, P., Vialettes, B., Gin, H., Timsit, J., Bauduceau, B., et al.: The clinical variability of maternally inherited diabetes and deafness is associated with the degree of heteroplasmy in blood leukocytes. J. Clin. Endocrinol. Metab. **94**, 3025–3030 (2009)
104. Hattersley, A., Bruining, J., Shield, J., Njolstad, P., Donaghue, K., International Society for Pediatric and Adolescent Diabetes: ISPAD Clinical Practice Consensus Guidelines 2006-2007. The diagnosis and management of monogenic diabetes in children. Pediatr. Diabetes. **7**, 352–360 (2006)
105. Thomas, C.C., Philipson, L.H.: Update on diabetes classification. Med. Clin. North Am. **99**, 1–16 (2015)
106. Pihoker, C., Gilliam, L.K., Ellard, S., Dabelea, D., Davis, C., Dolan, L.M., Greenbaum, C.J., Imperatore, G., Lawrence, J.M., Marcovina, S.M., et al.: Prevalence, characteristics and clinical diagnosis of maturity onset diabetes of the young due to mutations in HNF1A, HNF4A, and glucokinase: results from the SEARCH for diabetes in youth. J. Clin. Endocrinol. Metab. **98**, 4055–4062 (2013)
107. Schober, E., Rami, B., Grabert, M., Thon, A., Kapellen, T., Reinehr, T., Holl, R.W., DPV-Wiss Initiative of the German Working Group for Paediatric Diabetology: Phenotypical aspects of maturity-onset diabetes of the young (MODY diabetes) in comparison with type 2 diabetes mellitus (T2DM) in children and adolescents: experience from a large multicentre database. Diabet. Med. **26**, 466–473 (2009)
108. Fajans, S.S.: MODY–a model for understanding the pathogeneses and natural history of type II diabetes. Horm. Metab. Res. **19**, 591–599 (1987)
109. Fajans, S.S., Bell, G.I.: MODY: history, genetics, pathophysiology, and clinical decision making. Diabetes Care. **34**, 1878–1884 (2011)
110. Vaxillaire, M., Bonnefond, A., Froguel, P.: The lessons of early-onset monogenic diabetes for the understanding of diabetes pathogenesis. Best Pract. Res. Clin. Endocrinol. Metab. **26**, 171–187 (2012)
111. Galler, A., Stange, T., Müller, G., Näke, A., Vogel, C., Kapellen, T., Bartelt, H., Kunath, H., Koch, R., Kiess, W., et al.: Incidence of childhood diabetes in children aged less than 15 years

and its clinical and metabolic characteristics at the time of diagnosis: data from the childhood diabetes registry of Saxony, Germany. Horm. Res. Paediatr. **74**, 285–291 (2010)
112. Neu, A., Feldhahn, L., Ehehalt, S., Hub, R., Ranke, M.B., DIARY Group Baden-Württemberg: Type 2 diabetes mellitus in children and adolescents is still a rare disease in Germany: a population-based assessment of the prevalence of type 2 diabetes and MODY in patients aged 0–20 years. Pediatr. Diabetes. **10**, 468–473 (2009)
113. Byrne, M.M., Sturis, J., Menzel, S., Yamagata, K., Fajans, S.S., Dronsfield, M.J., Bain, S.C., Hattersley, A.T., Velho, G., Froguel, P., et al.: Altered insulin secretory responses to glucose in diabetic and nondiabetic subjects with mutations in the diabetes susceptibility gene MODY3 on chromosome 12. Diabetes. **45**, 1503–1510 (1996)
114. Frayling, T.M., Bulamn, M.P., Ellard, S., Appleton, M., Dronsfield, M.J., Mackie, A.D., Baird, J.D., Kaisaki, P.J., Yamagata, K., Bell, G.I., et al.: Mutations in the hepatocyte nuclear factor-1alpha gene are a common cause of maturity-onset diabetes of the young in the U.K. Diabetes. **46**, 720–725 (1997)
115. Lango Allen, H., Johansson, S., Ellard, S., Shields, B., Hertel, J.K., Raeder, H., Colclough, K., Molven, A., Frayling, T.M., Njølstad, P.R., et al.: Polygenic risk variants for type 2 diabetes susceptibility modify age at diagnosis in monogenic HNF1A diabetes. Diabetes. **59**, 266–271 (2010)
116. Wang, H., Maechler, P., Antinozzi, P.A., Hagenfeldt, K.A., Wollheim, C.B.: Hepatocyte nuclear factor 4alpha regulates the expression of pancreatic beta -cell genes implicated in glucose metabolism and nutrient-induced insulin secretion. J. Biol. Chem. **275**, 35953–35959 (2000)
117. Edghill, E.L., Bingham, C., Ellard, S., Hattersley, A.T.: Mutations in hepatocyte nuclear factor-1beta and their related phenotypes. J. Med. Genet. **43**, 84–90 (2006)
118. Menzel, R., Kaisaki, P.J., Rjasanowski, I., Heinke, P., Kerner, W., Menzel, S.: A low renal threshold for glucose in diabetic patients with a mutation in the hepatocyte nuclear factor-1alpha (HNF-1alpha) gene. Diabet. Med. **15**, 816–820 (1998)
119. Shih, D.Q., Dansky, H.M., Fleisher, M., Assmann, G., Fajans, S.S., Stoffel, M.: Genotype/phenotype relationships in HNF-4alpha/MODY1: haploinsufficiency is associated with reduced apolipoprotein (AII), apolipoprotein (CIII), lipoprotein(a), and triglyceride levels. Diabetes. **49**, 832–837 (2000)
120. Bingham, C., Bulman, M.P., Ellard, S., Allen, L.I., Lipkin, G.W., Hoff, W.G., Woolf, A.S., Rizzoni, G., Novelli, G., Nicholls, A.J., et al.: Mutations in the hepatocyte nuclear factor-1beta gene are associated with familial hypoplastic glomerulocystic kidney disease. Am. J. Hum. Genet. **68**, 219–224 (2001)
121. Horikawa, Y., Iwasaki, N., Hara, M., Furuta, H., Hinokio, Y., Cockburn, B.N., Lindner, T., Yamagata, K., Ogata, M., Tomonaga, O., et al.: Mutation in hepatocyte nuclear factor-1 beta gene (TCF2) associated with MODY. Nat. Genet. **17**, 384–385 (1997)
122. Bellanné-Chantelot, C., Chauveau, D., Gautier, J.F., Dubois-Laforgue, D., Clauin, S., Beaufils, S., Wilhelm, J.M., Boitard, C., Noël, L.H., Velho, G., et al.: Clinical spectrum associated with hepatocyte nuclear factor-1beta mutations. Ann. Intern. Med. **140**, 510–517 (2004)
123. Lindner, T.H., Njolstad, P.R., Horikawa, Y., Bostad, L., Bell, G.I., Sovik, O.: A novel syndrome of diabetes mellitus, renal dysfunction and genital malformation associated with a partial deletion of the pseudo-POU domain of hepatocyte nuclear factor-1beta. Hum. Mol. Genet. **8**, 2001–2008 (1999)
124. Osbak, K.K., Colclough, K., Saint-Martin, C., Beer, N.L., Bellanné-Chantelot, C., Ellard, S., Gloyn, A.L.: Update on mutations in glucokinase (GCK), which cause maturity-onset diabetes of the young, permanent neonatal diabetes, and hyperinsulinemic hypoglycemia. Hum. Mutat. **30**, 1512–1526 (2009)
125. Ellard, S., Beards, F., Allen, L.I., Shepherd, M., Ballantyne, E., Harvey, R., Hattersley, A.T.: A high prevalence of glucokinase mutations in gestational diabetic subjects selected by clinical criteria. Diabetologia. **43**, 250–253 (2000)

126. Schnyder, S., Mullis, P.E., Ellard, S., Hattersley, A.T., Flück, C.E.: Genetic testing for glucokinase mutations in clinically selected patients with MODY: a worthwhile investment. Swiss Med. Wkly. **135**, 352–356 (2005)
127. Meek, C.L., Lewis, H.B., Patient, C., Murphy, H.R., Simmons, D.: Diagnosis of gestational diabetes mellitus: falling through the net. Diabetologia. **58**(9), 2003–2012 (2015)
128. Chakera, A.J., Carleton, V.L., Ellard, S., Wong, J., Yue, D.K., Pinner, J., Hattersley, A.T., Ross, G.P.: Antenatal diagnosis of fetal genotype determines if maternal hyperglycemia due to a glucokinase mutation requires treatment. Diabetes Care. **35**, 1832–1834 (2012)
129. Spyer, G., Macleod, K.M., Shepherd, M., Ellard, S., Hattersley, A.T.: Pregnancy outcome in patients with raised blood glucose due to a heterozygous glucokinase gene mutation. Diabet. Med. **26**, 14–18 (2009)
130. Schwitzgebel, V.M., Mamin, A., Brun, T., Ritz-Laser, B., Zaiko, M., Maret, A., Jornayvaz, F.R., Theintz, G.E., Michielin, O., Melloul, D., et al.: Agenesis of human pancreas due to decreased half-life of insulin promoter factor 1. J. Clin. Endocrinol. Metab. **88**, 4398–4406 (2003)
131. Fernandez-Zapico, M.E., van Velkinburgh, J.C., Gutiérrez-Aguilar, R., Neve, B., Froguel, P., Urrutia, R., Stein, R.: MODY7 gene, KLF11, is a novel p300-dependent regulator of Pdx-1 (MODY4) transcription in pancreatic islet beta cells. J. Biol. Chem. **284**, 36482–36490 (2009)
132. Plengvidhya, N., Kooptiwut, S., Songtawee, N., Doi, A., Furuta, H., Nishi, M., Nanjo, K., Tantibhedhyangkul, W., Boonyasrisawat, W., Yenchitsomanus, P.T., et al.: PAX4 mutations in Thais with maturity onset diabetes of the young. J. Clin. Endocrinol. Metab. **92**, 2821–2826 (2007)
133. Borowiec, M., Liew, C.W., Thompson, R., Boonyasrisawat, W., Hu, J., Mlynarski, W.M., El Khattabi, I., Kim, S.H., Marselli, L., Rich, S.S., et al.: Mutations at the BLK locus linked to maturity onset diabetes of the young and beta-cell dysfunction. Proc. Natl. Acad. Sci. U. S. A. **106**, 14460–14465 (2009)
134. Molven, A., Ringdal, M., Nordbø, A.M., Raeder, H., Støy, J., Lipkind, G.M., Steiner, D.F., Philipson, L.H., Bergmann, I., Aarskog, D., et al.: Mutations in the insulin gene can cause MODY and autoantibody-negative type 1 diabetes. Diabetes. **57**, 1131–1135 (2008)
135. Bonnefond, A., Philippe, J., Durand, E., Dechaume, A., Huyvaert, M., Montagne, L., Marre, M., Balkau, B., Fajardy, I., Vambergue, A., et al.: Whole-exome sequencing and high throughput genotyping identified KCNJ11 as the thirteenth MODY gene. PLoS One. **7**, e37423 (2012)
136. Naylor, R., Philipson, L.H.: Who should have genetic testing for maturity-onset diabetes of the young? Clin. Endocrinol. **75**, 422–426 (2011)
137. Stanley, C.A.: Hyperinsulinism in infants and children. Pediatr. Clin. N. Am. **44**, 363–374 (1997)
138. Rahman, S.A., Nessa, A., Hussain, K.: Molecular mechanisms of congenital hyperinsulinism. J. Mol. Endocrinol. **54**, R119–R129 (2015)
139. Glaser, B., Blech, I., Krakinovsky, Y., Ekstein, J., Gillis, D., Mazor-Aronovitch, K., Landau, H., Abeliovich, D.: ABCC8 mutation allele frequency in the Ashkenazi Jewish population and risk of focal hyperinsulinemic hypoglycemia. Genet. Med. **13**, 891–894 (2011)
140. Kapoor, R.R., Flanagan, S.E., Arya, V.B., Shield, J.P., Ellard, S., Hussain, K.: Clinical and molecular characterisation of 300 patients with congenital hyperinsulinism. Eur. J. Endocrinol. **168**, 557–564 (2013)
141. Arya, V.B., Mohammed, Z., Blankenstein, O., De Lonlay, P., Hussain, K.: Hyperinsulinaemic hypoglycaemia. Horm. Metab. Res. **46**, 157–170 (2014)
142. Gloyn, A.L., Diatloff-Zito, C., Edghill, E.L., Bellanné-Chantelot, C., Nivot, S., Coutant, R., Ellard, S., Hattersley, A.T., Robert, J.J.: KCNJ11 activating mutations are associated with developmental delay, epilepsy and neonatal diabetes syndrome and other neurological features. Eur. J. Hum. Genet. **14**, 824–830 (2006)
143. Kane, C., Shepherd, R.M., Squires, P.E., Johnson, P.R., James, R.F., Milla, P.J., Aynsley-Green, A., Lindley, K.J., Dunne, M.J.: Loss of functional KATP channels in pancreatic beta-

cells causes persistent hyperinsulinemic hypoglycemia of infancy. Nat. Med. **2**, 1344–1347 (1996)
144. Fournet, J.C., Mayaud, C., de Lonlay, P., Gross-Morand, M.S., Verkarre, V., Castanet, M., Devillers, M., Rahier, J., Brunelle, F., Robert, J.J., et al.: Unbalanced expression of 11p15 imprinted genes in focal forms of congenital hyperinsulinism: association with a reduction to homozygosity of a mutation in ABCC8 or KCNJ11. Am. J. Pathol. **158**, 2177–2184 (2001)
145. Pinney, S.E., MacMullen, C., Becker, S., Lin, Y.W., Hanna, C., Thornton, P., Ganguly, A., Shyng, S.L., Stanley, C.A.: Clinical characteristics and biochemical mechanisms of congenital hyperinsulinism associated with dominant KATP channel mutations. J. Clin. Invest. **118**, 2877–2886 (2008)
146. Huopio, H., Reimann, F., Ashfield, R., Komulainen, J., Lenko, H.L., Rahier, J., Vauhkonen, I., Kere, J., Laakso, M., Ashcroft, F., et al.: Dominantly inherited hyperinsulinism caused by a mutation in the sulfonylurea receptor type 1. J. Clin. Invest. **106**, 897–906 (2000)
147. Kapoor, R.R., Flanagan, S.E., Fulton, P., Chakrapani, A., Chadefaux, B., Ben-Omran, T., Banerjee, I., Shield, J.P., Ellard, S., Hussain, K.: Hyperinsulinism-hyperammonaemia syndrome: novel mutations in the GLUD1 gene and genotype-phenotype correlations. Eur. J. Endocrinol. **161**, 731–735 (2009)
148. Flanagan, S.E., Kapoor, R.R., Mali, G., Cody, D., Murphy, N., Schwahn, B., Siahanidou, T., Banerjee, I., Akcay, T., Rubio-Cabezas, O., et al.: Diazoxide-responsive hyperinsulinemic hypoglycemia caused by HNF4A gene mutations. Eur. J. Endocrinol. **162**, 987–992 (2010)
149. Kapoor, R.R., Locke, J., Colclough, K., Wales, J., Conn, J.J., Hattersley, A.T., Ellard, S., Hussain, K.: Persistent hyperinsulinemic hypoglycemia and maturity-onset diabetes of the young due to heterozygous HNF4A mutations. Diabetes. **57**, 1659–1663 (2008)
150. Christesen, H.B., Tribble, N.D., Molven, A., Siddiqui, J., Sandal, T., Brusgaard, K., Ellard, S., Njølstad, P.R., Alm, J., Brock Jacobsen, B., et al.: Activating glucokinase (GCK) mutations as a cause of medically responsive congenital hyperinsulinism: prevalence in children and characterisation of a novel GCK mutation. Eur. J. Endocrinol. **159**, 27–34 (2008)
151. Snider, K.E., Becker, S., Boyajian, L., Shyng, S.L., MacMullen, C., Hughes, N., Ganapathy, K., Bhatti, T., Stanley, C.A., Ganguly, A.: Genotype and phenotype correlations in 417 children with congenital hyperinsulinism. J. Clin. Endocrinol. Metab. **98**, E355–E363 (2013)
152. Stanescu, D.E., Hughes, N., Kaplan, B., Stanley, C.A., De León, D.D.: Novel presentations of congenital hyperinsulinism due to mutations in the MODY genes: HNF1A and HNF4A. J. Clin. Endocrinol. Metab. **97**, E2026–E2030 (2012)
153. Højlund, K., Hansen, T., Lajer, M., Henriksen, J.E., Levin, K., Lindholm, J., Pedersen, O., Beck-Nielsen, H.: A novel syndrome of autosomal-dominant hyperinsulinemic hypoglycemia linked to a mutation in the human insulin receptor gene. Diabetes. **53**, 1592–1598 (2004)
154. Flanagan, S.E., Mackay, D.J., Greeley, S.A., McDonald, T.J., Mericq, V., Hassing, J., Richmond, E.J., Martin, W.R., Acerini, C., Kaulfers, A.M., et al.: Hypoglycaemia following diabetes remission in patients with 6q24 methylation defects: expanding the clinical phenotype. Diabetologia. **56**, 218–221 (2013)
155. Giurgea, I., Missirian, C., Cacciagli, P., Whalen, S., Fredriksen, T., Gaillon, T., Rankin, J., Mathieu-Dramard, M., Morin, G., Martin-Coignard, D., et al.: TCF4 deletions in Pitt-Hopkins Syndrome. Hum. Mutat. **29**, E242–E251 (2008)
156. Støy, J., Greeley, S.A.W., Paz, V.P., Ye, H., Pastore, A.N., Skowron, K.B., Lipton, R.B., Cogen, F.R., Bell, G.I., Philipson, L.H., et al.: Diagnosis and treatment of neonatal diabetes: a United States experience. Pediatr. Diabetes. **9**(5), 450–459 (2008)
157. Fuld, K., Conrad, B., Buckingham, B., Wilson, D.M.: Insulin pumps in young children. Diabetes Technol. Ther. **12**(Suppl 1), S67–S71 (2010)
158. Suzuki, S., Koga, M., Niizeki, N., Furuya, A., Matsuo, K., Tanahashi, Y., Tsuchida, E., Nohara, F., Okamoto, T., Nagaya, K., et al.: Evaluation of glycated hemoglobin and fetal hemoglobin-adjusted HbA1c measurements in infants. Pediatr. Diabetes. **14**(4), 267–272 (2013)
159. De Franco, E., Flanagan, S.E., Houghton, J.A., Allen, H.L., Mackay, D.J., Temple, I.K., Ellard, S., Hattersley, A.T.: The effect of early, comprehensive genomic testing on clinical care in neonatal diabetes: an international cohort study. Lancet. **386**(9997), 957–963 (2015)

160. Rubio-Cabezas, O., Flanagan, S.E., Damhuis, A., Hattersley, A.T., Ellard, S.: KATP channel mutations in infants with permanent diabetes diagnosed after 6 months of life. Pediatr. Diabetes. **13**(4), 322–325 (2011)
161. Greeley, S.A.W., John, P.M., Winn, A.N., Ornelas, J., Lipton, R.B., Philipson, L.H., Bell, G.I., Huang, E.S.: The cost-effectiveness of personalized genetic medicine: the case of genetic testing in neonatal diabetes. Diabetes Care. **34**(3), 622–627 (2011)
162. Carmody, D., Bell, C.D., Hwang, J.L., Dickens, J.T., Sima, D.I., Felipe, D.L., Zimmer, C.A., Davis, A.O., Kotlyarevska, K., Naylor, R.N., et al.: Sulfonylurea treatment before genetic testing in neonatal diabetes: pros and cons. J. Clin. Endocrinol. Metab. **99**(12), E2709–E2714 (2014)
163. Pearson, E.R., Flechtner, I., Njølstad, P.R., Malecki, M.T., Flanagan, S.E., Larkin, B., Ashcroft, F.M., Klimes, I., Codner, E., Iotova, V., et al.: Switching from insulin to oral sulfonylureas in patients with diabetes due to Kir6.2 mutations. N. Engl. J. Med. **355**(5), 467–477 (2006)
164. Thurber, B.W., Carmody, D., Tadie, E.C., Pastore, A.N., Dickens, J.T., Wroblewski, K.E., Naylor, R.N., Philipson, L.H., Greeley, S.A.W., The United States Neonatal Diabetes Working Group: Age at the time of sulfonylurea initiation influences treatment outcomes in KCNJ11-related neonatal diabetes. Diabetologia. **58**(7), 1430–1435 (2015)
165. Seedher, N., Kanojia, M.: Co-solvent solubilization of some poorly-soluble antidiabetic drugs. Pharm. Dev. Technol. **14**(2), 185–192 (2009)
166. Slingerland, A.S., Nuboer, R., Hadders-Algra, M., Hattersley, A.T., Bruining, G.J.: Improved motor development and good long-term glycaemic control with sulfonylurea treatment in a patient with the syndrome of intermediate developmental delay, early-onset generalised epilepsy and neonatal diabetes associated with the V59M mutation in the KCNJ11 gene. Diabetologia. **49**(11), 2559–2563 (2006)
167. Mlynarski, W., Tarasov, A.I., Gach, A., Girard, C.A., Pietrzak, I., Zubcevic, L., Kusmierek, J., Klupa, T., Malecki, M.T., Ashcroft, F.M.: Sulfonylurea improves CNS function in a case of intermediate DEND syndrome caused by a mutation in KCNJ11. Nat. Clin. Pract. Neurol. **3**(11), 640–645 (2007)
168. Slingerland, A.S., Hurkx, W., Noordam, K., Flanagan, S.E., Jukema, J.W., Meiners, L.C., Bruining, G.J., Hattersley, A.T., Hadders-Algra, M.: Sulphonylurea therapy improves cognition in a patient with the V59M KCNJ11 mutation. Diabet. Med. **25**(3), 277–281 (2008)
169. Koster, J.C., Cadario, F., Peruzzi, C., Colombo, C., Nichols, C.G., Barbetti, F.: The G53D mutation in Kir6.2 (KCNJ11) is associated with neonatal diabetes and motor dysfunction in adulthood that is improved with sulfonylurea therapy. J. Clin. Endocrinol. Metab. **93**(3), 1054–1061 (2008)
170. Mohamadi, A., Clark, L.M., Lipkin, P.H., Mahone, E.M., Wodka, E.L., Plotnick, L.P.: Medical and developmental impact of transition from subcutaneous insulin to oral glyburide in a 15-yr-old boy with neonatal diabetes mellitus and intermediate DEND syndrome: extending the age of KCNJ11 mutation testing in neonatal DM. Pediatr. Diabetes. **11**(3), 203–207 (2010)
171. Shah, R.P., Spruyt, K., Kragie, B.C., Greeley, S.A.W., Msall, M.E.: Visuomotor performance in KCNJ11-related neonatal diabetes is impaired in children with DEND-associated mutations and may be improved by early treatment with sulfonylureas. Diabetes Care. **35**(10), 2086–2088 (2012)
172. Beltrand, J., Elie, C., Busiah, K., Fournier, E., Boddaert, N., Bahi-Buisson, N., Vera, M., Bui-Quoc, E., Ingster-Moati, I., Berdugo, M., et al.: Sulfonylurea therapy benefits neurological and psychomotor functions in patients with neonatal diabetes owing to Potassium Channel mutations. Diabetes Care. **38**(11), 2033–2041 (2015)
173. Lahmann, C., Kramer, H.B., Ashcroft, F.M.: Systemic Administration of Glibenclamide Fails to achieve therapeutic levels in the brain and cerebrospinal fluid of rodents. PLoS One. **10**(7), e0134476 (2015)
174. Lehto, M., Tuomi, T., Mahtani, M.M., Widén, E., Forsblom, C., Sarelin, L., Gullström, M., Isomaa, B., Lehtovirta, M., Hyrkkö, A., et al.: Characterization of the MODY3 phenotype.

Early-onset diabetes caused by an insulin secretion defect. J. Clin. Invest. **99**(4), 582–591 (1997)
175. Pearson, E.R., Velho, G., Clark, P., Stride, A., Shepherd, M., Frayling, T.M., Bulman, M.P., Ellard, S., Froguel, P., Hattersley, A.T.: Beta-cell genes and diabetes: quantitative and qualitative differences in the pathophysiology of hepatic nuclear factor-1alpha and glucokinase mutations. Diabetes. **50**(Suppl 1), S101–S107 (2001)
176. Stride, A., Vaxillaire, M., Tuomi, T., Barbetti, F., Njolstad, P.R., Hansen, T., Costa, A., Conget, I., Pedersen, O., Sovik, O., et al.: The genetic abnormality in the beta cell determines the response to an oral glucose load. Diabetologia. **45**(3), 427–435 (2002)
177. Fajans, S.S., Brown, M.B.: Administration of sulfonylureas can increase glucose-induced insulin secretion for decades in patients with maturity-onset diabetes of the young. Diabetes Care. **16**(9), 1254–1261 (1993)
178. Pearson, E.R., Starkey, B.J., Powell, R.J., Gribble, F.M., Clark, P.M., Hattersley, A.T.: Genetic cause of hyperglycaemia and response to treatment in diabetes. Lancet. **362**(9392), 1275–1281 (2003)
179. Pearson, E.R., Liddell, W.G., Shepherd, M., Corrall, R.J., Hattersley, A.T.: Sensitivity to sulphonylureas in patients with hepatocyte nuclear factor-1alpha gene mutations: evidence for pharmacogenetics in diabetes. Diabet. Med. **17**(7), 543–545 (2000)
180. Shepherd, M., Pearson, E.R., Houghton, J., Salt, G., Ellard, S., Hattersley, A.T.: No deterioration in glycemic control in HNF-1alpha maturity-onset diabetes of the young following transfer from long-term insulin to sulphonylureas. Diabetes Care. **26**(11), 3191–3192 (2003)
181. Naylor, R.N., John, P.M., Winn, A.N., Carmody, D., Greeley, S.A.W., Philipson, L.H., Bell, G.I., Huang, E.S.: Cost-effectiveness of MODY genetic testing: translating genomic advances into practical health applications. Diabetes Care. **37**(1), 202–209 (2014)
182. Ostoft, S.H., Bagger, J.I., Hansen, T., Pedersen, O., Faber, J., Holst, J.J., Knop, F.K., Vilsboll, T.: Glucose-lowering effects and low risk of hypoglycemia in patients with maturity-onset diabetes of the young when treated with a GLP-1 receptor agonist: a double-blind, randomized, crossover trial. Diabetes Care. **37**(7), 1797–1805 (2014)
183. Chakera, A.J., Steele, A.M., Gloyn, A.L., Shepherd, M.H., Shields, B., Ellard, S., Hattersley, A.T.: Recognition and Management of Individuals with Hyperglycemia because of a heterozygous glucokinase mutation. Diabetes Care. **38**, 1383–1392 (2015)
184. Steele, A.M., Shields, B.M., Wensley, K.J., Colclough, K., Ellard, S., Hattersley, A.T.: Prevalence of vascular complications among patients with glucokinase mutations and prolonged, mild hyperglycemia. JAMA. **311**(3), 279–286 (2014)
185. Stride, A., Shields, B., Gill-Carey, O., Chakera, A.J., Colclough, K., Ellard, S., Hattersley, A.T.: Cross-sectional and longitudinal studies suggest pharmacological treatment used in patients with glucokinase mutations does not alter glycaemia. Diabetologia. **57**, 54–56 (2014)
186. Carmody, D., Lindauer, K.L., Naylor, R.N.: Adolescent non-adherence reveals a genetic cause for diabetes. Diabet. Med. **32**, e20–e23 (2015)
187. Thornton, P.S., Stanley, C.A., De Leon, D.D., Harris, D., Haymond, M.W., Hussain, K., Levitsky, L.L., Murad, M.H., Rozance, P.J., Simmons, R.A., et al.: Recommendations from the pediatric Endocrine Society for evaluation and Management of Persistent Hypoglycemia in neonates, infants, and children. J. Pediatr. **167**, 238–245 (2015)
188. Lord, K., Radcliffe, J., Gallagher, P.R., Adzick, N.S., Stanley, C.A., De Leon, D.D.: High risk of diabetes and neurobehavioral deficits in individuals with surgically treated hyperinsulinism. J. Clin. Endocrinol. Metab. **100**, 4133–4139 (2015)
189. Stanley, C.A., Rozance, P.J., Thornton, P.S., De Leon, D.D., Harris, D., Haymond, M.W., Hussain, K., Levitsky, L.L., Murad, M.H., Simmons, R.A., et al.: Re-evaluating "transitional neonatal hypoglycemia": mechanism and implications for management. J. Pediatr. **166**(1520–1525), e1521 (2015)
190. Palladino, A.A., Bennett, M.J., Stanley, C.A.: Hyperinsulinism in infancy and childhood: when an insulin level is not always enough. Clin. Chem. **54**, 256–263 (2008)
191. De Leon, D.D., Stanley, C.A.: Mechanisms of disease: advances in diagnosis and treatment of hyperinsulinism in neonates. Nat. Clin. Pract. Endocrinol. Metab. **3**, 57–68 (2007)

192. Lord, K., De Leon, D.D.: Monogenic hyperinsulinemic hypoglycemia: current insights into the pathogenesis and management. Int. J. Pediatr. Endocrinol. **2013**, 3 (2013)
193. Kapoor, R.R., James, C., Hussain, K.: Advances in the diagnosis and management of hyperinsulinemic hypoglycemia. Nat. Clin. Pract. Endocrinol. Metab. **5**, 101–112 (2009)
194. Shah, P., Rahman, S.A., McElroy, S., Gilbert, C., Morgan, K., Hinchey, L., Senniappan, S., Levy, H., Amin, R., Hussain, K.: Use of long-acting somatostatin analogue (lanreotide) in an adolescent with diazoxide-responsive congenital hyperinsulinism and its psychological impact. Horm. Res. Paediatr. **84**, 355–360 (2015)
195. Senniappan, S., Alexandrescu, S., Tatevian, N., Shah, P., Arya, V., Flanagan, S., Ellard, S., Rampling, D., Ashworth, M., Brown, R.E., et al.: Sirolimus therapy in infants with severe hyperinsulinemic hypoglycemia. N. Engl. J. Med. **370**, 1131–1137 (2014)
196. Khawash, P., Hussain, K., Flanagan, S.E., Chatterjee, S., Basak, D.: Nifedipine in congenital hyperinsulinism - a case report. J. Clin. Res. Pediatr. Endocrinol. **7**, 151–154 (2015)
197. Stein, S.A., Maloney, K.L., Pollin, T.I.: Genetic counseling for diabetes mellitus. Curr. Genet. Med. Rep. **2**, 56–67 (2014)
198. Palladino, A.A., Stanley, C.A.: The hyperinsulinism/hyperammonemia syndrome. Rev. Endocr. Metab. Disord. **11**, 171–178 (2010)
199. de Lonlay, P., Giurgea, I., Sempoux, C., Touati, G., Jaubert, F., Rahier, J., Ribeiro, M., Brunelle, F., Nihoul-Fékété, C., Robert, J.J., et al.: Dominantly inherited hyperinsulinaemic hypoglycaemia. J. Inherit. Metab. Dis. **28**, 267–276 (2005)
200. Yorifuji, T.: Congenital hyperinsulinism: current status and future perspectives. Ann. Pediatr. Endocrinol. Metab. **19**, 57–68 (2014)
201. Julier, C., Nicolino, M.: Wolcott-Rallison syndrome. Orphanet J. Rare Dis. **5**, 29 (2010)
202. Flanagan, S.E., Xie, W., Caswell, R., Damhuis, A., Vianey-Saban, C., Akcay, T., Darendeliler, F., Bas, F., Guven, A., Siklar, Z., et al.: Next-generation sequencing reveals deep intronic cryptic ABCC8 and HADH splicing founder mutations causing hyperinsulinism by pseudoexon activation. Am. J. Hum. Genet. **92**, 131–136 (2013)
203. Hartemann-Heurtier, A., Simon, A., Bellanne-Chantelot, C., Reynaud, R., Cave, H., Polak, M., Vaxillaire, M., Grimaldi, A.: Mutations in the ABCC8 gene can cause autoantibody-negative insulin-dependent diabetes. Diabetes Metab. **35**, 233–235 (2009)
204. Barroso, I., Gurnell, M., Crowley, V.E., Agostini, M., Schwabe, J.W., Soos, M.A., Maslen, G.L., Williams, T.D., Lewis, H., Schafer, A.J., et al.: Dominant negative mutations in human PPARgamma associated with severe insulin resistance, diabetes mellitus and hypertension. Nature. **402**, 880–883 (1999)
205. Agostini, M., Schoenmakers, E., Mitchell, C., Szatmari, I., Savage, D., Smith, A., Rajanayagam, O., Semple, R., Luan, J., Bath, L., et al.: Non-DNA binding, dominant-negative, human PPARgamma mutations cause lipodystrophic insulin resistance. Cell Metab. **4**, 303–311 (2006)
206. Gething, M.J.: Role and regulation of the ER chaperone BiP. Semin. Cell Dev. Biol. **10**, 465–472 (1999)
207. Ladiges, W.C., Knoblaugh, S.E., Morton, J.F., Korth, M.J., Sopher, B.L., Baskin, C.R., MacAuley, A., Goodman, A.G., LeBoeuf, R.C., Katze, M.G.: Pancreatic beta-cell failure and diabetes in mice with a deletion mutation of the endoplasmic reticulum molecular chaperone gene P58IPK. Diabetes. **54**, 1074–1081 (2005)
208. Shashi, V., McConkie-Rosell, A., Rosell, B., Schoch, K., Vellore, K., McDonald, M., Jiang, Y.H., Xie, P., Need, A., Goldstein, D.B.: The utility of the traditional medical genetics diagnostic evaluation in the context of next-generation sequencing for undiagnosed genetic disorders. Genet. Med. **16**, 176–182 (2014)

Application of NGS in the Diagnosis of Cardiovascular Genetic Diseases

Guoliang Wang, Vivan Niewiadonski, Jianping Li, Ruirui Ji, Wenxin Zou, Daniel J. Penny, and Yuxin Fan

Abstract Cardiovascular diseases (CVDs) are the leading global cause of death and encompass a broad range of disorders, including diseases of the vasculature, the myocardium, and the heart's electrical circuit, and congenital heart disease (CHD). In the etiology of most CVDs, a clear hereditary component has been demonstrated. CVDs can be divided in two major categories: the monogenic and the polygenic/multifactorial forms and have long been at the forefront of gene testing in the clinic. The advent of next-generation sequencing (NGS) technologies has led to increasingly comprehensive testing for CVDs in both the monogenic and the polygenic/multifactorial forms, although the interpretation of the NGS data is still a challenge at this time. This chapter describes the genetic background of CVDs including inherited cardiomyopathy, inherited primary arrhythmia syndromes, CHD and inherited aortopathy, as well as the utility of NGS in the detection of CVDs-related genetic alterations.

Keywords NGS • Inherited cardiovascular diseases • Molecular diagnosis

1 Introduction

Cardiovascular diseases (CVDs) are the leading global cause of death, accounting for 17.3 million deaths per year, a number that is expected to grow to more than 23.6 million by 2030. In the United States, an estimated 85.6 million adults have CVDs, and the estimated direct and indirect costs of CVDs are more than $300 billion. CVD encompasses a broad range of disorders, including diseases of the vasculature, the myocardium, and the heart's electrical circuit, and congenital heart disease (CHD) [1]. In the etiology of most CVDs, a clear hereditary component has been

G. Wang, M.D., Ph.D. • V. Niewiadonski, M.S. • J. Li, M.D., M.S. • R. Ji, M.D., Ph.D.
W. Zou, Ph.D. • D.J. Penny, M.D., Ph.D. • Y. Fan, M.D., Ph.D., FACMG, FAAP (✉)
John Welsh Cardiovascular Diagnostic Laboratory, Section of Cardiology,
Department of Pediatrics, Texas Children's Hospital, Baylor College of Medicine,
1102 Bates Avenue, Feigin Center, Suite 430.09, Houston, TX 77030, USA
e-mail: yuxinf@bcm.edu; http://www.bcm.edu/pediatrics/welsh

L.-J.C. Wong (ed.), *Next Generation Sequencing Based Clinical Molecular Diagnosis of Human Genetic Disorders*, DOI 10.1007/978-3-319-56418-0_12

demonstrated. CVDs can be divided in two major categories: the monogenic (rare) and the polygenic/multifactorial forms. In the monogenic diseases, the mutation in a single gene causes the pathology. These diseases have Mendelian traits that show the classical inheritance patterns: autosomal dominant, autosomal recessive, X-linked, or mitochondrial (maternally inherited). Examples of these traits in cardiovascular medicine include structural cardiomyopathies (e.g., hypertrophic or dilated cardiomyopathy) and channelopathies (e.g., Brugada syndrome [BrS] and long QT syndrome [LQTS]), familial dyslipidemias, familial pulmonary hypertension, and Marfan syndrome (MFS). However, the most common CVDs (e.g., coronary artery disease [CAD]) have complex traits that arise from elaborate gene-gene and gene-environmental interactions that confer risk for disease in a probabilistic manner. In these cases, a series of polymorphic variants in several genes increases the risk of developing the disease. Examples of this category of CVDs include coronary heart disease and hypertension [2, 3]. CVDs have long been at the forefront of gene testing in the clinic, and this trend is likely to continue. For many of these Mendelian forms of CVD, direct DNA sequencing and/or linkage analysis have successfully yielded the causal gene and mutation. However, mapping gene loci associated with complex traits requires substantial levels of information and analysis [4]. The advent of next-generation sequencing (NGS) technologies has led to increasingly comprehensive testing for CVDs in both the monogenic and the polygenic/multifactorial forms. Understanding the genetic etiology for these disorders has improved their clinical recognition and management and led to new guidelines for treatment and family-based diagnosis and surveillance [5].

In this chapter, we introduce genetics in human CVDs including inherited cardiomyopathy, inherited cardiac arrhythmias, congenital heart disease and inherited aortopathy. Then, we focus on using NGS for the detection of CVDs related DNA changes.

2 Inherited Cardiomyopathy

Inherited cardiomyopathies are a group of cardiovascular disorders and a major cause of heart disease in all age groups, often with an onset in adolescence or early adult life. Based on ventricular morphological and functional features, inherited cardiomyopathies are classified into hypertrophic cardiomyopathy (HCM), dilated cardiomyopathy (DCM), arrhythmogenic right ventricular cardiomyopathy (ARVC), left ventricular noncompaction (LVNC), and restrictive cardiomyopathy (RCM) [6]. Since a mutation in the β-myosin heavy chain gene (*MYH7*) was first identified as being responsible for causing HCM [7], many advances have been made to define the genetic etiology of inherited cardiomyopathies. For instance, HCM is now viewed as a "disease of the sarcomere," as most of the genes associated with HCM encode proteins of the myofilaments or Z-disc of the sarcomeres [8]. In contrast, DCM is far more genetically heterogeneous, with mutations in genes encoding cytoskeletal, nucleoskeletal, mitochondrial, and calcium-handling proteins [9]. Many genes have been identified for causing ARVC, most resulting in disturbed desmosome/intercalated disk function. Although understanding the genetic basis for the development of RCM and LVNC has been more elusive, genes for both groups of disorders have been identified and appear to include sarcomere dysfunction as a critical factor [10].

2.1 Hypertrophic Cardiomyopathy

HCM (HCM, OMIM: 612,098) is characterized by asymmetric or concentric wall thickening in the absence of an underlying systemic condition or other cardiac disease. With an estimated prevalence of 1 in 500 in the general population [11], HCM is the most common inherited heart condition. Although the age at onset of HCM can range from infancy to old age, manifestations usually do not appear before adolescence in carriers of a pathogenic variant. HCM is inherited primarily in an autosomal-dominant pattern, although reduced penetrance and clinical variability are common [12]. Clinically, most patients with HCM are asymptomatic or mildly symptomatic [13, 14]. The major effects of this disorder on human health are its predilection to be inherited, its reputation as the most common cause of sudden cardiac death (SCD) in young, healthy individuals, and its potential to develop heart failure (HF) because of diastolic factors or development of systolic dysfunction[10].

2.1.1 Genetics of Hypertrophic Cardiomyopathy

Since the pathogenic missense mutation in the *MYH7* (*MYH7* R403Q) was revealed two decades ago, hundreds of mutations have been identified in at least 29 putative HCM-susceptibility genes(Table 1) [7, 10, 15]. The most common genetic subtypes of HCM are sarcomeric- or myofilament-HCM, caused by mutations in eight genes encoding proteins of the myofilaments of the cardiac sarcomere [15]. In those patients with positive genetic tests, myosin-binding protein C (*MYBPC3*) and *MYH7* are, by far, the two most commonly identified HCM-associated genes, with an estimated prevalence of 25–35% for each. Other genes, including *TNNT2, TNNI3,TPM1, and ACTN2*, are known to account each for a small proportion of patients (1–5%) [16]. Collectively, sarcomere variants are identified in as many as 60% of patients with HCM who also have a family history and in 40% of patients with sporadic HCM [17]. In addition to defects in the sarcomere-encoding genes, patients with HCM have been identified hosting mutations in Z-disk and other nonsarcomere-encoding genes. The giant protein titin (TTN) and its interactive Z-disc proteins, including muscle LIM protein (MLP), Z-band alternatively spliced PDZ-motif (ZASP), telethonin, nexilin, myopalladin, myozenin-2, α-actinin 2, cardiac ankyrin repeat protein (CARP), and vinculin, have been identified as causes of HCM when the respective gene is mutated [18]. Although almost 1000 variants for HCM have been identified in all known HCM-associated genes to date, most are private and can, therefore, be detected only through comprehensive genetic testing [17].

Table 1 Inherited cardiomyopathy-associated genes

Gene	OMIM	Locus	Protein	HCM	DCM	RCM	ARVDC	LVNC
AARS2	612,035	6p21.1	Alanyl-tRNA Synthetase 2	X				
ABCC9	601,439	12p12.1	Sulfonylurea Receptor 2		X			
ACTC1	102,540	15q14	Alpha-Cardiac Actin	X	X	X		X
ACTN2	102,573	1q43	Alpha-Actinin-2	X	X			
ANKRD1	609,599	10q23.31	Ankyrin Repeat Domain 1	X	X			
BAG3	603,883	10q26.11	BAG3	X	X	X		
CASQ2	114,251	1p13.1	Calsequestrin 2					X
CAV3	601,253	3p25.3	Caveolin 3	X	X			
CRYAB	123,590	11q23.1	Crystallin, Alpha B	X	X			
COX15	603,646	10q24.2	Cytochrome C Oxidase Subunit 15	X				
CSRP3	600,824	11p15.1	Cysteine and Glycine-Rich Protein 3	X	X			
CTF1	600,435	16p11.2	Cardiotrophin 1		X			
CTNNA3	116,806	3p22.1	Catenin, Alpha 3				X	
DES	125,660	2q35	Desmin		X	X	X	
DMD	300,377	Xp21.2-p21.1	Dystrophin		X			X
DNAJC19	608,977	3q26.33	DnaJ Homolog Subfamily C Member 19				X	
DSC2	125,645	18q12.1	Desmocollin 2		X		X	
DSG2	125,671	18q12.1	Desmoglein 2		X		X	
DSP	125,647	6p24.3	Desmoplakin		X		X	
DTNA	601,239	18q12.1	Dystrobrevin, Alpha					X
EMD	300,384	Xq28	Emerin		X			
EYA4	603,550	6q23.2	EYA Transcriptional Coactivator and Phosphatase 4		X			
FHL2	602,633	2q12.2	Four and A Half LIM Domains 2		X			
FKTN	607,440	9q31.2	Fukutin		X			

FOXD4	611,085	9q21.11	Forkhead Box D4		X			
FXN	606,829	9q21.11	Frataxin	X				
GAA	606,800	17q25.3	Glucosidase, Alpha; Acid	X				
GATAD1	614,518	7q21.2	GATA Zinc Finger Domain Containing 1		X			
GLA	300,644	Xq22.1	Galactosidase, Alpha	X		X		X
ILK	602,336	11p15.4	Integrin-Linked Kinase		X			
JPH2	605,267	20q13.12	Junctophilin 2	X				
JUP	173,325	17q21.2	Junction Plakoglobin				X	
KRAS	190,070	12p12.1	Kirsten Rat Sarcoma Viral Oncogene Homolog	X				
LAMA4	600,133	6q21	Laminin, Alpha 4		X			
LAMP2	309,060	Xq24	Lysosomal-Associated Membrane Protein 2	X	X			
LDB3	605,906	10q23.2	LIM Domain Binding 3	X	X			X
LMNA	150,330	1q22	Lamin A/C		X		X	X
MIB1	608,677	18q11.2	Mindbomb E3 Ubiquitin Protein Ligase 1					X
MYBPC3	600,958	11p11.2	Myosin Binding Protein C, Cardiac	X	X			X
MYF6	159,991	12q21.31	Myogenic Factor 6 (Herculin)					
MYH6	160,710	14q11.2	Myosin Heavy Chain 6	X	X			
MYH7	160,760	14q11.2	Myosin Heavy Chain 7	X	X	X		X
MYL2	160,781	12q24.11	Myosin Light Chain 2	X		X		
MYL3	160,790	3p21.31	Myosin Light Chain 3	X		X		
MYLK2	606,566	20q11.21	Myosin Light Chain Kinase 2	X				
MYO6	600,970	6q14.1	Myosin VI	X				
MYOM1	603,508	18p11.31	Myomesin 1	X				
MYOZ2	605,602	4q26	Myozenin 2	X				
MYPN	608,517	10q21.3	Myopalladin		X	X		

(continued)

Table 1 (continued)

Gene	OMIM	Locus	Protein	HCM	DCM	RCM	ARVDC	LVNC
NEBL	605,491	10p12.31	Nebulette		X			
NEXN	613,121	1p31.1	Nexilin	X	X			
PDLIM3	605,889	4q35.1	PDZ And LIM Domain 3		X			
PKP2	602,861	12p11.21	Plakophilin 2		X		X	
PKP4	604,276	2q24.1	Plakophilin 4				X	
PLN	172,405	6q22.31	Phospholamban	X	X		X	
PRDM16	605,557	1p36.32	PR Domain Containing 16		X			X
PRKAG2	602,743	7q36.1	AMP-Activated Protein Kinase, Gamma 2 Non-Catalytic Subunit	X				
PRKAR1A	188,830	17q24.2	CAMP-Dependent Regulatory Protein Kinase, Type I-Alpha	X				
PSEN1	104,311	14q24.2	Presenilin 1		X			
PSEN2	600,759	14q24.2	Presenilin 2		X			
PTPN11	176,876	12q24.13	Protein Tyrosine Phosphatase 2C	X				
RAF1	164,760	Xp11.3	Raf Proto-Oncogene Serine/Threonine Kinase	X				
RBM20	613,717	10q25.2	RNA Binding Motif Protein 20		X			
RYR2	180,902	1q43	Ryanodine Receptor 2	X			X	
SCN5A	600,163	3p22.2	Voltage-Gated Sodium Channel Subunit Alpha Nav1.5		X			
SCO2	604,272	22q13.33	SCO2 Cytochrome C Oxidase Assembly Protein	X				
SDHA	600,857	5p15.33	Flavoprotein	X	X			
SGCD	601,411	5q33.2-q33.3	Sarcoglycan, Delta		X			
SPEG	615,950	2q35	Striated Muscle Preferentially Expressed Protein Kinase		X			
STRN	614,765	2p22.2	Striatin				X	
SURF1	185,620	9q34.2	Surfeit 1	X				
SYNE1	608,441	6q25.2	Synaptic Nuclear Envelope Protein 1		X			

SYNE2	608,442	14q23.2	Synaptic Nuclear Envelope Protein 2		X			
TAZ	300,394	Xq28	Protein G4.5	X	X			X
TBX20	606,061	Xq21.1	T-Box Protein 20		X			
TCAP	604,488	17q12	Titin Cap Protein	X	X			
TGFB3	190,230	14q24.3	Transforming Growth Factor Beta-3				X	
TMEM43	612,048	3p25.1	Transmembrane Protein 43				X	
TMPO	188,380	12q23.1	Thymopoietin		X			
TNNC1	191,040	3p21.1	Troponin C Type 1	X	X			
TNNI3	191,044	19q13.42	Troponin I Type 3	X	X	X		
TNNT2	191,045	1q32.1	Troponin T Type 2	X	X	X		X
TPM1	191,010	15q22.2	Tropomyosin 1	X	X	X		X
TTN	188,840	2q31.2	Titin	X	X	X	X	
TTR	176,300	1`8q12.1	Transthyretin	X	X	X		X
VCL	193,065	10q22.2	Vinculin	X	X			X
TOTAL	86			44	55	13	16	15

Mutations identified in sarcomeric genes typically are single nucleotide substitutions and, in most instances, the mutant protein is thought to incorporate into the sarcomere and act in a dominant-negative manner. However, about a half of the reported *MYBPC3* mutations are truncations caused by nonsense and frameshift mutations; these and some *MYBPC3* missense mutations can result in haploinsufficiency, a condition in which the gene product of the wild-type allele cannot compensate for the decreased product from the mutant allele [19]. For Z-disk and calcium modulator genes, the specific mechanism has not been clearly elucidated. Genetic testing for HCM has been commercially available for almost a decade, but a low mutation detection rate and costs have hindered uptake [20].

2.1.2 NGS of Hypertrophic Cardiomyopathy

Currently, genetic testing has been recommended for any patient with an established clinical diagnosis of HCM and for family members and appropriate relatives after the identification of the HCM-causative mutation in an index case [21]. Including genetic testing in the diagnostic strategy is more likely to be cost-effective than are clinical tests alone when considering family screening and consequences for the prevention of SCD [22–24]. Genetic testing identifies mutation carriers who will benefit from regular clinical investigations or early discussion of implantable cardioverter-defibrillator (ICD), as well as relatives who do not have the causal mutation and, therefore, can be released without long-term follow-up [25]. With the rapid development in genetic testing technology, an entire exome or a panel of HCM-related genes can now be tested simultaneously by the NGS, providing an opportunity to detect numerous mutations in same or different genes that are responsible for HCM. Several groups have developed NGS-based approaches for a comprehensive and cost-efficient genetic diagnosis of cardiomyopathies and have been commercialized. For example, NGS panels testing services offered by Baylor College of Medicine (BCM)/ John Welsh Cardiovascular Diagnostic Laboratory are listed in Table 2. Other laboratories including Ambry Genetics, GeneDx, etc. provide similar tests. These approaches demonstrate the feasibility of using NGS techniques in targeted sequencing of cardiomyopathy associated genes [26, 27].

2.2 *Dilated Cardiomyopathy*

DCM (DCM, OMIM: 613,694) is characterized by left ventricular dilation and systolic dysfunction (a reduction in myocardial force generation) and is the most common indication for cardiac transplantation [6]. The annual incidence is 2–8/10,000 (0.57/100,000/year in children), and the estimated prevalence is 1/2500 population, although this figure may be underestimated [6, 28, 29]. The age at onset includes newborn through late adulthood, but most patients are diagnosed between 20–50 years of age [30]. Clinical manifestations include HF, thromboembolism, and SCD.

Table 2 Cardiomyopathy NGS testing panels (www.bcm.edu/pediatrics/welsh).

Cardiomyopathy NGS panels	Targeted genes	Coverage
Hypertrophic Cardiomyopathy–HCM (41 genes)	*AARS2, ACTC1, ACTN2, ANKRD1, BAG3, CAV3, CSRP3, FXN, GAA, GLA, JPH2, KRAS, LAMP2, LDB3, MYBPC3, MYH6, MYH7, MYL2, MYL3, MYLK2, MYO6, MYOM1, MYOZ2, NEXN, PLN, PRKAG2, PRKAR1A, PTPN11, RAF1, RYR2, SCO2, SDHA, SURF1, TCAP, TNNC1, TNNI3, TNNT2, TPM1, TTN, TTR, VCL*	1. NGS only: 97.9% 2. NGS + Sanger fill-in:100%
Dilated Cardiomyopathy–DCM (52 genes)	*ABCC9, ACTC1, ACTN2, ANKRD1, BAG3, CAV3, CRYAB, CSRP3, CTF1, DES, DMD, DSC2, DSG2, DSP, EMD, EYA4, FHL2, FKTN, GATAD1, ILK, LAMA4, LAMP2, LDB3, LMNA, MYBPC3, MYH6, MYH7, MYPN, NEBL, NEXN, PDLIM3, PKP2, PLN, PRDM16, PSEN1, PSEN2, RBM20, SCN5A, SDHA, SGCD, SPEG, SYNE2, TAZ, TBX20, TCAP, TMPO, TNNC1, TNNI3, TNNT2, TPM1, TTN, VCL*	
Restrictive Cardiomyopathy–RCM (6 genes)	*ACTC1, BAG3, DES, MYH7, TNNI3, TNNT2*	
Arrhythmogenic Right Ventricular Dysplasia/Cardiomyopathy–ARVD/C (12 genes)	*DES, DSC2, DSG2, DSP, JUP, PKP2, PKP4, PLN, RYR2, TGFB3, TMEM43, TTN*	
Left Ventricular Noncompaction–LVNC (13 genes)	*ACTC1, CASQ2, DTNA, LDB3, LMNA, MIB1, MYBPC3, MYH7, PRDM16, TAZ, TNNT2, TPM1, VCL*	

DCM can also present with muscular involvement and may be the presenting or primary clinical feature of several multisystem conditions, including Emery-Dreifuss muscular dystrophy (EDMD), Barth syndrome, myofibrillar myopathy, limb-girdle muscular dystrophy (LGMD), and Duchenne or Becker muscular dystrophy (DMD/BMD) [31].

2.2.1 Genetics of Dilated Cardiomyopathy

DCM cases with a genetic etiology represent approximately 30–50%, based on the presence of a family history. More than 50 disease genes have been identified (Table 1); the most common mode of inheritance is autosomal-dominant transmission, although X-linked, autosomal-recessive, and mitochondrial inheritance forms have been described [9]. Most of these genes encode cytoskeletal, sarcomeric, or Z-disk proteins, but mutations in a small number of ion channel-encoding and desmosome-encoding genes also have been identified. They include genes primarily encoding cytoskeletal δ-sarcoglycan (SGCD), β-sarcoglycan (SGCB), desmin (DES), lamin A/C (LMNA), vinculin, sarcomeric/myofibrillar (α-cardiac actin [ACTC], troponin T [TNNT2], troponin I [TNNI3]), MYH7, myosin-binding protein C, α-tropomyosin (TPM1), and Z-disk proteins MLP/ cysteine and glycine-rich protein 3 (CSRP3), TTN, telethonin/TCAP, α-actinin-2 (ACTN2), nebulette (NEBL), myopalladin (MYPN), ANKRD1/CARP, and ZASP/LIM- domain binding 3 (LBD3)). Ion channel encoding genes identified to date include the cardiac sodium channel gene *SCN5A* and calcium homeostasis regulator phospholamban (*PLN*), and associated desmosome-encoding genes including desmoplakin (*DSP*), desmoglein-2 (*DSG2*), and desmocolin-2 (*DSC2*) have also been shown to result in a DCM phenotype [10]. New genes for DCM are continually being discovered, with recent additions including the gene encoding BCL2-associated athanogene 3 (*BAG3*) [32], *RBM20* [33], and *TTN* [34]. *TTN* may contribute to as many as 25% of familial and 18% of sporadic DCM cases, rendering it by far the most commonly mutated gene in DCM [34].

Given the diversity of affected cellular processes, numerous proximal factors probably contribute to contractile dysfunction in DCM. Molecular mechanisms of DCM-causing mutations include diminished force generation and transmission, alterations of energy production and regulation, and intracellular calcium defects [35]. Sarcomere mutations may cause DCM through two mechanisms: deficits of force production and deficits of force transmission [36]. Mutations in cytoskeletal and Z-disc proteins cause mainly defects of force transmission. The altered desmosomal proteins appear to disrupt the links among the intercalated disk, Z-disk, and sarcomere [10]. Mitochondiral DNA mutations may alter energy production and regulation in affected cardiomyocyte [37], and *PLN* mutations cause intracellular calcium defects [38].

2.2.2 NGS of Dilated Cardiomyopathy

Because of significant locus and allelic heterogeneity, genetic testing for DCM has been of limited utility, with pathogenic variants identified in 17–40% of cases using current 40 gene panels, and most genes contributing only a small percentage of pathogenic variants [10]. Clear genotype-phenotype correlations are rare. Exceptions include variants in the *LMNA* and *SCN5A* genes, which typically are associated with DCM and conduction system disease [39, 40]. With the application of NGS techniques, *TTN*-targeted sequencing revealed that *TTN* truncating mutations are a common cause of DCM [34]. Whole exome sequencing (WES) identified novel *GATA* Zinc Finger Domain Containing 1 (*GATAD1*) and *BAG3* mutations in patients with DCM [32, 41]. Due to the high level of complexity of genotype-phenotype associations and advanced genetic testing efficiency in DCM, genetic testing by NGS may be useful for the identification of non-carriers and asymptomatic carriers, as well as for prevention strategies, sport recommendations, and defibrillator implantation. It can also guide reproductive decision-making including utilization of pre-implantation genetic diagnostic strategies [42]. Table 2 summarized the NGS panel for dilated cardiomyopathy developed in Baylor College of Medicine (BCM)/ John Welsh Cardiovascular Diagnostic Laboratory.

2.3 Restrictive Cardiomyopathy

Restrictive cardiomyopathy (RCM) (RCM1,OMIM# 115210; RCM3, OMOM#612422) is a rare disease of the myocardium characterized by increased stiffness of the ventricles leading to compromised diastolic filling with preserved systolic function. These changes may develop in association with local inflammatory or systemic, infiltrative, or storage disease [43]. RCM accounts for approximately 5% of all cases of primary heart muscle disease. In people with familial RCM, the heart muscle is stiff and cannot fully relax after each contraction. Most affected individuals have severe signs and symptoms of HF. Adult patients with RCM present with dyspnea, fatigue, and limited exercise capacity. In children, RCM may present with failure to thrive, fatigue and even syncope [44, 45]. RCM carries a poor prognosis, particularly in children, despite optimal medical treatment. Several studies have reported that 66–100% die or receive a cardiac transplant within a few years of diagnosis [46, 47].

2.3.1 Genetics of Restrictive Cardiomyopathy

RCM may be associated with systemic disease but is most often idiopathic. The results of recent molecular genetic investigations have revealed that a substantial proportion of RCM without associated systemic disease is caused by mutations in sarcomeric disease genes that have been associated with HCM, DCM, and noncompaction cardiomyopathy. Mutations in several genes have been found to cause familial RCM. Mutations in the cardiac troponin I gene (*TNNI3*) are the major causes of this condition. Mogensen et al. reported a large family in which individuals were affected by either idiopathic RCM or HCM. Linkage analysis to selected sarcomeric contractile protein genes identified *TNNI3* as the likely disease-causing gene [43]. The fact that *TNNI3* mutations were identified in a significant proportion of such patients indicates that idopathic RCM is part of the clinical expression of sarcomeric contractile protein disease and of HCM [48]. Chen et al. also reported a *TNNI3* missense mutation (R192H) in idiopathic RCM in a 12-year-old Chinese girl [49]. The case further improves the knowledge of the causes of cardiomyopathy disease and shows that the spectrum of sarcomeric gene mutations may be involved in pediatric RCM. Mutations in several sarcomeric genes have been reported subsequently in patients with RCM. Peddy et al. reported a novel presentation of a mutation in the *TNNT2* gene causing RCM in a child [50]. These findings, together with a possibly aggravating mutation in *MYBPC3*, further expand the phenotypic spectrum caused by *TNNT2* and *MYBPC3* mutations [50]. Mutations involving cardiac troponin T likely lead to altered calcium sensitivity of the troponin complex, contributing to altered relaxation of the cardiac muscle, which is the hallmark of RCM [50]. Ware et al. first described the *MYH7* mutation in a child with RCM. These findings expand the phenotypic presentation of mutations in this sarcomeric protein, previously well known to cause both HCM and DCM [51]. RCM can also be a manifestation of desmin-related cardiomyopathy, caused by mutations in *DES* [52]. A point mutation in *BAG3* is known to cause fulminant skeletal myopathy and early death in knockout mice and myofibrillar myopathies with RCM or HCM in humans [32]. Familial RCM also can be caused by a mutation in *ACTC1* [47].

2.3.2 NGS of Restrictive Cardiomyopathy

Genetic testing may be useful to confirm the diagnosis of familial RCM [53]. NGS is perhaps one of the most exciting advances in the last decade in the field of life sciences and biomedical research. With the availability of massive parallel sequencing, a human DNA blueprint can be decoded to explore hidden information, with reduced time and cost. On the basis of previous studies, cardiomyopathy gene panels have been developed whereby a selected number of genes can be targeted and sequenced on NGS platforms commercially and are available for patients on demand (Table 2) [54]. The massively parallel approach of NGS technology allows the simultaneous sequencing of many genes; it is, therefore, an ideal technology for molecular genetic diagnosis of RCM and other genetically heterogeneous disorders.

2.4 Arrhythmogenic Right Ventricular Dysplasia/ Cardiomyopathy (ARVD/C)

ARVD/C (ARVD/C9, OMIM# 609040) is a rare, inherited cardiomyopathy characterized by ventricular arrhythmias, SCD, and abnormalities of the right (and less commonly left) ventricular structure and function. It is an inherited condition with an estimated prevalence of 1 per 5000. Patients usually present during the second to fifth decades of life with palpitations, lightheadedness, syncope, or sudden death [55].

2.4.1 Genetics of ARVD/C

ARVD/C is inherited as an autosomal-dominant trait, meaning that the risk of a family member inheriting an abnormal gene is 50% for all offspring of the genetically affected proband, whether male or female [56]. Approximately 50–60% of patients with ARVD/C are estimated to have a mutation in genes associated with cardiac desmosomes [57]. The desmosomal proteins involved are desmoplakin (encoded by the *DSP* gene), plakophilin 2 (PKP2), DSG2, DSC-2, and junctional plakoglobin (JUP). The nondesmosomal proteins related to ARVD/C are DES, transmembrane protein 43 (TMEM43), transforming growth factor β-3 (TGFβ3), LMNA, TTN, PLN, and α-T-catenin (CTNNA3). A mutation in the cardiac ryanodine receptor, encoded by the ryanodine gene 2 (*RyR2*), was identified in a patient affected by ARVD/C[58]. To date, the majority of pathogenic mutations have been identified in genes coding for desmosomal proteins, with the *PKP2* gene being responsible for approximately 35–40% of cases. Mutations in the genes *DSP*, *DSG2*, and *DSC2* are responsible for nearly 15–20% of ARVD/C cases. The most prevalent form of ARVD/C (type 9) is caused by mutations in the *PKP2* gene, which encodes the PKP2, an essential armadillo repeat protein located in the outer dense plaque of cardiac desmosomes that interacts with numerous other cell adhesion proteins. How *PKP2* mutations perturb cardiac desmosome assembly and function in ARVD/C is unknown. It has been speculated that lack of PKP2 or incorporation of mutant PKP2 into cardiac desmosomes impairs cell-cell contacts and, as a consequence, disrupts adjacent cardiomyocytes, particularly in response to mechanical stress or stretch. Mura et al. reported the first case of copy number variation (CNV) identified in an ARVC family using high-density single nucleotide polymorphisms (SNPs) arrays. A heterozygous deletion of about 122.5 Kb on chromosome 12p11.21, encompassing the entire *PKP2* gene, was detected in all affected family members. Hence, *PKP2* deletions may arise by non-recurrent rearrangements due to replication-based mechanisms of DNA repair [58].

2.4.2 NGS of ARVD/C

For ARVD/C NGS panel (Table 2), all coding exons of the related genes, plus approximately 20 nucleotides of flanking DNA for each exon can be sequenced. Sequencing is accomplished by capturing specific regions with an optimized solution-based hybridization kit, followed by massively parallel sequencing of the captured DNA fragments. Additional Sanger sequencing is performed for any regions not captured or with insufficient number of sequence reads. All pathogenic and undocumented variants are confirmed by Sanger sequencing.

2.5 *Left Ventricular Noncompaction (LVNC)*

LVNC (LVNC OMIM#604169), also known as spongy myocardium, is a distinct form of cardiomyopathy occurring in-utero when segments of spongy myocardium fail to transform into compact, mature musculature, resulting in prominent myocardial trabeculae, deep intra-trabecular recesses, and decreased cardiac function [59]. It was first described in 1990 [60]. Prevalence of LVNC is estimated to be approximately 0.25% of adults referred for echocardiography [61]. The clinical manifestation of LVNC is highly variable, ranging from no symptoms to a progressive deterioration in cardiac function that results in congestive HF, arrhythmias, thromboembolic events, and SCD [62].

2.5.1 Genetics of LVNC

LVNC is a genetically heterogeneous cardiomyopathy, with both familial and sporadic forms. Autosomal dominant is the most common form of inheritance, but autosomal recessive, X-linked, and maternally inherited (matrilineal) mitochondrial inheritance have been reported. More than ten genes have been described in LVNC, and some of the genetic mutations are associated with overlapping phenotypes with HCM and DCM [63]. It is most commonly attributed to mutations in seven genes (*TAZ, DTNA, LDB3, LMNA, SCN5A, MYH7*, and *MYBPC3*). With additional contributing variants reported in rare instances (*ACTC1, TNNT2, MIB1, PRDM16*, and *TPM1*), all LVNC loci encode proteins involved in cellular energy, muscle development, and ion channel formation, or are components of the muscle filaments. Despite these advances, many of the causative LVNC genes have yet to be identified. Mutations in *MYH7* are the single most common cause of LVNC, accounting for 8–13% of cases, with the remainder of the genes reported to be mutated in rare cases.

2.5.2 NGS of LVNC

LVNC is predominantly a genetic cardiomyopathy with variable presentation ranging from asymptomatic to severe. Accordingly, the diagnosis of LVNC requires genetic counseling, DNA diagnostics, and cardiological family screening. Molecular diagnosis offers reliable identification of asymptomatic relatives at risk. In the absence of an identified genetic cause for LVNC, or when relatives decline DNA testing, cardiological screening remains the appropriate method to identify familial disease. Combining genetic testing with clinical screening of family members can greatly enhance the detection rate of familial LVNC to 67% [64]. LVNC genetic testing has no prognostic and therapeutic implications, as clear genotype–phenotype correlations have not been identified [53].

Multi-gene testing using NGS is a highly accurate and reproducible approach to the routine molecular genetic testing of patients with cardiomyopathies (Table 2). The method has led to dramatic improvement in efficiency and speed of gene sequencing, as well as markedly reduced costs for clinical genetic testing [65]. Schaefer et al. reported the case of an infant who died at 6 months with a LVNC. Molecular analysis was performed using a NGS sequencing strategy that leads to identify compound heterozygous pathogenic mutations in the *MYBPC3* gene. The new approach using NGS strategy allows a rapid molecular diagnosis for families presenting with cardiomyopathies with a broad coverage of the known disease-causing genes at a reasonable cost. Gene panel size could include a large number of genes in order to identify gene variants that could explain more LNVC cases than present [66].

3 Inherited Primary Arrhythmia Syndromes

Inherited primary arrhythmia syndromes comprise a group of syndromes with unique genetic abnormalities and presentations but with very similar clinical outcomes and complications, the most terrifying of which are life-threatening arrhythmias and SCD. Such diseases usually are of autosomal-dominant inheritance, have a structurally normal heart, often affect otherwise healthy persons, and can generally be treated successfully if recognized early. Patients commonly have affected relatives who are still asymptomatic. Inherited primary arrhythmia syndromes are rare, but they often occur in patients of young age. Early diagnosis can markedly reduce the risk of SCD. The common heritable arrhythmias include LQTS, short QT syndrome (SQTS), catecholaminergic polymorphic ventricular tachycardia (CPVT), and BrS. These disorders are thought to be responsible for 10–15% of cases of sudden unexplained death in young adults and children [67].

Heart rhythm is generated by an elegant interplay of ions at the cellular level. Ion channels for sodium (Na+), potassium (K+), and calcium (Ca2+) in the myocardial cellular membrane are responsible for allowing this interplay across the membrane. The inherited abnormalities of the genes encoding these ion channel proteins, or "accessory" proteins, essential to the ion channel functions cause dysfunction in the ion channels and life-threatening arrhythmias [68]. The current trend in the management of these potentially deadly disorders is to use pharmacotherapy (antiarrhythmic agents) and defibrillators for the prevention of sudden death; however, targeted therapy at a molecular level appears to be the path of the future.

Recent advances in DNA sequencing have led to genetic testing being applied widely in the clinical setting and in research of inherited primary arrhythmia syndromes. In particular, NGS allows the large-scale and rapid assessment of target gene panels, whole exome, and entire genome. These assessments have enabled us to define and pinpoint many such disorders, which previously were labeled as idiopathic, to specific genes on various chromosomes and have provided extraordinary help in making definite diagnoses and risk stratifications, as well as guiding management, of these diseases. Although WES (whole exome sequencing) and WGS (whole genome sequencing) approaches are beginning to compete, targeted genetic testing is currently the cornerstone of testing in the clinical setting [21, 67, 69]. Our understanding of the genetic architecture and causal mechanisms for many inherited primary arrhythmia syndromes remains limited. NGS has facilitated our exploration in inherited primary arrhythmia syndromes, though it has introduced numerous challenges with respect to the interpretation of genetic variations and has unleashed a flood of biological data of unknown clinical significance. Herein, we briefly summarize the application of NGS in the clinical setting of LQTS and BrS.

Table 3 Long QT syndrome by genetic subtype

LQTS subtype	Gene	OMIM	Protein	Functional effect of mutation	Frequency of cases (%)
LQT1	*KCNQ1*	192,500	Alpha-subunit of *I*Ks	Loss-of-function, reduced *I*Ks	30–35
LQT2	*KCNH2*	613,688	Alpha-subunit of *I*Kr	Loss-of-function, reduced *I*Kr	25–30
LQT3	*SCN5A*	603,830	Alpha-subunit of *I*Na	Gain-of-function, increased late *I*Na inward current	5–10
LQT4	*ANK2*	600,919	Ankyrin-B; links membrane proteins with underlying cytoskeleton	Loss-of-function, disrupts multiple ion channels	<1
LQT5	*KCNE1*	613,695	Beta-subunit of *I*Ks	Loss-of-function, reduced *I*Ks	<1
LQT6	*KCNE2*	613,693	Beta-subunit of *I*Kr	Loss-of-function, reduced *I*Kr	<1
LQT7	*KCNJ2*	600,681	Alpha-subunit of *I*K1	Loss-of-function, reduced *I*K1	<1
LQT8	*CACNA1c*	601,005	Alpha-subunit of *I*CaL	Gain-of-function, increased *I*CaL	Rare
LQT9	*CAV3*	611,818	Caveolin-3; a scaffolding protein in caveolae	Increased late *I*Na inward current	<1
LQT10	*SCN4B*	611,819	Beta 4-subunit of *I*Na	Gain-of-function, increased late *I*Na inward current	Rare
LQT11	*AKAP9*	611,820	A kinase-anchor protein-9; sympathetic *I*Ks activation	Loss-of-function, reduced *I*Ks	Rare
LQT12	*SNTA1*	612,955	Alpha1-syntrophin; regulation of *I*Na	Increased late *I*Na inward current	Rare
LQT13	*KCNJ5*	613,485	Kir 3.4	Loss-of-function, reduced *I*KACh	Rare
LQT14	*CALM1*	616,247	Calmodulin-1	Altered calcium signaling	<1
LQT15	*CALM2*	616,249	Calmodulin-2	Altered calcium signaling	<1

Abbreviation: *LQTS* long QT syndrome

3.1 *Long QT Syndrome (LQTS)*

LQTS is a genetic disease characterized by its hallmark electrocardiographic feature of QT prolongation and T wave abnormalities, its trademark arrhythmia of *torsades de pointes* (TdP), and its predisposition for syncope, "seizures," and SCD in young individuals with structurally normal hearts. Most LQTS index cases manifest diagnostic QT prolongation on their resting 12-lead ECG, whereas approximately 10–40% of patients with LQTS (index cases and relatives) have non-diagnostic QT intervals at rest and what are referred to as "normal QT interval" or "concealed LQTS" [70–73]. With an estimated incidence of at least 1 in 2500 people [74], LQTS is underscored by marked clinical heterogeneity ranging from a lifelong asymptomatic state to sudden death during infancy. LQTS is more likely to express itself before puberty in males and after puberty in females [73, 75, 76]. Besides age and sex, the degree of QTc prolongation is associated with likelihood of a first LQT-triggered cardiac event (syncope or aborted cardiac arrest), and occurrence of such cardiac events, particularly while on therapy, is a strong predictor of recurrences [77–79]. Among symptomatic index cases, the untreated 10-year mortality rate is approximately 50% [78–81].

3.1.1 Genetics of LQTS

Since the sentinel discovery of the primary LQTS-causative genes in 1995, genetic studies have identified 15 genes for which pathogenic mutations cause LQTS. These genes encode potassium-channel proteins, sodium-channel proteins, calcium channel-related factors, and membrane adaptor proteins on chromosomes 3, 4, 7, 11, 12, 17, 20, and 21 (Table 3) [68]. Patients with LQT1, LQT2, and LQT3 genotypes with mutations in *KCNQ1*, *KCNH2*, and *SCN5A*, respectively, account for more than 92% of patients with genetically confirmed LQTS. As many as 15–20% of patients with LQTS remain genetically elusive [69]. The majority of LQTS is inherited as an autosomal-dominant trait, the Romano-Ward syndrome. Sporadic (or *de novo*) alterations occur in less than 5–10% of LQTS cases. The autosomal-recessive form of LQTS, also known as Jervell and Lange-Nielsen syndrome, which probably affects fewer than 1 in a million people and involves the same (homozygous) or different (compound heterozygous) *KCNQ1* mutations from both parents, is more virulent and is associated with deafness [82, 83]. Mutations in *KCNJ2* (Kir2.1, LQT7) result in the neurologic musculoskeletal Andersen-Tawil syndrome, with associated QT prolongation. The other LQTS genotypes (LQT4 and LQT8–13) have each been identified in only a few families or in single individuals. Common variants in the LQTS genes (e.g., SNPs), and in some cases unrelated genes, are thought to contribute to the variable penetrance of LQTS within affected family members having the same gene mutation [84].

3.1.2 NGS of LQTS

The clinical utility of NGS for inherited arrhythmia syndromes is to detect mutations of the genes that are associated with the diseases for confirmation of a suspected condition, or cascade screening in relatives of probands (index affected family members) who carry a disease-causing mutation. The identification of the genetic basis of Mendelian diseases associated with arrhythmia has allowed the integration of this information into the diagnosis and clinical management of patients and at-risk family members. The rapid expansion of genetic testing options and the increasing complexity involved in the interpretation of results create unique opportunities and challenges. Competency to incorporate genetics into clinical management and to provide appropriate family-based risk assessment and information is needed.

Four NGS platforms are currently employed in genetic testing. The choice of NGS platform depends on the aim and the type of study being performed. Different NGS platforms have been used for the genetic screening of inherited primary arrhythmia syndromes. Li et al. [85] used two different platforms, MiSeq (Illumina, San Diego, CA, USA) and Ion Torrent PGM (Life Technologies Ldt, Paisley, UK), to investigate five genes (*KCNQ1, KCNH2, SCN5A, KCNE1, KCNE2*) associated with LQTS. Their results showed that Ion PGM is less expensive and fast, whereas MiSeq has a lower run time and higher capacity. In any case, both platforms are valid instruments for a molecular diagnosis of inherited cardiac diseases because they are faster, less expensive, and more comprehensive than traditional genetic diagnostic tests. A literature search in PubMed with the search query "next generation sequencing and inherited arrhythmia syndromes" has found that 11 of 15 studies (73%) used Illumina sequencing system to screen LQTS genes, indicating that the NGS platform is more valid and practical in the genetic testing of inherited primary arrhythmia syndromes.

Targeted gene panels, WES, and WGS have been applied in LQTS genetic testing. Targeted gene panel is currently the main approach for genetic diagnosis. In 2011, the HRS/EHRA Expert Consensus Statement recommended comprehensive or LQT1–3 targeted LQTS genetic testing for any patients with strong clinical index of suspicion for LQTS, and mutation-specific genetic testing for family members and other appropriate relatives [21]. The initial testing for LQTS likely does not need to include a broad sequencing panel, considering that most affected individuals have variations in only one of a small number of known genes [86]. However, the currently-available panels for diagnosis of LQTS are able to identify a causative variant in only 72% of tested patients [87]. For accurate diagnosis of patients with rare diseases, especially those with atypical clinical manifestation such as atypical Timothy syndrome (TS2), comprehensive and unbiased mutation screening such as WES and WGS may be required [88, 89]., The advantage of the more comprehensive testing is the potential for increased sensitivity of mutation detection. The major drawbacks of more comprehensive testing include potentially increased costs and increased detection of genetic variants of unknown clinical significance.

Numerous genotype–phenotype relationships in LQTS have been discovered and include genotype-suggestive ECG patterns, genotype-suggestive arrhythmogenic triggers, genotype-based natural histories, and genotype-specific responses to pharmacotherapy [90–93]. Most of these relationships pertain to the major LQTS genotypes: LQT1, LQT2, and LQT3. Compared to those with the more common potassium channel loss-of-function subtypes (LQT1 and LQT2), patients with LQT3 appear to have the highest mortality per event [91]. Within each of the two major LQTS genotypes (LQT1 and LQT2), the mutation's location within the protein and its functional sequelae have been proposed as independent risk factors [94, 95]. Specific genetic variants, such as the Jervell and Lange-Nielsen syndrome[83] and the extremely rare Timothy syndrome (LQT8) [96], are highly malignant, manifest with major arrhythmic events very early, and respond poorly to therapies. Within the most common genetic groups, specific locations, types of mutations, and degree of mutation dysfunction are associated with different risks. Mutations in the cytoplasmic loops of LQT1 [70, 71], LQT1 mutations with dominant-negative ion current effects, and mutations in the pore region of LQT2 are associated with higher risks[67, 95], and the same is true even for some specific mutations with an apparently mild electrophysiological effect [74]. By contrast, mutations in the C-terminal region tend to be associated with a mild phenotype [97]. Concealed mutation-positive patients are at low, but not zero, risk for spontaneous arrhythmic events. A major risk factor for patients with asymptomatic, genetically diagnosed LQTS comes from drugs that block the I_{Kr} current and by conditions that lower their levels of plasma potassium. Among genotyped patients, LQT1 males, who are asymptomatic at a young age [98], are at low risk of becoming symptomatic later in life, whereas females, and especially LQT2 females, remain at risk even after age 40. Thus, the genetic test result has joined traditional risk factors (i.e., gender, age at onset, QTc at rest, syncope) as independent prognostic risk factors [99, 100]. NGS has facilitated the gene sequencing, enabled application of genetic testing in diagnosis of LQTS, and continues to increase its sensitivity and accuracy.

Fifteen genetic forms of LQTS have been identified. Their risk factors and managements are different and largely depend on their genotype as aforementioned. The lifestyle of the patient should be managed according to the diagnosis. Patients with LQT1 should void strenuous exercise, and exposure to abrupt load noise should be reduced for patients with LQT2. Beta blocker pharmacotherapy is the primary treatment for the management of most patients with LQTS [99–102]. Among the three most common genotypes, beta blockers are extremely protective in patients with LQT1 and moderately protective in those with LQT2 [103]. In contrast, targeting of the pathologic, LQT3-associated late sodium current with propranolol (as the preferred beta blocker) and the possible addition of mexiletine, flecainide, or ranolazine represents the preferred pharmacotherapeutic option for management of LQT3 [104–106]. Prophylactic ICD therapy should be considered in very-high-risk patients such as symptomatic patients with two or more gene mutations, including those with the Jervell and Lange-Nielsen variant with congenital deafness [83]. Gene-specific LQTS therapies with sodium channel blockers have been utilized to a limited extent in high-risk patients with LQTS refractory to beta-blockers or in patients with recurrent events despite use of ICD and left cardiac sympathetic denervation (LCSD) therapies [104]. The use of sodium channel blockers has generally been limited to LQT3 patients. Therefore, the genetic diagnosis can guide the treatment of most LQTS patients.

3.2 *Brugada Syndrome (BrS)*

BrS is characterized by the presence of a typical electrocardiographic (ECG) pattern (right bundle branch block and persistent ST-segment elevation in right precordial leads) and is associated with a high risk of SCD [107]. Most such patients remain completely asymptomatic. The symptoms usually appear in patients who are around 40 years of age. Males are more often symptomatic than are females, probably from the influence of hormones and gender distribution of ion channels across the heart. The prevalence of the disease manifesting with clinical symptoms is estimated to be 1 in 5000 to 10,000 in Western countries and may be more prevalent in South Asia. Phenotypic expression of BrS rarely occurs in children. Inheritance of BrS is via an autosomal-dominant mode of transmission.

3.2.1 Genetics of Brugada Syndrome

Mutations in 19 genes, most of which are located in the SCN5A gene, encoding the alpha-subunit of the Na(+) cardiac channel have been identified as associated with the Brugada phenotype (Table 4) [108]. These mutations cause either a decrease in inward sodium or calcium current or an increase in outward potassium currents, resulting in an outward shift in the balance of current active during the early phases of the action potential [108]. However, genetic bases of BrS are only partially understood. More than 70% BrS patients still remain genetically undiagnosed.

3.2.2 NGS of Brugada Syndrome

A search in the PubMed database with the query "next generation sequencing and Brugada syndrome" found that 77% (7/9) of studies employed the Illumina sequencing system to perform DNA sequencing and 89% of these studies (8/9) used targeted gene panels to screen known and suspicious genes associated with BrS, indicating that the Illumina sequencing system is a valid platform for genetic testing of BrS and that a targeted gene panel is currently the main approach for the genetic diagnosis of BrS. These studies either confirmed BrS-causing mutation genes or discovered new potential BrS candidate genes, demonstrating that clinical genetic diagnosis of BrS is justified. As genetic diagnoses of more than 70% of BrS cases still remain elusive, more comprehensive genetic testing such as WES or WGS may be required. With advances in technology, NGS will continue to provide more data and information for clinical decision-making in the era of personalized medicine.

Table 4 Brugada syndrome associated genes

Type	Gene	OMIM	Locus	Protein	Ion Channel	% of Probands
BrS1	*SCN5A*	601,144	3p21	Nav1.5, type V alpha subunit of *I*na	↓ *I*Na	11–28%
BrS2	*GPD1-L*	611,777	3p24	glycerol-3-phosphate dehydrogenase 1-like	↓ *I*Na	Rare
BrS3	*CACNA1C*	114,205	12p13.3	Cav1.2, alpha 1C subunit of *ICa*	↓ *I*Ca	6.60%
BrS4	*CACNB2b*	611,876	10p12.33	Cavß2b, beta 2 subunit of *I*Ca	↓ *I*Ca	4.80%
BrS5	*SCN1B*	612,838	19q13.1	Navß1, type I beta subunit of *I*Na	↓ *I*Na	1.10%
BrS6	*KCNE3, MiRP2*	613,119	11q13–14	beta subunit 3 of *Ikr*	↑ *I*to	Rare
BrS7	*SCN3B*	613,120	11q23.3	Navß3, type III beta subunit of *I*Na	↓ *I*Na	Rare
BrS8	*KCNJ8*	613,123	12p11.23	Kir6.1, subfamily J, member 8 of *I*kr	↑ *I*K-ATP	2%
BrS9	*CACNA2D1*	616,399	7q21.11	Cavα2δ, alpha2/delta subunit 1 of *I*Ca	↓ *I*Ca	1.80%
BrS10	*KCND3*	605,411	1p13.2	Kv4.3, subfamily D, member 3 of *I*Ca	↑ *I*to	Rare
BrS11	*RANGRF, MOG1*	607,954	17p13.1	RAN guanine nucleotide release fact	↓ *I*Na	Rare
BrS12	*SLMAP*	602,701	3p21.2-p14.3	sarcolemma associated protein	↓ *I*Na	Rare
BrS13	*ABCC9, SUR2A*	601,439	12p12.1	ATP binding cassette subfamily C member 9	↑ *I*K-ATP	Rare
BrS14	*SCN2B*	601,327	11q23	Navß2, type II beta subunit of *I*na	↓ *I*Na	Rare
BrS15	*PKP2*	602,861	12p11	Plakophillin-2	↓ *I*Na	Rare
BrS16	*FGF12*	601,513	3q28	fibroblast growth factor 12	↓ *I*Na	Rare
BrS17	*SCN10A*	604,427	3p22.2	Nav1.8, type X alpha subunit of *I*na	↓ *I*Na	16.70%
BrS18	*HEY2*	604,674	6q22.31	hes-related family bHLH transcription factor with YRPW motif 2	↓*I*Na	Rare
BrS19	*SEMA3A*	603,961	7p12.11	semaphorin 3A	↑ *I*to	Rare

Abbreviation: *BrS* Brugada syndrome

The diagnosis of BrS is a clinically made and requires the signature type 1 Brugada ECG pattern in combination with one or more clinical variables such as unexplained syncope and family history of premature, unexplained sudden death [109]. Genetic testing is not involved in the diagnosis, but the identification of a causative mutation may help confirm a clinically uncertain diagnosis. Genetic testing in families with an underlying causal gene defect may play a decisive role in whom should precautions be taken and who should be monitored [21].

Genetic testing is recommended for support of the clinical diagnosis, for early detection of relatives at potential risk, and particularly for the purpose of advancing research and, in turn, our understanding of genotype-phenotype relations. The role of genetic markers in risk stratification of BrS remains a matter of debate [110]. According to the latest guidelines (HRS/EHRA consensus statement) [21], genetic testing is recommended (Class I) for relatives of an index case with an identified BrS-causative mutation. Genetic testing can be useful (Class II) for any patient in whom a cardiologist has established a clinical index of suspicion for BrS based on examination of the patient's clinical history, family history, and expressed ECG (resting 12-lead ECGs or provocative drug challenge testing) phenotype [21]. On the contrary, the latest joint CCS/CHRS position paper [111] and other studies [112] suggest that a Type 1 BrS ECG alone should be sufficient indication (Class I recommendation) for genetic testing. In an individual expressing an isolated Type 2 or 3 BrS pattern, genetic testing has a Class III recommendation [21]. Although identification of a genotype may not be helpful in the approach to therapy at present, it could be argued that with additional evidence, some genotypes may offer innovative therapeutic strategies (e.g., use of $I_{K\text{-}ATP}$ blockers in cases involving a gain of function of $I_{K\text{-}ATP}$ or use of I_{Kr} blockers in cases of I_{Kr} gain of function) [108].

4 Congenital Heart Disease (CHD)

CHD, the most common human congenital defect, includes a large set of structural and functional deficits that arise during cardiac embryogenesis and is a leading cause of death in infants. Reller et al. demonstrated that in North America, CHD occurs in 8.1 per 1000 live births [113]. CHD is also identified in 10% of stillbirths and is presumed to be a substantive cause of early fetal demise [114]. Additionally, undiagnosed mild malformations of the heart often appear later in adulthood or remain undiagnosed for life. The number of adults with some form of CHD is also growing rapidly due to the advances in diagnostics, therapeutic methods, and surgical treatments that are more effective, allowing patients with CHD to achieve adulthood [115]. The improvement of molecular genetic technologies has led to important advances in understanding the pathogenesis of CHDs and has also provided evidence of how common genetic variation can influence the risk of certain types of CHD.

CHD is a complex disease that often displays genetic heterogeneity and variable penetrance and expression; it encompasses a broad category of anatomic malformations that can range from a small septal defect or leaky valve to a severe malformation such as a single ventricle, requiring extensive surgical repair or leading to death. Genetically, CHD mutations may occur as autosomal-dominant, autosomal-recessive, or X-linked traits that are expressed with high penetrance and with variable clinical manifestations. Several classification systems exist for describing CHD, the most common of which is purely clinical: *cyanotic* if the malformation results in deoxygenated blood bypassing the lung and causes cyanosis (blue patient) or *non-cyanotic* if the malformation does not result in cyanosis [115]. Table 5 shows the categories of CHD classifications according to IPCCC (International Pediatric and Congenital Cardiac Code) and the most common diagnoses within each category.

CHD may be categorized as syndromic or non-syndromic depending on the coexistence of non-cardiac anomalies. Among the many well-known examples of syndromic CHD are Holt-Oram syndrome, Alagille syndrome, and Noonan syndrome. Many of these syndromes have a monogenic mode of inheritance, and genetic causes are well established. In contrast, non-syndromic or isolated CHD is related only to heart defects and accounts for the majority of CHD cases. It occurs sporadically, and families with clear monogenic inheritance of non-syndromic CHD are scarce. The high heritability of CHD suggests a strong genetic component, and numerous genes have been linked to both syndromic and non-syndromic forms of CHD [116].

4.1 Syndromic CHD

Cardiac malformations are among the most prevalent malformations in congenital syndromes. Epidemiologic and population-based studies estimate that syndromic cardiovascular malformations account for approximately 25% of cases [5]. CHD syndromes can be due to chromosome dosage disorders, large chromosomal deletions, small micro-deletions, CNVs, or single gene defects [115]. Table 6 summarizes the most common syndromes that include CHD as a primary manifestation as well as their known genetic causes.

With regard to syndromic CHD, the alterations categorized as single genes defects are the ones that can be detected using NGS technology because causing variants are comprised of single-point mutations or small insertions/deletions that are easily detected using this methodology. For example, mutations in the genes related to the RAS/mitogen activated protein kinase (MAPK) pathway, which is important for control of cell proliferation and differentiation, cause dysregulation of this pathway and result in a spectrum of disorders known as "RASopathies," including Noonan and Costello syndromes [5]. Other molecular mechanisms responsible for syndromic CHD have also been identified; an example is Holt-Oram syndrome, characterized by atrial and ventricular septal defects, progressive atrioventricular conduction system disease, and upper limb malformations associated with mutations in *TBX5*, a member of the T-box gene family that encodes transcription factors that contain a conserved DNA-binding motif. T-box proteins

Table 5 IPCCC classification of congenital heart disease and most common diagnoses

Classification category	Diagnoses
Abnormalities of position and connection of the heart	Dextrocardia Atrial Situs Inversus Double Inlet Left Ventricle (DILV) Double Inlet Right Ventricle (DIRV) Transposition of the Great Arteries (TGA) Double Outlet Left Ventricle (DOLV) Double Outlet Right Ventricle (DORV) Common Arterial Trunk (CAT), aka Truncus Arteriosus (TA)
Tetralogy of Fallot and variants	Tetralogy of Fallot (TOF) Pulmonary Atresia (PA) and Venticular Septal Defect (VSD)
Abnormalities of great veins	Supervior Vena Cava (SVC) abnormality Inferior Vena Cava (SVC) abnormality Coronary sinus abnormality Total Anomalous Pulmonary Venous Connection (TAPVC) Partially Anomalous Pulmonary Venous Connection (PAPVC)
Abnormalities of atriums and atrial septum	Atrial Septal Defect (ASD) Patent Foramen Ovale (PFO)
Abnormalities of AV valves and AV septal defect	Tricuspid Regurgitation (TR) Tricuspid Stenosis (TS) Ebstein's anomaly Mitral Regurgitation (MR) Mitral Stenosis (MS) Mitral Valve Proplapse (MVP) Atrioventricular Septal Defect (AVSD)
Abnormalities of ventricles and ventricular septum	Single Ventricle Ventricular imbalance: dominant LV + hypoplastic RV, or dominant RV + hypoplastic RV Aneurysm (RV, LV, or septal) Hypoplastic Left Heart Syndrome (HLHS) Double Chambered Right Ventricle (DCRV) Ventricular Septal Defect (VSD)
Abnormalities of VA valves and great arteries	Aortopulmonary Window (AP Window) Pulmonary Stenosis (PS), valvar or subvalvar Pulmonary Artery Stenosis (PAS) Aortic Stenosis (AS), valvar or subvalvar Aortic Insufficiency (AI) Bicuspid Aortic Valve (BAV) Supravalvar Aortic Stenosis (SVS) Coarctation of the Aorta (COA) Interrupted Aortic Arach (IAA)
Abnormalities of coronary arteries, arterial duct and pericardium; AV fistulae	Anomalous Origin of Coronary Artery from Pulmonary Artery (ALCAPA) Patent Ductus Arteriosus (PDA)

Table 6 Examples of syndromic CHD and genetic causes [115]

Syndrome with CHD	Genetic cause for CHD	% with CHD
Disorders of chromosome dosage		
Edward Syndrome (Trisomy 13)	Unknown	80–100%
Patau Syndrome (Trisomy 18)	Unknown	80–100%
Down Syndrome (Trisomy 21)	Unknown	40–50%
Turner (Monosomy X)	Unknown	20–50%
Chromosomal microdeletions		
Di Georges Syndrome	22q11.2 deletion resulting in absent *TBX1* gene	80–100%
Williams-Beuren Syndrome	Microdeletion of *ELN* gene; Mutations in *ELN* gene	80–100%
Single gene defects		
Holt-Oram Syndrome	*TBX5* mutations	85%
Alagille Syndome	*JAG1* or *Notch1/2* mutations; Microdeletion or rearrangement at 20p12 resulting in absent *JAG1* gene	>90%
Noonan Syndrome	Mutations in *PTPN11, SOS1, RAF1, KRAS, BRAF, MEK1, MEK2*, and *HRAS*.	80%
CHARGE Association	Mutations in *CHD7* and *SEMA3E*; Microdeletion at 22q11.2	85%
Char Syndrome	Mutations in *TFAP2B*	100%
Ellis-can Creveld Syndrome	Mutations in *EVC* or *EVC2*	60%
Cardiofaciocutaneous Syndrome	Mutations in *KRAS, BRAK, MEK1*, or *MEK2*; Microdeletion at 12q21.2-q22	71%
Costello Syndrome	Mutations in *HRAS* (overlap with Noonan and Cardiofaciocutaneous Syndrome)	63%

function in regulating cell-fate decisions and early pattern formation, and different gene family members contribute to organogenesis [117]. Another molecular mechanism implicated in syndromic CHD is related to ligands and receptor molecules; for example, Alagille syndrome, a multisystem disorder with heart, skeletal, liver, eye, and facial features, is caused by dominant mutations in the *JAG1*, a gene encoding a ligand in the Notch signaling pathway, or in the *NOTCH2* gene. The presence of mutations (frameshifts, nonsense, disrupted or cryptic splice signals, missense) in either gene reduces Notch signaling, a highly conserved pathway involved in lineage specification and cell-fate decision during development [118]. Other syndromes caused by alterations in single genes, related to embryonic development, can also display CHDs among their clinical features. Table 7 provides a list of genes and associated phenotypes that may also include CHD as one of the clinical characteristics.

Table 7 Single genes related to syndromic CHD

Gene	OMIM	Locus	Protein	Associated phenotype
B3GAT3	606,374	11q12.3	Beta-1,3-Glucuronyltransferase 3	Multiple joint dislocations, short stature, craniofacial dysmorphism, and CHD.
BCOR	300,485	Xp11.4	BCL6 Corepressor	Microphthalmia, syndromic 2, ASD
DTNA	601,239	18q12.1	Dystrobrevin, Alpha	Left ventricular noncompaction 1, with or without congenital heart defects
ESCO2	609,353	8p21.1	Establishment Of Sister Chromatid Cohesion N-Acetyltransferase 2	SC phocomelia syndrome
FANCA	607,139	16q24.3	Fanconi Anemia, Complementation Group A	Fanconi anemia
FANCC	613,899	9q22.32	Fanconi Anemia, Complementation Group C	Fanconi anemia
FANCD2	613,984	3p25.3	Fanconi Anemia, Complementation Group D2	Fanconi anemia
FANCE	613,976	6p21.31	Fanconi anemia, complementation group E	Fanconi anemia
FBN1	134,797	15q21.1	Fibrillin 1	Marfan syndrome
GPC3	300,037	Xq26.2	Glypican 3	Simpson-Golabi-Behmel syndrome, type 1
HOXA1	142,955	7p15.2	Homeobox A1	Athabaskan brainstem dysgenesis syndrome / Bosley-Salih-Alorainy syndrome
IGBP1	300,139	Xq13.1	Immunoglobulin (CD79A) Binding Protein 1	Corpus callosum, agenesis of, with mental retardation, ocular coloboma, micrognathia, VSD and PDA
MAP2K2	176,872	15q22.31	Mitogen-Activated Protein Kinase Kinase 2	Cardiofaciocutaneous syndrome 3
MID1	300,552	Xp22.2	Midline 1	Opitz GBBB syndrome, type I
MKKS	604,896	20p12.2	McKusick-Kaufman Syndrome	McKusick-Kaufman syndrome
MYCN	164,840	2p24.3	V-Myc Avian Myelocytomatosis Viral Oncogene Neuroblastoma Derived Homolog	Feingold syndrome

(continued)

Table 7 (continued)

Gene	OMIM	Locus	Protein	Associated phenotype
NF1	162,200	17q11.2	Neurofibromin 1	Neurofibromatosis, type 1
NIPBL	608,667	5p13.2	Nipped-B-Like Protein	Cornelia de Lange syndrome
NPHP3	608,002	3q22.1	Nephrocystin-3	Meckel syndrome, type 7
NUBPL	613,621	14q12	Nucleotide Binding Protein-Like	Mitochondrial complex I deficiency
PRKG1	176,894	10q11.2-q21.1	Protein Kinase, CGMP-Dependent, Type I	Familial thoracic aortic aneurysm
RAI1	607,642	17p11.2	Retinoic Acid Induced 1	Smith-Magenis syndrome
RBM10	300,080	Xp11.3	RNA Binding Motif Protein 10	TARP syndrome
RPSA	150,370	3p22.1	Ribosomal Protein SA	Isolated congenital asplenia
SALL1	602,218	16q12.1	Spalt-Like Transcription Factor 1	Townes-Brocks syndrome
SKI	164,780	1p36.33-p36.32	V-Ski Avian Sarcoma Viral Oncogene Homolog	Shprintzen-Goldberg syndrome
TCTN3	613,847	10q24.1	Tectonic Family Member 3	Joubert syndrome
TGFB2	190,220	1q41	Transforming Growth Factor, Beta 2	Loeys-Dietz syndrome
TGFBR2	190,182	3p24.1	Transforming growth factor-beta receptor type 2	MFS-like syndrome, Loeys-Dietz syndrome
ZMPSTE24	606,480	1p34.2	Zinc Metallopeptidase STE24	Restrictive dermopathy

4.2 *Non-syndromic CHD*

Non-syndromic or isolated CHD refers to conditions in individuals who do not present with either a genetic syndrome or extra cardiac anomalies and is the most prevalent form of CHD. Mutations in many genes have been associated with several CHD phenotypes, yet the evidence varies for each gene. Only a minority of CHD seems to be due to monogenetic mutations, and the majority occur sporadically [116]. Gene mutations can be classified as highly penetrant mutations in disease-causing genes, low-penetrance mutations in susceptibility genes, and common variants in CHD risk-genes. Transcription factor genes are the most common group of genes implicated in CHD. Other genes are part of signaling transduction pathways and structural components of the heart [115]. Additionally, recent studies have found a variety of CNVs, large deletions or duplications of DNA segments, that also contribute to non-syndromic CHD [119–121]. Table 8 presents a list some important disease-causing genes related to non-syndromic CHD.

Non-syndromic CHD occurs from low frequency (LOF) mutations in a variety of genes that encode molecules that participate in developmental signaling pathways. Many of the genes implicated in non-syndromic CHD (e.g., *GATA4*, *FOG2*, *NKX2.5*, *NKX2.6*, *ZIC3*, *CITED2*, *TBX1*, and *TBX20)* are transcriptional regulators of heart morphogenesis, Others such as *ZIC3*, *NODAL*, and *LEFTY2* are receptors and ligands involved in signaling pathways, which encode molecules that restrict the expression of Nodal-responsive genes to the left side of embryos [116]. Mutations in these genes disrupt normal signals that direct cardiac looping and cause a spectrum of heart malformations [122–124]. Additionally, the Notch signaling pathway is implicated in numerous developmental processes and participates in epithelial-to-mesenchymal transformation, a process that is critical for normal valvulogenesis. Mutations in *NOTCH1* typically cause malformations of the aortic valve [125]. Genes that encode cardiac structural proteins comprise another category of monogenic causes of non-syndromic CHD. Rare mutations in genes such as *MYH6*, *MYH7* (encoding the α and β cardiac myosin heavy chains, respectively), and *ACTC* (a cardiac actin) have been reported as rare causes of autosomal-dominant atrial septal defects, among other disorders [126].

Although families exhibiting autosomal-dominant inheritance of isolated cardiovascular malformations have been reported, many cases of non-syndromic CHD are unlikely to result from simple, single-gene disorders. Instead, many cases of CHD are likely the result of multiple genetic alterations that increase susceptibility to CHD and interact with environmental factors. High throughput analysis by NGS technology can allow comprehensive and simultaneous detection of several pathogenic variants in genes involved in developmental pathways important for cardiac morphogenesis, thereby providing insights into these more complex inheritance models in which the cumulative effect of numerous genetic risk factors leads to disease.

4.3 NGS of Congenital Heart Disease

NGS enables rapid analysis of large amounts of genetic information and is well suited to the study of complex diseases such as CHD. The techniques applied can be targeted (exome sequencing and disease-specific gene panels) or non-targeted (WGS) depending on the purpose of the study. Although these techniques use the same sequencing processes, their capture methods are different and each one presents a set of advantages and disadvantages [127]. Although WES has the technical advantage of screening variants in all genes, in a relatively short–time, it also involves considerable expense at the analytical level due to the many incidental findings that may be generated. Also, WES is inferior to targeted gene panels in

Table 8 Single genes related to non-syndromic CHD

Gene	OMIM	Locus	Protein	Associated phenotype
Transcription factors and cofactors				
ANKRD1	609,599	10q23.31	Ankyrin repeat domain	TAPVR
CITED2	602,937		c-AMP responsive element-binding protein	ASD; VSD
GATA4	600,576	8p23.1	GATA4 transcription factor	ASD, PS, VSD, TOF, AVSD, PAPVR
GATA6	601,656	18q11.2	GATA6 transcription factor	ASD, TOF, PS, AVSD, PDA, OFT defects, VSD
HAND2	602,407	4q34.1	Helix-loop-helix transcription factor	TOF
IRX4	606,199	5p15.33	Iroquois homeobox 4	VSD
MED13L	608,771	12q24.21	Mediator complex subunit 13-like	TGA
NKX2.5	600,584	5q35.1	Homeobox containing transcription factor	ASD, VSD, TOF, HLH, CoA, TGA, DORV, IAA, OFT defects
NKX2–6	611,770	8p21.2	Homeobox containing transcription factor	PTA
TBX1	602,054	22q11.21	T-Box 1 transcription factor	TOF, (22q11 deletion syndromes)
TBX5	601,620	12q24.21	T-Box 5 transcription factor	AVSD, ASD, VSD, (Holt Oram syndrome)
TBX20	606,061	7p14.2	T-Box 20 transcription factor	ASD, MS, VSD
TFAP2B	601,601		Transcription factor AP-2 beta	PDA, (Char syndrome)
ZIC3	300,265	Xq26.3	Zinc finger transcription factor	TGA, PS, DORV, TAPVR, ASD, HLH, VSD, Dextrocardia, L-R axis defects, Heterotaxy
ZFPM2 (FOG2)	603,693	8q23.1	Zinc Finger Protein, FOG Family Member 2	TOF, DORV
Receptors, ligands, and signaling				
ACVR1	102,576	2q24.1	BMP receptor	AVSD
ACVR2B	602,730	3p22.2	Activin receptor	PS, DORV, TGA, dextrocardia
ALDH1A2	603,687	15q21.3	Retinaldehyde dehydrogenase	TOF
CFC1	605,194	2q21.1	Cryptic protein	TOF; TGA; AVSD; ASD; VSD; IAA; DORV
CRELD1	607,170	3p25.3	Epidermal growth factor-related proteins	ASD; AVSD
FOXH1	603,621	8q24.3	Forkhead activin signal transducer	TOF, TGA
GDF1	602,880	19p13.11	Growth differentiation factor-1	Heterotaxy, TOF, TGA, DORV

(continued)

Table 8 (continued)

Gene	OMIM	Locus	Protein	Associated phenotype
GJA1	121,014	6q22.31	Connexin 43	ASD, HLH, TAPVR, (Oculodentodigital dysplasia)
JAG1	601,920	20p12.2	Jagged-1 ligand	PAS, TOF, (Alagille syndrome)
LEFTY2	601,877	1q42.12	Left-right determination factor	TGA, AVSD, IAA, CoA, L-R axis defects, IVC defects
NODAL	601,265	10q22.1	Nodal homolog (TGF-beta superfamily)	TGA, PA, TOF, DORV, dextrocardia, IVC defect, TAPVR, AVSD
NOTCH1	190,198	9q34.3	NOTCH1 (Ligand of JAG1)	BAV, AS, CoA, HLH
PDGFRA	173,490	4q12	Platelet-derived growth factor receptor alpha	TAPVR
PTPN11	176,876	12q24.13	Protein Tyrosine Phosphatase, Non-Receptor Type 11	AVSD
SMAD6	602,931	15q22.31	MAD-related protein	BAV, CoA, AS
TAB2	605,101	6q25.1	TGF-beta activated kinase	OFT defects
TDGF1	187,395	3p21.31	Teratocarcinoma-derived growth factor 1	TOF, VSD
VEGF	192,240	6p21.1	Vascular endothelial growth factor	CoA, OFT defects
Structural proteins				
ACTC1	102,540	15q14	Alpha cardiac actin	ASD
ELN	130,160	7q11.23	Elastin	SVAS, PAS, PS, AS, (Williams-Beuren syndrome)
MYH6	160,710	14q11.2	Alpha myosin heavy chain	ASD, TA, AS, PFO, TGA
MYH7	160,760	14q11.2	Beta myosin heavy chain	Ebstein anomaly, ASD, NVM
TLL1	606,742	4q32.3	Tolloid-Like Protein 1	ASD
FLNA	300,017	Xq28	Filamin A, Alpha	Cardiac valvular dysplasia, X-linked dilated

AS aortic valve stenosis, *ASD* atrial septal defect, *AV* atrioventricular, *AVSD* atrioventricular septal defect, *BAV* bicuspid aortic valve, *CoA* coarctation of the aorta, *DORV* double outlet right ventricle, *HLHS* hypoplastic left heart syndrome, *HLV* hypoplastic left ventricle, *HRV* hypoplastic right ventricle, *IAA* interrupted aortic arch, *MS* mitral valve stenosis, *NMV* Noncompaction of the Ventricular Myocardium, *PA* pulmonary atresia, *PAPVR* partial anomalous pulmonary venous retour, *PAS* pulmonary artery stenosis, *PDA* patent ductus arteriosus, *PS* pulmonary valve stenosis, *PTA* persistent truncus arteriosus, *RV* right ventricle, *SVAS* supravalvular aortic stenosis, *TAPVR* total anomalous pulmonary venous retour, *TGA* transposition of the great arteries, *TOF* tetralogy of Fallot, *VSD* ventricular septal defect

Table 9 Studies performed using NGS technology

Analysis type	Cohort Size	Likely causal variant identified	Sequencing platform	Study
WES	2	*MYH6* (p.Ala290Pro)	Illumina HiSeq2000	Arrington et al. [129]
WES	362	*SMAD2* (p.Trp244Cys) among others	Illumina HiSeq2000	Zaidi et al. [130]
WES	1	*SHROOM3* (p.Gly60Val)	Illumina GAII	Tariq et al. [132]
WES	1	*PLXND1* (p.Arg1299Cys)	Illumina HiSeq2000	Ta-Shma et al. [133]
WES	2	*ACTC1* (p.Met178Leu)	Illumina HiSeq2000	Greenway et al. [134]
WES	17	Not found	Illumina HiSeq2000	Martin et al. [135]
Target Panel (57 genes)	16 families	*TFAP2b* (p.Arg285Gln) among others	Illumina HiSeq2000	Blue et al. [127, 131]

WES whole exome sequencing

terms of depth of coverage because NGS technologies require that multiple sequencing reads overlap to provide greater confidence of genotype calls. Therefore, WES is currently recommended only as follow-up when other investigations, including targeted gene panels, fail to yield results [128]. Some studies have been published regarding the application of NGS techniques to CHD. For instance, Arrington et al. applied the exome sequencing to screen family members with diverse CHD and identified one likely causal variant (*MYH6* p.Ala290Pro) [129]. Zaid et al. applied WES in a cohort of 362 patients with CHD and their parents and revealed *de novo* mutations in histone-modifying genes [130]. Blue et al. developed a NGS gene panel for 57 genes previously implicated in CHD and screened 16 families with strong CHD histories, with an average read depth of 1,873X. The 57 genes of this panel were selected based on their known involvement in human CHD or suggestive evidence from mouse studies. The investigators thus identified a pathogenic variant in 5 of 16 (31%) of the families studied [131]. Other studies have focused on exome sequencing of other genes related to CHD [132–135]. Table 9 summarizes the available literature using NGS for CHD research up to date.

5 Inherited Aortopathy

Aortopathies are a group of overlapping and heterogeneous aorta disorders characterized by dilation, tortuosity, aneurysm and dissection of the aorta [136]. Aortic aneurysm and dissections account for 1–2% of all deaths in the Western countries, became the 15th leading cause of mortality in individuals over 55 years of age, and can be categorized into two main groups based on their locations: thoracic aortic aneurysm (TAA) and abdominal aortic aneurysms (AAA), both demonstrate a strong genetic component in their aetiology [137–139]. While AAA is a complex disorder that integrates predisposing genetic influence with lifestyle-associated risk factors, TAA is highly associated with hereditary factors and does not exhibit an obligate association with cardiovascular risk factors [140]. The thoracic aorta is anatomically segmented into the aortic root (sinuses of Valsalva), sinotubular junction, tubular ascending aorta, aortic arch, and descending thoracic aorta [141, 142]. Thoracic aortic aneurysms leading to type A dissections (TAAD) is the most common fatal condition of aortopathy, and can be sporadic, inherited in isolation (familial nonsyndromic) or in association with genetic syndromes [143, 144]. More than 20% of thoracic aortic aneurysms are inherited disorders [136].

Syndromic connective tissue diseases such as Marfan syndrome (MFS), Loeys-Dietz syndrome (LDS), Ehlers-Danlos syndrome vascular type (EDS type 4), and other syndromes including Bicuspid aortic valve (BAV), Noonan syndrome and Turner syndrome all involve aortopathies resulted from mutations in different genes (Table 10) [145, 146]. When TAAD occurs in the absence of syndromic features, it is inherited in an autosomal dominant manner with decreased penetrance and variable expression, the disease is referred to as familial TAAD (FTAAD). Many genes are found to be associated with this FTAAD. Given the overlapping phenotypes presented by these disorders, genetic sequencing is often needed to make accurate diagnosis and appropriate clinical intervention. Recently emerged high throughput NGS has many advantages over the traditional Sanger sequencing methods, making it feasible for a rapid, cost-effective, comprehensive detection of mutations in many genes associated with these aortopathies [136, 147].

5.1 Syndromic Aortopathy in Marfan Syndrome

Marfan syndrome (MFS, OMIM 154700) was first described in 1896 by the French pediatrician Antoine Bernard-Jean Marfan, who reported the association of long slender digits and other skeletal abnormalities in a 5-year-old girl, Gabrielle [148]. MFS is an autosomal dominant disorder of connective tissue with pleiotropic manifestations that occur in the skeletal, ocular and cardiovascular systems. All cases of the MFS appear to be caused by mutations in the fibrillin-1 gene (*FBN1*) on chromosome 15q21.1. The involved skeletal system manifestation is characterized by bone overgrowth and joint laxity. The extremities are disproportionately long for the size of the trunk (dolichostenomelia). Overgrowth of the ribs can push the sternum in (pectus excavatum) or out (pectus carinatum). Scoliosis is common and can be mild or severe and progressive. Myopia is the most common ocular feature seen in approximately 60% of affected individuals, with the displacement of the lens from the center of the pupil as a hallmark. People with MFS are also at increased risk for retinal detachment, glaucoma, and early cataract formation. However, the major sources of morbidity and early mortality in the MFS relate to the cardiovascular system. The main life-threatening cardiovascular manifestations include

Table 10 Genes involved in inherited aortopahies

Gene	OMIM	Locus	Protein	Associated phenotype
ACTA2	102,620	10q22-q24	Smooth muscle actin, alpha 2	TAAD4
BGN	301,870	Xq28	Biglycan	Aortic rupture and dissection
CBS	613,381	21q22.3	Cystathionine beta-synthase	MFS-like symptoms
COL3A1	120,180	2q31	Collagen type III alpha 1	EDS4
COL4A5	303,630	Xq22.3	Collagen type IV alpha 5	Alport syndrome aortic arch aneurysm
COL5A1	120,215	9q34.3	Collagen type V alpha 1	Classic type EDS, aortic root dilation
COL5A2	120,190	2q32.2	Collagen type V alpha 2	Classic type EDS, aortic root dilation
DCN	125,255	12q21.33	Decorin	Aortic dissection
EFEMP2 (FBLN-4)	604,633	11q13.1	EGF-Containing fibulin-like extracellular matrix protein 2	AR cutis laxa
ELN	130,160	7q11.23	Elastin	AD cutis laxa
ENG	111,395	9q34.11	Endoglin	BAV, HHT
FBN1	134,797	15q21.1	Fibrillin 1	MFS
FBN2	612,570	5q23-q31	Fibrillin 2	BAV-TAAD syndrome
FLNA	300,017	Xq28	Filamin A	Aortic valve disease, aortic coarctation
JAG1	601,920	20p12.1	Jagged 1	Alagille syndrome, aortic coarctation and aneurysm
MAT2A	601,468	2p11.2	Methionine adenosyltransferase II, alpha	Familial TAAD
MFAP5	601,103	12p13.31	Microfibrillar-associated protein 5	Familial TAAD
MYH11	160,745	16p13.13-p13.12	Myosin heavy chain 11	TAAD-patent ductus arteriosus
MYLK	600,922	3q21.1	Myosine light chain kinase	Familial TAAD
NOTCH1	190,198	9q34.3	Homolog of drosophila Notch 1	BAV-TAAD syndrome
NOTCH2	600,275	1p12	Homolog of drosophila Notch 2	Alagille syndrome, aortic coarctation and aneurysm
PLOD1	153,454	1p36.22	Procolllagen-lysine, 2-oxoglutarate 5-dioxygenase	EDSVI, arterial rupture

(continued)

Table 10 (continued)

Gene	OMIM	Locus	Protein	Associated phenotype
PRKG1	176,894	10q11.2-q21.1	Regulatory cGMP-dependent protein kinase type I	Familial TAAD
PTPN11	176,876	12q24.1	PTPN11(SHP2)	TAAD
SKI	164,780	1p36.33-p36.32	V-SKI avian sarcoma viral oncogene homolog	Shprintzen-Goldberg craniosynostosis syndrome
SLC2A10	606,145	20q13.1	Solute carrier family 2, member 10	Arterial tortuosity syndrome
SMAD3	603,109	15q22.33	MAD homolog 3	LDS3, Aneurysms-osteoarthritis syndrome
SMAD4	600,993	18q21.1	MAD homolog 4	Thoracic aortic diseases with JPS-HHT
TGFB2	190,220	1q41	TGF-beta 2	LDS4
TGFB3	190,230	14q24.3	TGF-beta 3	LDS5
TGFBR1	190,181	9q22.33	Transforming growth factor-beta receptor type 1	LDS1, TAAD
TGFBR2	190,182	3p24.1	Transforming growth factor-beta receptor type 2	MFS-like syndrome, LDS2, TAAD

ascending aortic aneurysm, dilatation of the aorta at the level of the sinuses of Valsalva, a predisposition for aortic tear and rupture, mitral valve prolapse with or without regurgitation, tricuspid valve prolapse, and enlargement of the proximal pulmonary artery. Histological aspects of MFS aortopathy include a medial degeneration from disarray and fragmentation of elastic fibers and accumulation of basophilic ground substance areas depleted of smooth muscle cells (SMCs).

Approximately three quarters of individuals with MFS are inherited from the affected parents, the remaining quarter of probands with MFS have a *de novo* mutation. The children of an individual with MFS are at 50% risk of inheriting the mutant allele and the disorder. Prenatal testing for pregnancies at increased risk is possible if the pathogenic variant in the family is known [148]. The estimated prevalence of MFS is 1:5000–1:10,000 and there is no apparent enrichment in any ethnic or racial group and no gender preference. With proper management, the life expectancy of someone with MFS approximates that of the general population. Reduced or mutated forms of fibrillin-1 prejudice the homeostasis of the extracellular matrix leading to alteration of the aortic mechanical properties, increased TGF-β bioavailability and TGF-β-related signaling, medial remodeling, and changes of SMC phenotype [149].

5.2 Syndromic Aortopathy in Loeys-Dietz Syndrome

In 2005, the Loeys-Dietz syndrome (LDS, OMIM 609192 and OMIM 608967) is defined by Loeys et al. as a new MFS related syndrome [150, 151], characterized by the triad of hypertelorism, cleft palate or bifid uvula, and arterial tortuosity, combined with widespread aneurysms. The aortic aneurysms tend to be more aggressive than in patients with MFS, leading to dissection and rupture at smaller diameters and at younger ages. Additional MFS-like cardiovascular, craniofacial, and skeletal symptoms are often observed. Distinguishing features between LDS and MFS include hypertelorism, cleft palate/bifid uvula, craniosynostosis, cervical spine instability, arterial tortuosity, and aneurysm beyond the aortic root. Although initially 2 types of LDS were distinguished, they are now thought to be part of a continuum of disease. LDS1 is caused by mutations in the *TGFBR1* gene, whereas LDS2 is caused by mutations in the *TGFBR2* gene. LDS3, which is associated with early-onset osteoarthritis, is caused by mutations in the *SMAD3* gene, while LDS4 and LDS5 are caused by mutations in the *TGFB2* and *TGFB3* genes respectively.

5.3 Syndromic Aortopathy in Ehlers-Danlos Syndrome

Ehlers-Danlos syndrome type 4 (also known as the vascular type EDS), results from mutations in the type III procollagen (*COL3A1*) gene, is characterized by thin, translucent skin; easy bruising; arterial, intestinal, and/or uterine fragility and characteristic facial appearance. Vascular dissection or rupture, gastrointestinal perforation, or organ rupture are the presenting signs in the majority of adults with EDS type 4. Affected patients are at risk for aneurysms and rupture or dissection, especially of medium-sized arteries including coronary, splanchnic, and uterine artery [141]. Although EDS type 4 patients can present with aortic aneurysms and dissections, the frequency of aortic disease versus others is however not well established [141, 152].

Arterial rupture may occur spontaneously or may be preceded by aneurysm, arteriovenous fistulae or dissection. Attempts to surgically repair arteries are often complicated owing to the presence of friable tissue that does not heal well. The diffuse vascular disease and the difficulties encountered during repair lead to premature deaths. Interestingly, these patients also have thin, translucent skin, wounds that heal poorly and cause atrophic scars, and a high risk for bowel and uterine rupture, being similar to complications observed in patients with LDS [149].

5.4 BAV Aortopathy Syndrome

Bicuspid aortic valve (BAV) is one of the most common congenital heart anomalies (both in children and in adults), occurring in 0.5–2% of the population based on echocardiography and autopsy series. BAV is the cause of aortic stenosis in 70–85%

of pediatric cases and about 50% of adult cases. The tissue pathology in BAV is not limited to the valves' leaflets but extends from the left ventricular outflow tract to the ascending thoracic aorta. Acute thoracic aortic emergencies such as aneurysm and dissection are associated with BAVs although its etiology is still in controversy [153–155].

5.5 *Other Aortopathies*

Other genetic syndromes associated with a slightly increased risk for thoracic aortic disease include Turner syndrome [139], Noonan syndrome [146], arterial tortuosity syndrome [156, 157], adult polycystic kidney disease [158], osteogenesis imperfecta [141], and Alagille syndrome [159].

Sporadic young-onset thoracic aortic aneurysm and/or dissection (TAAD) in the absence of a genetic syndrome is a complex multifactorial disease mainly involving the ascending aorta [160]. Approximately 20% of these non-syndromic TAAD cases have an affected family member. Autosomal dominant inheritance with variable expression and reduced penetrance is described in FTAAD. FTAAD exhibits significant clinical and genetic heterogeneity. Mutations in the genes related to aortic wall structure and function, including *MYH11*, *ACTA2*, *SLC2A10*, *MYLK*, *SMAD3*, *NOTCH1*, have also been linked to non-syndromic familial forms of TAA [143, 144]. And more recently, *MAT2A* [161] and *MFAP5* [162]genes were added into this list.

Filamin A (*FLNA*) encodes a non-muscle actin binding protein that has an important role in cross-linking cortical actin filaments into a dynamic three-dimensional structure. *FLNA* mutations result in X-linked inheritance of a brain malformation known as periventricular heterotopia [142]. The disorder occurs mostly in females and affected women have an increased number of miscarriages of male fetuses, suggesting that hemizygous males die perinatally. In addition, *FLNA* mutations also cause an Ehlers-Danlos syndrome with joint and skin hyperextensibility and aortic dissections [141].

5.6 *NGS Analysis of Genes for Aortopathies*

As mentioned above, inherited aortopathies are very complicated disorders with overlapping manifestations between various syndromes and the vast candidate genes (available and more emerging) involved. It's difficult to differentiate the diagnosis without analysis of all potential candidate genes by sequencing as well as cytogenetic analyses. Traditional Sanger sequencing is impossible to fulfil this huge task in consideration of cost and time to be taken. However, the high throughput, cost-effective next generation sequencing (NGS) techniques provided a feasible solution to clinical applications for inherited aortopathy diagnosis. In 2012,

Table 11 NGS panel for inherited aortopathy

Inherited aortopathy NGS panel	Targeted genes	Coverage
Marfan Syndrome, Loeys-Dietz Syndrome, Aortopathy and Related Disorders (18 genes)	*ACTA2, CBS, COL3A1, FBN1, FBN2, MAT2A, MFAP5, MYH11, MYLK, NOTCH1, PRKG1, SLC2A10, SMAD3, SMAD4, TGFB2, TGFB3, TGFBR1, TGFBR2*	1. NGS only: 97.9% 2. NGS + Sanger fill-in:100%

Wooderchak-Donahue et al. [136](ARUP, Utah) first compared different enrichment methods for targeted NGS analysis of a panel of 10 genes related to aortopathy. And recently they reported a study of 175 cases using NGS and microarray-based comparative genomic hybridization (aCGH) analysis for this panel of genes, and found high negative rate suggesting more genes exist for aortopathies and should be included into the panel [147]. Barbier et al. [162] identified the *MFAP5* gene associated with FTAAD using exome sequencing screen of the proband and related family members. Blinc et al. [163]recently confirmed the diagnosis of a previously healthy woman with Loeys-Dietz syndrome (LDS) type 3, after clinical exome sequencing detected a novel missense mutation in the evolutionary conserved region of *SMAD3* and helped the right management performed for this patient. Currently, commercial NGS panels for inherited aortopathy are available in many genetic diagnostic laboratories. Table 11 listed the genes in the inherited aortopathy NGS panel offered by BCM/John Welsh Cardiology Diagnostic Laboratory as a reference for potential use by clinicians and patients.

6 Conclusion

Cardiovascular diseases (CVDs) are the leading global cause of death, which encompass a broad range of disorders, including diseases of the vasculature, the myocardium, and the heart's electrical circuit, and congenital heart disease (CHD). Most CVDs have clear hereditary components. Understanding the genetic etiology for these disorders has improved their clinical recognition and management and led to new guidelines for treatment and family-based diagnosis and surveillance [5]. Genetic testing has long been used in clinical cardiology and a large number of genes and gene mutations have been identified to be related to specific cardiovascular diseases. Basically, genetic testing can be utilized in (1) confirmation of clinical diagnosis of a proband with inherited CVD; (2) risk assessment and stratification of family members of patients with inherited CVDs; (3) prenatal testing and prevention of inherited CVDs; and (4) pre-symptomatic assessment and prediction of inherited CVDs. The advent of NGS technologies has led to increasingly comprehensive testing for CVDs in both the monogenic and the polygenic/multifactorial forms. It is foreseeable that NGS technologies will be used increasingly. However, one of the main challenges of NGS technologies is the analysis of the data. There is still variability concerning sensitivity and specificity within NGS platforms and different software, and Sanger sequencing still continues to be the gold standard for validation in 'clinical sequencing' experiments.

References

1. Mozaffarian, D., et al.: Heart disease and stroke statistics – 2015 update: a report from the American Heart Association. Circulation. **131**, e535 (2015)
2. Faita, F., Vecoli, C., Foffa, I., Andreassi, M.G.: Next generation sequencing in cardiovascular diseases. World J. Cardiol. **4**, 288–295 (2012)
3. Hershberger, R.E.: Cardiovascular genetic medicine: evolving concepts, rationale, and implementation. J. Cardiovasc. Transl. Res. **1**, 137–143 (2008)
4. Kathiresan, S., Srivastava, D.: Genetics of human cardiovascular disease. Cell. **148**, 1242–1257 (2012)
5. Ware, S.M., Jefferies, J.L.: New genetic insights into congenital heart disease. J. Clin. Exp. Cardiolog. **S8**, 003 (2012). doi:10.4172/2155-9880
6. Maron, B.J.: Contemporary definitions and classification of the cardiomyopathies: an American Heart Association Scientific Statement from the Council on Clinical Cardiology, Heart Failure and Transplantation Committee; Quality of Care and Outcomes Research and Functional Genomics and Translational Biology Interdisciplinary Working Groups; and Council on Epidemiology and Prevention. *Circulation*. **113**, 1807–1816 (2006)
7. Geisterfer-Lowrance, A.A., et al.: A molecular basis for familial hypertrophic cardiomyopathy: a beta cardiac myosin heavy chain gene missense mutation. Cell. **62**, 999–1006 (1990)
8. Ho, C.Y., et al.: Genetic advances in sarcomeric cardiomyopathies: state of the art. Cardiovasc. Res. **105**, 397–408 (2015)
9. Mcnally, E.M., Golbus, J.R., Puckelwartz, M.J.: Genetic mutations and mechanisms in dilated cardiomyopathy. J. Clin. Invest. **123**, 19–26 (2013)
10. Towbin, J.A.: Inherited cardiomyopathies. Circ. J. **78**, 2347–2356 (2014)
11. Maron, B.J., et al.: Prevalence of hypertrophic cardiomyopathy in a general population of young Adults : echocardiographic analysis of 4111 subjects in the CARDIA study. Circulation. **92**, 785–789 (1995)
12. Teekakirikul, P., Kelly, M.A., Rehm, H.L., Lakdawala, N.K., Funke, B.H.: Inherited cardiomyopathies. J. Mol. Diagnostics. **15**, 158–170 (2013)
13. Profiles, P., Roberts, W.C., Mueller, F.O.: Sudden death in young competitive athletes. JAMA. **276**, 199–204 (1996)
14. Maron, B.J., Doerer, J.J., Haas, T.S., Tierney, D.M., Mueller, F.O.: Sudden deaths in young competitive athletes: analysis of 1866 deaths in the United States, 1980-2006. Circulation. **119**, 1085–1092 (2009)
15. Tester, D.J., Ackerman, M.J.: Genetic testing for potentially lethal, highly treatable inherited cardiomyopathies/channelopathies in clinical practice. Circulation. **123**, 1021–1037 (2012)
16. Maron, B.J., Maron, M.S., Semsarian, C.: Genetics of hypertrophic cardiomyopathy after 20 years: clinical perspectives. J. Am. Coll. Cardiol. **60**, 705–715 (2012)
17. Ho, C.Y.: Genetics and clinical destiny: improving care in hypertrophic cardiomyopathy. Circulation. **122**, 2430–2440 (2010)
18. Lopes, L.R., et al.: Genetic complexity in hypertrophic cardiomyopathy revealed by high-throughput sequencing. J. Med. Genet. **50**, 228–239 (2013)
19. Marston, S., et al.: Evidence from human Myectomy samples that MYBPC3 mutations cause hypertrophic cardiomyopathy through haploinsufficiency. Circ. Res. **105**, 219–222 (2009)
20. Ingles, J., et al.: Clinical predictors of genetic testing outcomes in hypertrophic cardiomyopathy. Genet. Med. **15**, 972–977 (2013)

21. Ackerman, M.J., et al.: HRS/EHRA expert consensus statement on the state of genetic testing for the channelopathies and cardiomyopathies this document was developed as a partnership between the Heart Rhythm Society (HRS) and the European Heart Rhythm Association (EHRA). Hear. Rhythm. **8**, 1308–1339 (2011)
22. Wordsworth, S., et al.: DNA testing for hypertrophic cardiomyopathy: a cost-effectiveness model. Eur. Heart J. **31**, 926–935 (2010)
23. Ingles, J., McGaughran, J., Scuffham, P.A., Atherton, J., Semsarian, C.: A cost-effectiveness model of genetic testing for the evaluation of families with hypertrophic cardiomyopathy. Heart. **98**, 625–630 (2012)
24. Zhang, L., et al.: Hypertrophic cardiomyopathy: can the noninvasive diagnostic testing identify high risk patients? World J. Cardiol. **6**, 764–770 (2014)
25. Charron, P.: Genetic analysis for predictive screening in hypertrophic cardiomyopathy. Heart. **98**, 603–604 (2012)
26. Meder, B., et al.: Targeted next-generation sequencing for the molecular genetic diagnostics of cardiomyopathies. Circ. Cardiovasc. Genet. **4**, 110–122 (2011)
27. Dames, S., Durtschi, J., Geiersbach, K., Stephens, J., Voelkerding, K.V.: Comparison of the illumina genome analyzer and roche 454 GS FLX for resequencing of hypertrophic cardiomyopathy-associated genes. J. Biomol. Tech. **21**, 73–80 (2010)
28. Towbin, J.A., et al.: Incidence, causes, and outcomes of dilated cardiomyopathy in children. JAMA. **296**, 1867–1876 (2006)
29. Hershberger, R.E., Hedges, D.J., Morales, A.: Dilated cardiomyopathy: the complexity of a diverse genetic architecture. Nat. Rev. Cardiol. **10**, 531–547 (2013)
30. Deck, G.W., Fuster, V.: Idiopathic dilated cardiomyopathy. N. Engl. J. Med. **331**, 1564–1575 (1994)
31. Finsterer, J., Stöllberger, C., Wahbi, K.: Cardiomyopathy in neurological disorders. Cardiovasc. Pathol. **22**, 389–400 (2013)
32. Norton, N., et al.: Genome-wide studies of copy number variation and exome sequencing identify rare variants in BAG3 as a cause of dilated cardiomyopathy. Am. J. Hum. Genet. **88**, 273–282 (2011)
33. Li, D., et al.: Identification of novel mutations in RBM20 in patients with dilated cardiomyopathy. Clin. Transl. Sci. **3**, 90–97 (2010)
34. Herman, D.S., et al.: Truncations of titin causing dilated cardiomyopathy. N. Engl. J. Med. **366**, 619–628 (2012)
35. Watkins, H., Ashrafian, H., Redwood, C.: Inherited cardiomyopathies. N. Engl. J. Med. **364**, 1643–1656 (2011)
36. Morita, H., Seidman, J., Seidman, C.E.: Genetic causes of human heart failure. J. Clin. Invest. **115**, 518–526 (2005)
37. Arbustini, E., et al.: Mitochondrial DNA mutations and mitochondrial abnormalities in dilated cardiomyopathy. Am. J. Pathol. **153**, 1501–1510 (1998)
38. DeWitt, M.M., MacLeod, H.M., Soliven, B., McNally, E.M.: Phospholamban R14 deletion results in late-onset, mild, hereditary dilated cardiomyopathy. J. Am. Coll. Cardiol. **48**, 1396–1398 (2006)
39. Van Rijsingen, I.A.W., et al.: Risk factors for malignant ventricular arrhythmias in Lamin A/C mutation carriers: a European cohort study. J. Am. Coll. Cardiol. **59**, 493–500 (2012)
40. McNair, W.P., et al.: SCN5A mutations associate with arrhythmic dilated cardiomyopathy and commonly localize to the voltage-sensing mechanism. J. Am. Coll. Cardiol. **57**, 2160–2168 (2011)
41. Theis, J.L., et al.: Homozygosity mapping and exome sequencing reveal GATAD1 mutation in autosomal recessive dilated cardiomyopathy. Circ. Cardiovasc. Genet. **4**, 585–594 (2011)
42. Mestroni, L., Taylor, M.R.G.: Genetics and genetic testing of dilated cardiomyopathy: a new perspective. Discov. Med. **15**, 43–49 (2013)
43. Mogensen, J., Arbustini, E.: Restrictive cardiomyopathy. Curr. Opin. Cardiol. **24**, 214–220 (2009)

44. Chen, S., Balfour, I.C., Jureidini, S.: Clinical spectrum of restrictive cardiomyopathy in children. J. Heart Lung Transplant. **2498**, 90–92 (2001)
45. Russo, L.M., Webber, S.A.: Idiopathic restrictive cardiomyopathy in children. Heart. **91**, 1199–1202 (2005). doi:10.1136/hrt.2004.043869
46. Rivenes, S.M., Kearney, D.L., Smith, E.O.B., Towbin, J.A., Denfield, S.W.: Sudden death and cardiovascular collapse in children with restrictive cardiomyopathy. Circulation. **102**, 876–883 (2015)
47. Kaski, J.P., et al.: Idiopathic restrictive cardiomyopathy in children is caused by mutations in cardiac sarcomere protein genes. Heart. **94**, 1478–1484 (2008)
48. Mogensen, J., et al.: Idiopathic restrictive cardiomyopathy is part of the clinical expression of cardiac troponin I mutations. J. Clin. Invest. **111**, 209–216 (2003)
49. Chen, Y., et al.: Pediatric restrictive cardiomyopathy due to a heterozygous mutation of the TNNI3 gene. J. Biomed. Res. **28**, 59–63 (2014)
50. Peddy, S.B., et al.: Infantile restrictive cardiomyopathy resulting from a mutation in the cardiac troponin T gene. Pediatrics. **117**, 1830–1833 (2006)
51. Sm, W., et al.: Pediatric restrictive cardiomyopathy associated with a mutation in b -myosin heavy chain. Clin. Genet. **134**, 165–170 (2008)
52. Lopes, L.R., Elliott, P.M.: Biochimica et Biophysica Acta genetics of heart failure ☆. BBA – Mol. Basis Dis. **1832**, 2451–2461 (2013)
53. Teo, L.Y.L., Moran, R.T., Tang, W.H.W.: Evolving approaches to genetic evaluation of specific cardiomyopathies. Curr. Heart Fail. Rep. **12**, 339–349 (2015)
54. Biswas, A., Sandeep, V.R.R.: Next generation sequencing in cardiomyopathy : towards personalized genomics and medicine. Mol. Biol. Rep. **41**, 4881–4888 (2014)
55. Reviews, F., Arrhythmia, O.N.: Arrhythmogenic right ventricular dysplasia/cardiomyopathy. Circ J. **79**, S57–S61 (2015)
56. Marcus, F.I., Edson, S., Towbin, J.A.: Genetics of arrhythmogenic right ventricular cardiomyopathy a practical guide for physicians. JAC. **61**, 1945–1948 (2013)
57. Campuzano, O., et al.: Genetics of arrhythmogenic right ventricular cardiomyopathy. J. Med. Genet. **50**, 280–289 (2013)
58. Egle, I., et al.: Identification of a PKP2 gene deletion in a family with arrhythmogenic right ventricular cardiomyopathy. Eur. J. Hum. Genet. **21**, 1226–1231 (2013)
59. Shemisa, K., Li, J., Tam, M., Barcena, J.: Left ventricular noncompaction cardiomyopathy. Cardiovasc. Diagn. Ther. **3**, 170–175 (2013)
60. Chin, T.K., Perloff, J.K., Williams, R.G., Jue, K., Mohrmann, R.: Isolated noncompaction of left ventricular myocardium a study of eight cases. Circulation. **82**, 507–514 (1990)
61. Bahler, R.C.: Prevalence and characteristics of left ventricular noncompaction in a community hospital cohort of patients with systolic dysfunction. Echocardiography. **25**, 8–12 (2008)
62. Ichida, F.: Left ventricular noncompaction. *Circ J*. **73**, 19–26 (2009)
63. Stress, O., With, I., Matter, W., After, R.: HHS public access. *Stroke*. **44**, 3516–3521 (2014)
64. Hoedemaekers, Y.M., et al.: The importance of genetic counseling, DNA diagnostics, and cardiologic family screening in left ventricular noncompaction cardiomyopathy. Circ. Cardiovasc. Genet. **3**, 232–239 (2010)
65. Waldmüller, S., et al.: Targeted 46-gene and clinical exome sequencing for mutations causing cardiomyopathies. Mol. Cell. Probes. **29**, 308–314 (2015)
66. Schaefer, E., et al.: Next-generation sequencing (NGS) as a fast molecular diagnosis tool for left ventricular noncompaction in an infant with compound mutations in the MYBPC3 gene. Eur. J. Med. Genet. **57**, 129–132 (2014)
67. Deo, R., Albert, C.M.: Epidemiology and genetics of sudden cardiac death. Circulation. **125**, 620–637 (2012)
68. Spears, D.A., Gollob, M.H.: Genetics of inherited primary arrhythmia disorders. Appl. Clin. Genet. **8**, 215–233 (2015)
69. Priori, S.G., et al.: HRS/EHRA/APHRS expert consensus statement on the diagnosis and management of patients with inherited primary arrhythmia syndromes. Hear. Rhythm. **10**, 1932–1963 (2013)

70. Priori, S.G., Napolitano, C., Schwartz, P.J.: Low penetrance in the long-QT syndrome: clinical impact. Circulation. **99**, 529–533 (1999)
71. Tester, D.J., Will, M.L., Haglund, C.M., Ackerman, M.J.: Effect of clinical phenotype on yield of long QT syndrome genetic testing. J. Am. Coll. Cardiol. **47**, 764–768 (2006)
72. Wei, J.Y.: The New England journal of medicine downloaded from nejm.Org at the TMC LIBRARY on June 13, 2014. For personal use only. No other uses without permission. From the NEJM archive. N. Engl. J. Med. **327**, 1735–1739 (2010)
73. Priori, S.G., et al.: Risk stratification in the long-QT syndrome. N. Engl. J. Med. **348**, 1866–1874 (2003)
74. Schwartz, P.J., et al.: Prevalence of the congenital long-QT syndrome. Circulation. **120**, 1761–1767 (2009)
75. Goldenberg, I., et al.: Risk factors for aborted cardiac arrest and sudden cardiac death in children with the congenital long-QT syndrome. Circulation. **117**, 2184–2191 (2008)
76. Hobbs, J.B., et al.: Risk of aborted cardiac arrest or sudden cardiac death during adolescence in the long-QT syndrome. JAMA. **296**, 1249–1254 (2006)
77. Mönnig, G., et al.: Electrocardiographic risk stratification in families with congenital long QT syndrome. Eur. Heart J. **27**, 2074–2080 (2006)
78. Sauer, A.J., et al.: Long QT syndrome in adults. J. Am. Coll. Cardiol. **49**, 329–337 (2007)
79. Jons, C., et al.: Risk of fatal arrhythmic events in long QT syndrome patients after syncope. J. Am. Coll. Cardiol. **55**, 783–788 (2010)
80. Wang, Q., et al.: SCN5A mutations associated with an inherited cardiac arrhythmia, long QT syndrome. Cell. **80**, 805–811 (1995)
81. Schwartz, P.J.: Idiopathic long QT syndrome: progress and questions. Am. Heart J. **109**, 399–411 (1985)
82. Goldenberg, I., et al.: Clinical course and risk stratification of patients affected with the Jervell and Lange-Nielsen syndrome. J. Cardiovasc. Electrophysiol. **17**, 1161–1168 (2006)
83. Schwartz, P., The, J.: Jervell and Lange-Nielsen syndrome: natural history, molecular basis, and clinical outcome. Circulation. **113**, 783–790 (2006)
84. Crotti, L., et al.: NOS1AP is a genetic modifier of the long-QT syndrome. Circulation. **120**, 1657–1663 (2009)
85. Li, X., et al.: Towards clinical molecular diagnosis of inherited cardiac conditions: a comparison of bench-top genome DNA sequencers. PLoS One. **8**, e67744 (2013)
86. Lubitz, S.A., Ellinor, P.T.: Next-generation sequencing for the diagnosis of cardiac arrhythmia syndromes. Heart Rhythm. **12**, 1062–1070 (2015)
87. Lieve, K.V., et al.: Results of genetic testing in 855 consecutive unrelated patients referred for long QT syndrome in a clinical laboratory. Genet. Test. Mol. Biomarkers. **17**, 553–561 (2013)
88. Fröhler, S., et al.: Exome sequencing helped the fine diagnosis of two siblings afflicted with atypical Timothy syndrome (TS2). BMC Med. Genet. **15**, 48 (2014)
89. Whole, R., Sequencing, G.: HHS public access. *Hear. Rhythm.* **11**, 1707–1713 (2014)
90. Zhang, L., et al.: Spectrum of ST-T-wave patterns and repolarization parameters in congenital long-QT syndrome: ECG findings identify genotypes. Circulation. **102**, 2849–2855 (2000)
91. Arrhythmias, G.T.L., et al.: Genotype-phenotype correlation in the long-QT syndrome. Circulation. **103**, 89–95 (2001)
92. Ackerman, M.J.: The long QT syndrome. Pediatr. Rev. **19**, 232–238 (1998)
93. Choi, G.: Spectrum and frequency of cardiac channel defects in swimming-triggered arrhythmia syndromes. Circulation. **110**, 2119–2124 (2004)
94. Moss, A.J., et al.: Clinical aspects of type-1 long-QT syndrome by location, coding type, and biophysical function of mutations involving the KCNQ1 gene. Circulation. **115**, 2481–2489 (2007)
95. Shimizu, W., et al.: Genotype-phenotype aspects of type 2 long QT syndrome. J. Am. Coll. Cardiol. **54**, 2052–2062 (2009)
96. Splawski, I., et al.: CaV1.2 calcium channel dysfunction causes a multisystem disorder including arrhythmia and autism. Cell. **119**, 19–31 (2004)

97. Donger, C., et al.: KVLQT1 C-terminal missense mutation causes a forme fruste long-QT syndrome. Circulation. **96**, 2778–2781 (1997)
98. Locati, E.H., et al.: Age- and sex-related differences in clinical manifestations in patients with congenital long-QT syndrome: findings from the international LQTS registry. Circulation. **97**, 2237–2244 (1998)
99. Villain, E., et al.: Low incidence of cardiac events with beta-blocking therapy in children with long QT syndrome. Eur. Heart J. **25**, 1405–1411 (2004)
100. Mönnig, G., et al.: Implantable cardioverter-defibrillator therapy in patients with congenital long-QT syndrome: a long-term follow-up. Hear. Rhythm. **2**, 497–504 (2005)
101. Schwartz, P.J., et al.: Who are the long-QT syndrome patients who receive an implantable cardioverter-defibrillator and what happens to them?: data from the European long-QT syndrome implantable cardioverter-defibrillator (LQTS ICD) registry. Circulation. **122**, 1272–1282 (2010)
102. Horner, J.M., et al.: Implantable cardioverter defibrillator therapy for congenital long QT syndrome: a single-center experience. Heart Rhythm. **7**, 1616–1622 (2010)
103. Moss, A.J., et al.: Effectiveness and limitations of beta-blocker therapy in congenital long-QT syndrome. Circulation. **101**, 616–623 (2000)
104. Schwartz, P.J., et al.: Long QT syndrome patients with mutations of the SCN5A and HERG genes have differential responses to Na+ channel blockade and to increases in heart rate. Implications for gene-specific therapy. *Circulation*. **92**, 3381–3386 (1995)
105. Ruan, Y., Liu, N., Bloise, R., Napolitano, C., Priori, S.G.: Gating properties of SCN5A mutations and the response to mexiletine in long-QT syndrome type 3 patients. Circulation. **116**, 1137–1144 (2007)
106. Moss, A.J., et al.: Safety and efficacy of flecainide in subjects with long QT-3 syndrome (DeltaKPQ mutation): a randomized, double-blind, placebo-controlled clinical trial. Ann. Noninvasive Electrocardiol. **10**, 59–66 (2005)
107. Sarquella-Brugada, G., Campuzano, O., Arbelo, E., Brugada, J., Brugada, R.: Brugada syndrome: clinical and genetic findings. Genet. Med. **18**, 3–12 (2016)
108. Antzelevitch, C.: Brugada syndrome clinical, genetic, molecular, cellular, and ionic aspects. Curr. Probl. Cardiol. **41**, 7–57 (2016)
109. Antzelevitch, C., et al.: Brugada syndrome: report of the second consensus conference. Hear. Rhythm. **2**, 429–440 (2005)
110. Mizusawa, Y., Wilde, A.A.M.: Brugada syndrome. *Circ. Arrhythmia Electrophysiol.* **5**, 606–616 (2012)
111. Gollob, M.H., et al.: Recommendations for the use of genetic testing in the clinical evaluation of inherited cardiac arrhythmias associated with sudden cardiac death: Canadian cardiovascular society/Canadian Heart Rhythm Society joint position paper. Can. J. Cardiol. **27**, 232–245 (2011)
112. Crotti, L., et al.: Spectrum and prevalence of mutations involving BrS1- through BrS12-susceptibility genes in a cohort of unrelated patients referred for Brugada syndrome genetic testing. J. Am. Coll. Cardiol. **60**, 1410–1418 (2012)
113. Reller, M.D., Strickland, M.J., Riehle-Colarusso, T., Mahle, W.T., Correa, A.: Prevalence of congenital heart defects in metropolitan Atlanta, 1998–2005. J. Pediatr. **153**, 807–813 (2008)
114. Hoffman, J.I., Kaplan, S.: The incidence of congenital heart disease. J. Am. Coll. Cardiol. **39**, 1890–1900 (2002)
115. Fahed, C.A., Nemer, M.G.: In: Cooper, D. (ed.) Mutations in Human Genetic Disease, vol. 52, pp. 119–148. InTech, Rijeka (2012)
116. Wessels, M., Willems, P.: Genetic factors in non-syndromic congenital heart malformations. Clin. Genet. **78**, 103–123 (2010)
117. Basson, C.T., et al.: Mutations in human TBX5 [corrected] cause limb and cardiac malformation in Holt-Oram syndrome. Nat. Genet. **15**, 30–35 (1997)
118. Fahed, A.C., Gelb, B.D., Seidman, J.G., Seidman, C.E.: Genetics of congenital heart disease: the glass half empty. Circ. Res. **112**, 707–720 (2013)

119. Erdogan, F., et al.: High frequency of submicroscopic genomic aberrations detected by tiling path array comparative genome hybridisation in patients with isolated congenital heart disease. J. Med. Genet. **45**, 704–709 (2008)
120. Silversides, C.K., et al.: Rare copy number variations in adults with tetralogy of Fallot implicate novel risk gene pathways. PLoS Genet. **8**, e1002843 (2012)
121. Soemedi, R., et al.: Contribution of global rare copy-number variants to the risk of sporadic congenital heart disease. Am. J. Hum. Genet. **91**, 489–501 (2012)
122. Kosaki, K., et al.: Characterization and mutation analysis of human LEFTY A and LEFTY B, homologues of murine genes implicated in left-right axis development. Am. J. Hum. Genet. **64**, 712–721 (1999)
123. Zhou, X., Sasaki, H., Lowe, L., Hogan, B.L., Kuehn, M.R.: Nodal is a novel TGF-beta-like gene expressed in the mouse node during gastrulation. Nature. **361**, 543–547 (1993)
124. De Luca, A., et al.: Familial transposition of the great arteries caused by multiple mutations in laterality genes. Heart. **96**, 673–677 (2010)
125. Garg, V., et al.: Mutations in NOTCH1 cause aortic valve disease. Nature. **437**, 270–274 (2005)
126. Ching, Y.-H., et al.: Mutation in myosin heavy chain 6 causes atrial septal defect. Nat. Genet. **37**, 423–428 (2005)
127. Blue, G.M., Winlaw, D.S.: Next generation sequencing in congenital heart disease: gene discovery and clinical application. J. Next Gener. Seq. Appl. **2**, 2–5 (2015)
128. Andelfinger, G.: Next-generation sequencing in congenital heart disease. J. Am. Coll. Cardiol. **64**, 2507–2509 (2014)
129. Arrington, C.B., et al.: Exome analysis of a family with pleiotropic congenital heart disease. Circ. Cardiovasc. Genet. **5**, 175–182 (2012)
130. Zaidi, S., et al.: De novo mutations in histone modifying genes in congenital heart disease. Nature. **498**, 220–223 (2013)
131. Blue, G.M., et al.: Targeted next-generation sequencing identifies pathogenic variants in familial congenital heart disease. J. Am. Coll. Cardiol. **64**, 2498–2506 (2014)
132. Tariq, M., Belmont, J.W., Lalani, S., Smolarek, T., Ware, S.M.: SHROOM3 is a novel candidate for heterotaxy identified by whole exome sequencing. Genome Biol. **12**, R91 (2011)
133. Ta-Shma, A., et al.: Isolated truncus arteriosus associated with a mutation in the plexin-D1 gene. Am. J. Med. Genet. A. **161A**, 3115–3120 (2013)
134. Greenway, S.C., et al.: Exome sequencing identifies a novel variant in ACTC1 associated with familial atrial septal defect. Can. J. Cardiol. **30**, 181–187 (2014)
135. Martin, L.J., et al.: Whole exome sequencing for familial bicuspid aortic valve identifies putative variants. Circ. Cardiovasc. Genet. **7**, 677–683 (2014)
136. Wooderchak-Donahue, W.L., et al.: A direct comparison of next generation sequencing enrichment methods using an aortopathy gene panel- clinical diagnostics perspective. BMC Med. Genet. **5**, 50 (2012)
137. Gillis, E., Laer, L.V., Loeys, B.L.: Genetics of thoracic aortic aneurysm. Circ. Res. **113**, 327–340 (2015)
138. Saratzis, A., Bown, M.J.: The genetic basis for aortic aneurysmal disease. Heart. **100**, 916–922 (2014)
139. Halushka, M.K.: Single gene disorders of the aortic wall. Cardiovasc. Pathol. **21**, 240–244 (2012)
140. Rea, G., Stewart, F.J.: Genetic biomarkers in aortopathy. Biomark. Med. **7**, 547–563 (2013)
141. Paterick, T.E., et al.: Aortopathies: etiologies, genetics, differential diagnosis, prognosis and management. *Am J Med.* **126**, 670–678 (2013)
142. Niwa, K.: Aortic root dilatation in tetralogy of Fallot long-term after repair – histology of the aorta in tetralogy of Fallot: evidence of intrinsic aortopathy. Int. J. Cardiol. **103**, 117–119 (2005)
143. Milewicz, D.M., et al.: Genetic basis of thoracic aortic aneurysms and dissections: focus on smooth muscle cell contractile dysfunction. Annu. Rev. Genomics Hum. Genet. **9**, 283–302 (2008)

144. Milewicz, D.M., Carlson, A.A., Regalado, E.S.: Genes predisposing to thoracic aortic aneurysms and dissections: associated phenotypes, gene-specific management, and genetic testing. Cardiol. Clin. **28**, 191–197 (2010)
145. Francois, K.: Aortopathy associated with congenital heart disease: a current literature review. Ann. Pediatr. Cardiol. **8**, 25–36 (2015)
146. Cornwall, J.W., Green, R.S., Nielsen, J.C., Gelb, B.D.: Frequency of aortic dilation in Noonan syndrome. Am. J. Cardiol. **113**, 368–371 (2014)
147. Wooderchak-Donahue, W., et al.: Clinical utility of a next generation sequencing panel assay for Marfan and Marfan-like syndromes featuring aortopathy. Am. J. Med. Genet. Part A. **167**, 1747–1757 (2015)
148. Judge, D.P., Dietz, H.C.: Marfan ' s syndrome. Lancet. **633**, 1966–1976 (2005)
149. Romaniello, F., et al.: Aortopathy in Marfan syndrome: an update. Cardiovasc. Pathol. **23**, 261–266 (2014)
150. Callewaert, B.L., et al.: Aneurysm syndromes caused by mutations in the TGF-β receptor. N. Engl. J. Med. **24**, 788–798 (2006)
151. Loeys, B.L., et al.: A syndrome of altered cardiovascular, craniofacial, neurocognitive and skeletal development caused by mutations in TGFBR1 or TGFBR2. Nat. Genet. **37**, 275–281 (2005)
152. Bradley, T.J., Bowdin, S.C.: Multidisciplinary aortopathy clinics should now Be the standard of Care in Canada. Can. J. Cardiol. **32**, 1–5 (2015)
153. Michelena, H.I., et al.: Bicuspid aortic valve aortopathy in adults: incidence, etiology, and clinical significance. Int. J. Cardiol. **201**, 400–407 (2015)
154. Bruckner, B.A., Reardon, M.J.: Bicuspid aortic valve and associated aortopathy: surgical considerations. Methodist Debakey Cardiovasc. J. **6**, 29–32 (2010)
155. Abdulkareem, N., Smelt, J., Jahangiri, M.: Bicuspid aortic valve aortopathy: genetics, pathophysiology and medical therapy. Interact. Cardiovasc. Thorac. Surg. **17**, 554–559 (2013)
156. Jeremy, R.W., Robertson, E., Lu, Y., Hambly, B.D.: Perturbations of mechanotransduction and aneurysm formation in heritable aortopathies. Int. J. Cardiol. **169**, 7–16 (2013)
157. Morris, S.A.: Arterial tortuosity in genetic arteriopathies. Curr. Opin. Cardiol. **30**, 587–593 (2015)
158. Westerland, O., et al.: Vascular manifestations of syndromic aortopathies: role of current and emerging imaging techniques. Clin. Radiol. **70**, 1344–1354 (2015)
159. Jones, J.A., Ikonomidis, J.S.: The pathogenesis of aortopathy in Marfan syndrome and related diseases. Curr. Cardiol. Rep. **12**, 99–107 (2010)
160. Cury, M., Zeidan, F., Lobato, A.C.: Aortic disease in the young: pictorial review of genetic aneurysm syndromes, connective tissue disorders and familial aortic aneurysms and dissections. *Int. J. Vasc. Med.* ***2013***, 1–7 (2013)
161. Guo, D.-C., et al.: MAT2A mutations predispose individuals to thoracic aortic aneurysms. Am. J. Hum. Genet. **96**, 170–177 (2015)
162. Barbier, M., et al.: MFAP5 loss-of-function mutations underscore the involvement of matrix alteration in the pathogenesis of familial thoracic aortic aneurysms and dissections. Am. J. Hum. Genet. **95**, 736–743 (2014)
163. Blinc, A., et al.: Clinical exome sequencing as a Novel tool for diagnosing Loeys-Dietz syndrome type 3. *Eur. J. Vasc. Endovasc. Surg.* ***50***, 816–821 (2015)

Comprehensive Analyses of the Mitochondrial Genome

Victor Wei Zhang and Lee-Jun C. Wong

Abstract Comprehensive molecular diagnoses of mitochondrial DNA (mtDNA) related disorders must include detection and quantification of every single nucleotide variant (SNV) across the entire coding regions, as well as structural variations such as large deletions with mapping of the breakpoints. Traditionally, diagnosis of mtDNA-related disorders is achieved by employing step-wise procedures, such as PCR based Sanger sequencing for SNV, real time quantitative PCR for heteroplasmy quantification, and array CGH or Southern blot for large structural variations. Although these assays together have good clinical utility, these procedures are tedious and have technical limitations. More importantly, they may yield underdiagnoses or mis-diagnoses for some patients. The clinically validated massively parallel next generation sequencing (NGS) with deep coverage and proper quality control, as described "Zhang-Wong method", can achieve a one-step cost-effective comprehensive diagnosis with greatly improved diagnostic yield, and is regarded as the "new gold standard" (Zhang et al. Clin. Chem. 58:1322–1331, 2012). The enhanced sensitivity, accuracy, and reproducibility of simultaneous detection and quantification of mtDNA SNVs, as well as the concurrent detection and junction characterization of single and multiple deletions offer extraordinary value in genetic counseling and patient management.

Keywords Mitochondrial disorders • mtDNA • mtDNA point mutations • mtDNA mutation heteroplasmy • mtDNA deletion • mtDNA multiple deletions • Next generation sequencing • Massively parallel sequencing

---

V.W. Zhang, M.D., Ph.D. (✉)
Department of Molecular and Human Genetics, Baylor College of Medicine, Houston, TX, USA

AmCare Genomics Laboratory, International BioIsland 2-4C-201, GuangZhou, China
e-mail: wzhang2@bcm.edu; victor.w.zhang@amcarelab.com

L.-J.C. Wong, Ph.D.
Department of Molecular and Human Genetics, Baylor College of Medicine, Houston, TX, USA

L.-J.C. Wong (ed.), *Next Generation Sequencing Based Clinical Molecular Diagnosis of Human Genetic Disorders*, DOI 10.1007/978-3-319-56418-0_13

1 Introduction

Mitochondrial diseases are clinically and genetically heterogeneous disorders due to defects in either the nuclear genome (three billion bp in 23 pairs of chromosomes) or the mitochondrial genome (16.6 kilo bp). Since the mitochondrial genome is small, traditional diagnosis of mitochondrial disorders focused on the Sanger analysis of the tiny circular double stranded mitochondrial DNA (mtDNA). This chapter describes NGS based molecular diagnosis of mitochondrial disorders caused by mtDNA defects.

Mitochondria are the only cellular organelles that contain their own genetic materials. Most human cell contains hundreds to thousands of mitochondria [2], each of which contains multiple copies of mtDNA molecules. The number of mitochondria and mtDNA molecules per cell depends on energy demand of the specific tissue. Since there are multiple copies of mtDNA, if mutation occurs, it often coexists with the wild type mtDNA, a phenomenon called "heteroplasmy". The degree of heteroplasmy of a pathogenic variant, nature of the specific variant, and its tissue distribution determine the clinical phenotype of the affected patient, including variable penetrance, expressivity, and age of onset [3, 4]. Phenotype may also be modified by genetic background and environmental factors.

Unlike nuclear genes, mtDNA contains no introns in the protein coding regions. The entire mtDNA, encoding a total of 37 genes, is efficiently utilized. Polycistronic messages are produced from both strands of the mtDNA. Two genes, *ATP6* and *ATP8* share part of their coding regions in different reading frames [5].

The mtDNA encodes two ribosomal RNAs, 22 tRNAs and 13 respiratory chain complex protein subunits. Beside protein coding regions, mutations in the rRNA and tRNA can also cause diseases (http://www.mitomap.org/MITOMAP). The most frequently occurring mtDNA pathogenic variants reside in the tRNA genes, such as m.3243A>G in $tRNA^{Leu(UUR)}$, the most common cause of mitochondrial encephalopathy lactic acidosis and stroke-like episodes (MELAS), and m.8344A>G in $tRNA^{Lys}$ associated with mitochondrial epilepsy and ragged red fibers (MERRF). In addition to the dense coding regions, there is also an approximately 1.1 kb noncoding displacement loop (D-loop) region where the origins of replication are located. The polymorphic nature of this non-coding region plays a critical role in forensic science and human evolution; however, pathogenic variants and large deletions in D-loop region have not been reported.

Traditional molecular diagnosis of mtDNA disorders to identify deleterious changes of mtDNA sequences was based on Sanger sequencing for single nucleotide variants (SNVs) and Southern blot analysis for large deletions [6]. However, these methods have several intrinsic technical limitations: (a) It does not detect low levels of heteroplasmy and does not provide reliable quantification; (b) Southern blot analysis is not sensitive enough to detect all deletions, and it alone does not determine deletion junction sequences; (c) It is difficult to distinguish single and multiple deletions unequivocally; (d) The presence of nuclear mtDNA homologs interferes with the accurate detection and quantification of SNVs; (e) The polymorphic variations across the entire mtDNA make it difficult to select multiple primer regions without any mtSNPs for amplifications. These pitfalls can be resolved by the application of the one-step comprehensive deep next generation sequencing (NGS) approach as described in the sections below.

2 The Design of One-step Amplification Coupled with High Coverage Depth Sequencing

2.1 *Rationales for the Enrichment of the Whole mtDNA in One Piece*

The absence or presence of various degrees of heteroplasmic pathogenic variants is critical in disease diagnosis, prognosis, and genetic counseling. The unambiguous detection of mtDNA single or multiple deletions in different tissues is also important in disease diagnosis [3]. However, traditional methods of amplicon based Sanger sequencing and Southern blot analysis have serious limitations in their application to molecular diagnosis (Table 1). The key to achieve comprehensive molecular diagnosis of mtDNA disorders that will solve these issues is to prepare the authentic entire mtDNA molecule for sequencing [1, 7]. The strategy uses a pair

Table 1 One-step NGS solutions to issues of traditional approaches in the diagnosis of mtDNA related disorders

Issues of conventional approaches	Issue/solution	NGS solutions
Detection of low levels of heteroplasmy	1	Deep and reliable coverage >5-10,000×, thus, a 1% variant heteroplasmy will provide 50-100 reads
Reliable quantification	2	A control sample with 1.1% variant heteroplasmy is spiked in with every batch of NGS analysis to ensure accurate determination of the limit of detection and to monitor variation across batches
Southern blot is not sensitive in the detection of low level deletions. It alone does not determine deletion junction sequences	3	The circular mtDNA molecule, with or without large deletion or duplication is amplified, as long as it contains the origin of replication. The deleted part is shown by sudden dip in read depth and the junction sequences can be read out directly from the NGS results
Hard to distinguish single deletion and multiple deletions unequivocally	4	The deep NGS is extremely sensitive to low levels of multiple deletions due to the preferential amplification of the deleted molecules. The pattern of coverage profile is distinct. A single deletion shows sharp drop in coverage depth at the deletion junction, while it shows continuous gradual drop in coverage depth within the range of multiple deletions
The interference of the NUMT	5	LR-PCR with a pair of back to back primers allow opposite direction amplification of only the circular form of mtDNA, not the linear form of the NUMT intersperse within the nuclear genome. Thus, NUMTs are not sequence analyzed

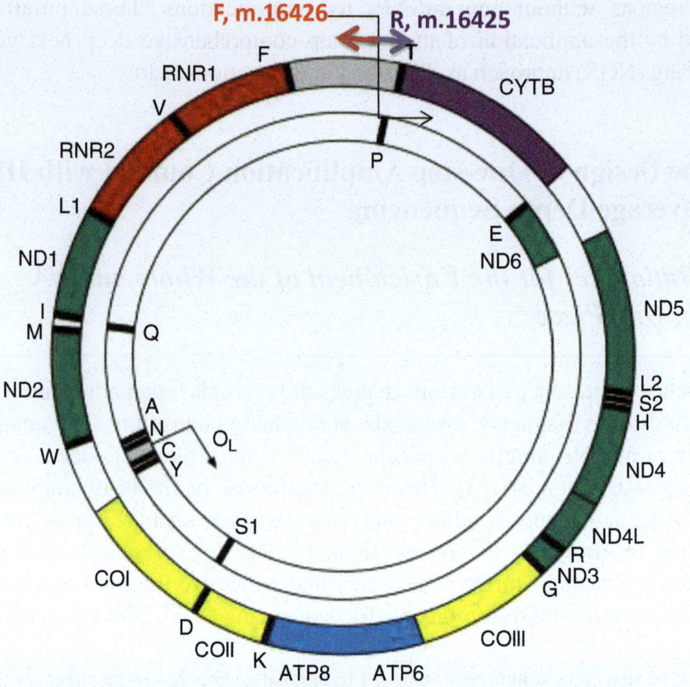

Fig. 1 The primer positions for the LR-PCR of the mitochondrial genome

of back-to-back primers near the origin of mtDNA replication to enrich the entire mtDNA in one piece (Fig. 1). Since the back-to-back primers (m.16426F and m.16425R) are oriented in opposite directions, only the genuine circular mtDNA molecules, not found in the nuclear genome, can be amplified into one piece of DNA template, thus avoiding interference of the linear embedded mtDNA homologs (NUMT) (Table 1, #5). The advantage of this approach is that the linearized mtDNA products will include molecules with AND without deletion or duplication (Table 1, #4) [1, 7]. In addition, every single nucleotide of coding regions of the entire 16,569 bp mtDNA is equally represented for uniform coverage, facilitating reliable detection and quantification of heteroplasmic variants, as well as the determination of deletions and breakpoint sequences (Table 1, #1, 2 and 3).

2.2 *Procedures for Single-Amplicon, LR-PCR Based mtDNA Enrichment*

Successful amplification of the entire circular mtDNA into one piece of LR-PCR product is essential for high quality NGS analysis [1, 7]. The PCR products need to be checked by agarose gel for its presence and quality before continuing on to next library preparation step to ensure fruitful results. Forward and reverse primers are mt16426F- 5'ccgcacaagagtgctactctcctc3' and mt16425R- 5'gatattgatttcacggaggatg-gtg3'. PCR was performed using TaKaRa LA Taq Hot Start polymerase kit (TaKaRa Bio Inc., Madison, WI, USA) and 100 ng of total genomic DNA isolated from blood or 15 ng from skeleton muscle as template in a 50 μl PCR system. After an initial 2 min incubation at 95 °C, it is followed by 30 cycles of PCR with 20 s of denaturation at 95 °C and 18 min of annealing and extension at 68°. The reaction is completed by one cycle of final extension at 68 °C for 20 min.1.5 μl PCR products was analyzed on 1.5% agarose gel with 1 kb plus DNA ladder (Invitrogen, Carlsbad, CA, USA).

3 The New Gold Standard for Molecular Diagnosis of mtDNA Related Disorders

3.1 Uniform Coverage: The Necessity of Amplification of the mtDNA in One Piece

The mtDNA must be enriched in one single LR-PCR step such that every nucleotide is equally represented and sequenced to ensure uniform coverage. The profiles of the coverage depth of the mtDNA using three different enrichment methods are compared in Fig. 2. The coverage profile of the mtDNA enriched by capture in solution using RNA or DNA [8–10] shows that different segments of the mtDNA are not captured and sequenced uniformly (Fig. 2a) [1, 9]. Therefore, it is not possible to detect large deletions or low level heteroplasmic variants with equal sensitivity from these sequence data. Multiple copies of mitochondrial pseudogenes/homologs are located on each of the nuclear chromosomes [11–13]. These nuclear mitochondrial sequences (NUMTs) are subject to genetic drift and therefore produce a significant background of sequence variants that must be contended with in order to discern the true mtDNA sequence. In addition, due to the abundance of NUMTs, in-solution probe hybridization /sequencing will co-capture NUMTs significantly even in the absence of mtDNA specific probes. Thus, interference from NUMT sequences may result in incorrect sequence information and/or errors in the quantification of mtDNA heteroplasmy [11–13].

Since the mtDNA is small (16.6 kb) and does not contain any introns, enrichment has historically been achieved by PCR, using24–36 pairs of primers [14–16] to amplify short overlapping regions, or 2–3 pairs of primers for long range PCR (LR-PCR) [8, 17, 18]. However, the coverage profile of the mtDNA enriched by multiplex amplification revealed uneven coverage depth of each amplicon due to different PCR efficiencies with spiky over-covered primer positions (Fig. 2b). The inconsistency in amplicon coverage depth prevents detection of large deletions and renders the detection sensitivity unreliable.

In contrast to capture based or multiple amplicon approaches, enrichment of the entire mtDNA using only one pair of back-to-back LR-PCR primers generates a

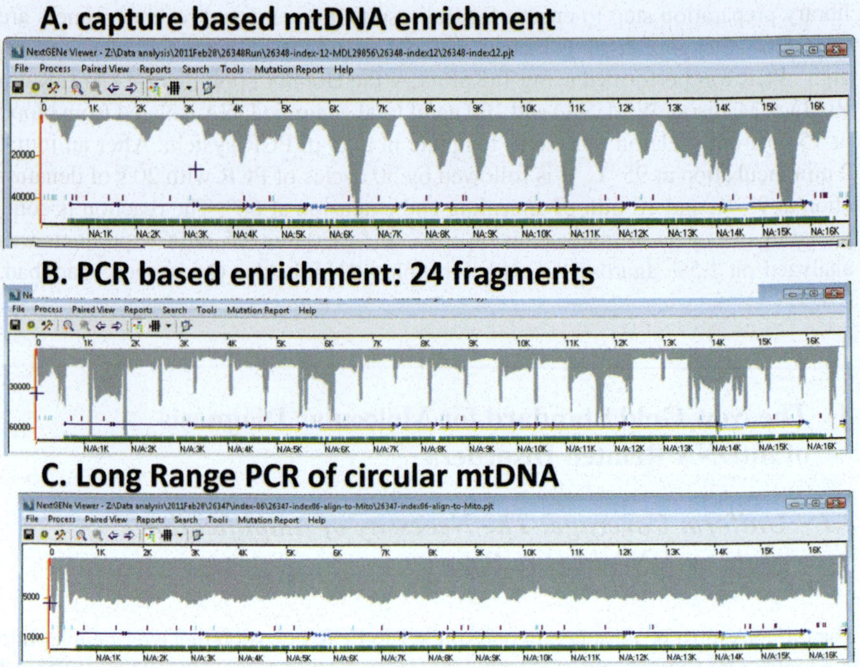

Fig. 2 Comparison of the coverage depth of the mitochondrial genome by different methods of enrichment methods. Enrichment of mtDNA by capture method (**a**), by PCR amplification (**b**), and the LR-PCR of the whole circular mitochondrial genome by one pair of back-to-back primers. The X axis is the nucleotide position from m.1 to m.16,569. The Y axis is the read depth at each nucleotide position

linearized mtDNA with every single nucleotide equally covered [1, 19], providing a uniform coverage profile (Fig. 2c) that can readily detect mtDNA deletions (Fig. 3) [1, 19]. Thus, it is clear that in order to detect large mtDNA deletions, the entire mtDNA must be enriched evenly throughout.

3.2 *Detection and Mapping of mtDNA Deletions*

Uniform coverage of the entire mtDNA by single amplicon LR-PCR/NGS allows the detection and mapping of mtDNA deletions. Sequencing of the mtDNA enriched by oligonucleotide probe-hybridization capture or multiplexed PCR does not provide uniform coverage throughout the genome [1]. Variable coverage depth among different regions or amplicons of the mtDNA makes it impossible to detect large mtDNA deletions reliably. In contrast, enrichment of the mtDNA with single amplicon LR-PCR amplification provides even coverage of the entire mtDNA (Figs. 2c and 3a), allowing accurate detection of large deletions.

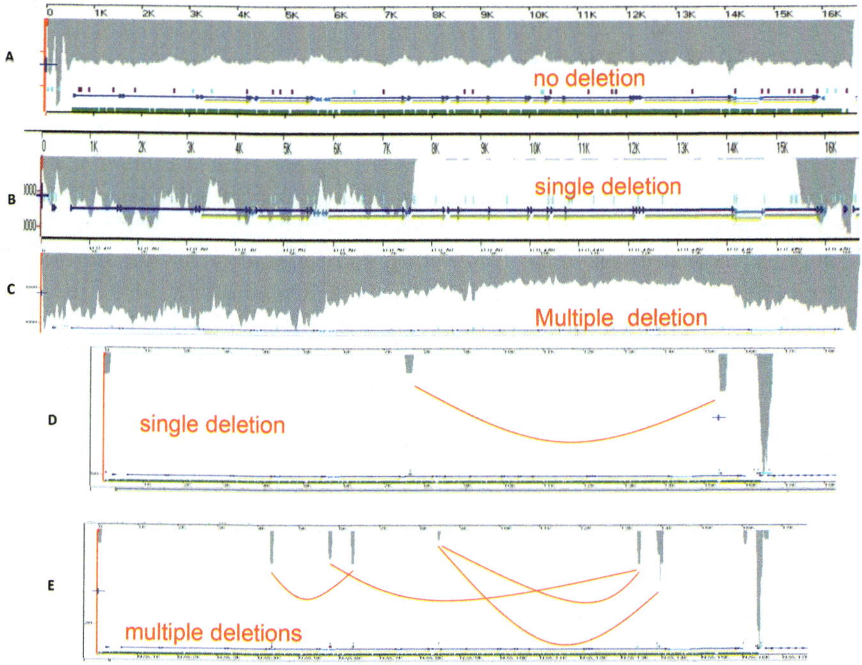

Fig. 3 Detection of single and multiple mtDNA large deletions and mapping of breakpoints. Coverage depth pattern of no (**a**), single (**b**), and multiple mtDNA large deletions (**c**) and mapping of corresponding breakpoints for single deletion (**d**), and multiple deletions (**e**)

Figure 3b shows LR-PCR/NGS analysis of a DNA sample from the blood specimen of a 10-year-old boy presenting with encephalopathy, exercise intolerance, easy fatigability, and sensorineural hearing loss. The sharp decrease in read coverage from m.7638 to m.15434 indicates a heteroplasmic large deletion of 7797 bp [7]. The degree of deletion heteroplasmy can be estimated by comparing the coverage of deleted versus non-deleted regions. The deletion breakpoint, at a single base resolution, is clearly revealed by this method (Fig. 3d). The deletion junction sequence can be confirmed by conventional PCR using primers flanking the deletion region followed by Sanger sequencing through the deletion junction [3].

The coverage profile of a muscle sample from a 70-year-old man with myopathy is shown in Fig. 3c. Instead of sharp deletion junctions, an arch-shaped coverage pattern was observed, suggesting multiple mtDNA deletions [7]. Realignment of the unmapped sequences to the reference sequences, with less stringent parameters, revealed multiple deletion junctions (Fig. 3e). A total of 48 junction sequences were identified. The majority of the deletion breakpoints that have been confirmed by targeted PCR followed by Sanger sequencing are shown in Fig. 3e.

3.3 Reliable Quantification of Heteroplasmy and Detection Limit

The mtDNA contains a number of homopolymeric stretches, high GC content regions, and short tandem repeats. Low heteroplasmy of deleterious mutations, including small indels in some repeat regions, can be clinically significant. Thus, it is important to assess the limit of detection of NGS-based assays [1, 7]. Different platforms may affect the depth of coverage and sequence error rate differently. Here, we focus on the analysis of mtDNA LR-PCR products by Illumina HiSeq sequencing. Several factors, such as DNA quality, amount of DNA template loaded to the sequencer, cluster generation, number of sequencing cycles, factors of multiplex, etc. can affect read depth. Therefore, proper controls should be included and analyzed together with each indexed specimen to ensure accuracy and reproducibility for reliable heteroplasmy sequence error rate quantification [1]. Different sequencing platforms provide different depth of read coverage, which may limit heteroplasmy detection [1, 8, 9, 17, 18].

3.3.1 Deep Coverage for More Accurate Quantification and Detection of Variants

The coverage depth depends on several factors including sequencing platforms, amount of DNA template loaded, and multiplexed samples, the coverage depth should be evaluated and optimized in individual laboratories. Although it seems that the deeper the coverage the better the detection sensitivity and accuracy, overly deep coverage may cause a concurrent increase in background noise, thus, worsening the limit of detection. The coverage depth should be optimized by balancing the sensitivity and specificity. In general, a coverage depth of greater than 10,000× will confidently detect the presence of 100× variant reads (1% heteroplasmy). However, the limit of detection also depends on the error rates.

3.3.2 Evaluation of Experimental and Analytical Errors

To assure that the quantification of heteroplasmy by NGS is reliable, a set of cloned synthetic 150 bp control DNAs (Fig. 4a) is spiked into each indexed sample as external quality controls ("ExQC"). To mimic a range of heteroplasmy, different proportions of control DNAs with different nucleotide changes at specific positions are mixed to form a series of synthetic heteroplasmies (Fig. 4b). To evaluate the reliability of quantitative measurements, the spiked-in control DNA was indexed together with each sample using the same barcode during library preparation. The control DNA sequences, except the nucleotide positions marked for heteroplasmy measurements, are used to calculate error rates. The instrument sequencing error is determined by the number of incorrect nucleotides of the control DNA sequences

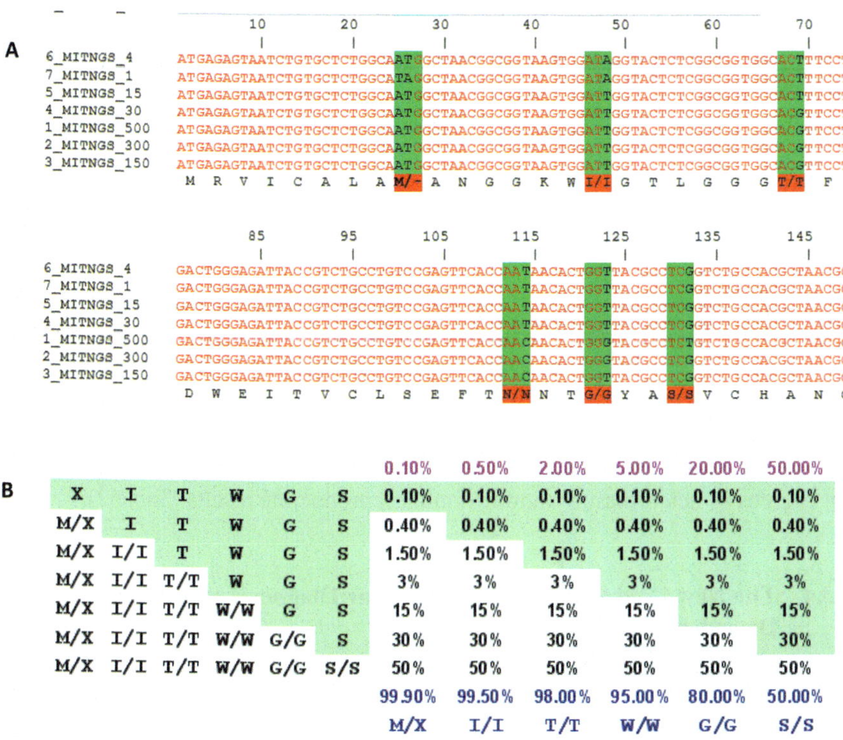

Fig. 4 Spiked-in quality control samples. (**a**) The synthetic controls with the variant positions indicated (**b**) the mixing in ratios indicated in the table to generate the desired percentage of artificial heteroplasmy

versus the total number of nucleotides mapped to the control DNA sequences [1]. The sample DNA has an analytical error rate of 0.326+/−0.335%, as compared to 0.151+/−0.394% for the control DNAs, indicating that the Illumina NGS platform has relatively low sequence error, but the library construction may introduce error significantly. Thus, the limit of detection, calculated as three standard deviations above the mean error ($0.326 + 3 \times 0.335$), is 1.33% under the described experimental, instrumental, and analytical conditions [1]. The experimental procedures can introduce errors, which, in general, are higher for mtDNA samples than for the synthetic ExQC samples. The observed and the expected percentages of the variants at specific positions exhibit an excellent correlation [1].

3.3.3 Reproducibility

In addition to the spike-in ExQC samples, to ensure the reproducibility of heteroplasmy quantification, a DNA sample with known percentage of heteroplasmy should always be used as an additional batch control. For example, a sample with a

Table 2 Reproducibility of heteroplasmy quantification

Variant	Number of independent measurements	% heteroplasmy (average ± STD)
m.3243	16	1.14 ± 0.09
m.4136	2	1.62 ± 0.02
m.7041	2	13.05 ± 0.25
m.9035	2	1.18 ± 0.03
m.15171	2	3.55 ± 0.07

1.1% heteroplasmic m.3243A>G variant is always included in every batch to make sure that the percentage heteroplasmy consistently reproduced within 2SD. As shown, in Table 2, the sample analyzed in 16 different batches over a period of 2 years yielded an average of these independent measurements of 1.14+/−0.09% with a tight range of 0.94–1.35, and errors within 2 SD, suggesting a reproducible quantification. Two independent measurements of four other different variants at different degrees of heteroplasmy showed similar reproducible results (Table 2).

3.3.4 The New Gold Standard for Molecular Diagnosis of mtDNA Disorders

Zhang and coworkers evaluated the clinical performance of the NGS analysis of the mtDNA generated by long range PCR with a set of non-overlapping back-to-back primers for the first time. This “Zhang-Wong method” enriches each nucleotide of the entire 16.6 kb mtDNA equally with an average coverage depth of 5,000–20,000×, providing uniform coverage and sufficient depth for the quantification of heteroplasmy. Analysis of a total of 12 samples that had been thoroughly analyzed by Sanger sequencing revealed that all variants identified by Sanger sequencing were also detected by the NGS approach. However, there were three samples with heteroplasmic variants missed by Sanger sequencing. One of them is a novel m.1630A>G (tRNA Val) variant at a level of 3.7% heteroplasmy, too low to be detected by Sanger sequencing, in the asymptomatic mother of a 2-year old affected child who harbored 33% heteroplasmy for the same variant. Studies of matrilineal family members and clinical correlation revealed that the m.1630A>G co-segregated with the disease and was likely to be a causative mutation. The detection of low level heteroplasmy of the m.1630A>G mutation in the mother indicates that this is an inherited case, thus, the recurrence risk is higher than a *de novo* case. In addition to the low heteroplasmy m.1630A>G, an m.16193insC variant present at 15% heteroplasmy and a 27% heteroplasmy of m.303insC/CC were identified by NGS, but detected as homoplasmy by Sanger sequencing. These results indicate that small insertions and deletions within a homopolymeric stretch, such as m.303_309insC, m.311_315insC and m.16193insC, may be difficult to detect by Sanger methods, but can be accurately identified by NGS. Thus, in the molecular diagnosis of mtDNA disorder, the NGS approach described here becomes the new gold standard to detect low heteroplasmy and small indels in homopolymer stretches, in addition to mtDNA large deletions as described in Sect. 3.2.

4 Clinical and Genetic Significance of Low Level Heteroplasmic Mutations

In general, Sanger sequencing does not reliably detect heteroplasmy less than 15% [20]. While specific primers or probes can be designed for the quantification PCR of target positions [21, 22], it is laborious to validate the method for every novel mtDNA variant. In addition, primers may contain nucleotide modifications and probes are specific for either the wild type or mutant allele [23–25], thus, difference in PCR amplification efficiency for these two alleles is expected, leading to inherent inaccuracies with the measurement of the degree of heteroplasmy. As described in Sect. 3 above, mtDNA analysis at 5–20,000× coverage, the experimental error rate of the NGS using Illumina HiSeq2000 platform was 0.326+/−0.335%, with a limit of detection of 1.33% [1]. The examples below demonstrate the importance of the ability of deep NGS to detect low levels of heteroplasmy accurately and reproducibly in precision risk assessment and genetic counselling.

4.1 Heteroplasmy May Be Low in Blood but High in Affected Tissues

The proband in Fig. 5a had a history of hearing loss and vision problems. Although screening for mtDNA common point mutations by allele-specific oligonucleotide (ASO) hybridization detected m.3243A>G in the tRNALeu(UUR) gene [26], the heteroplasmy was too low to be detected by Sanger sequencing. The LR-PCR/NGS approach detected a heteroplasmy of 7%. Her daughter carried the same mutation at 9.3% heteroplasmy in her blood sample, and showed symptoms of mitochondrial myopathy. Since there is a selection against the m.3243A>G mutation in the rapidly dividing blood cells [23, 27], the degree of heteroplasmy could be higher in her muscle causing the myopathy phenotype. Sanger sequencing would have missed the detection of this low heteroplasmic pathogenic variant in both the proband and the daughter. However, the LR-PCR NGS readily detected and accurately quantified the pathogenic variant and provided a quick, definitive diagnosis in one single step. This case illustrates the importance of employing the one-step NGS analysis for the diagnosis of mtDNA disorders. While the heteroplasmy levels are low in blood, they may be higher in affected tissues.

4.2 Sporadic Somatic Mutation in Affected Tissue

The proband of family in Fig. 5b is a 65-year-old woman with peripheral neuropathy, muscle weakness, ptosis, abnormal muscle histological findings and abnormal EMG. A heteroplasmic m.7222A>G (p.Y440C in COI) was detected in this patient's

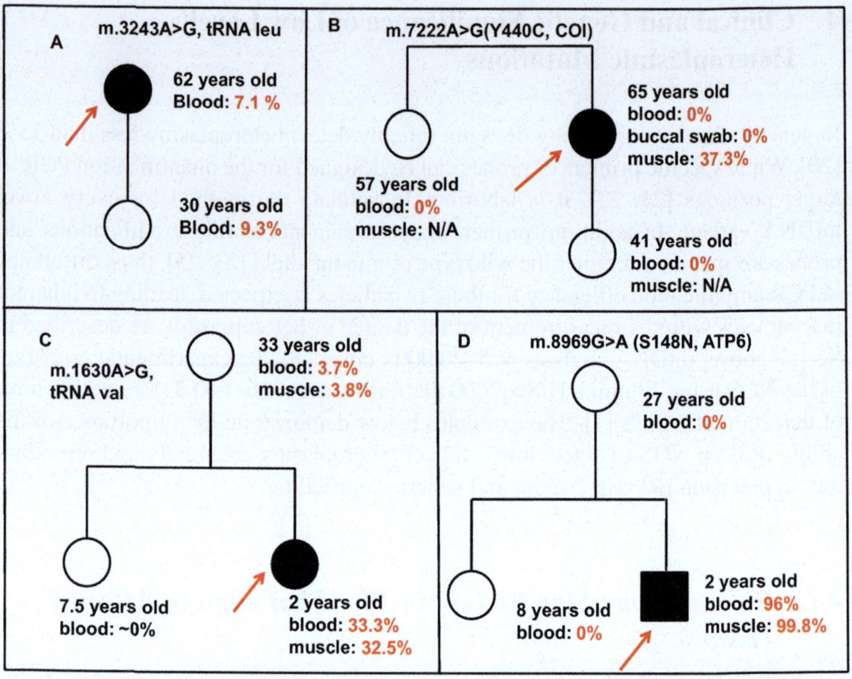

Fig. 5 Examples of pedigrees with implication of heteroplasmy in genetic counseling. (**a**) Definitive diagnosis may be missed by tradition Sanger method due to low heteroplasmy. (**b**) Somatic variant in affected tissue may not be transmitted to offspring. (**c**) Low heteroplasmy in asymptomatic mother implies high risk of intergenerational transmission. Low level of heteroplasmy is different from none at all. (**d**) Mother does not carry the pathogenic variant at all in her blood nor does the asymptomatic child. It is sporadic in the affected child

muscle specimen by Sanger sequencing, and subsequently verified and quantified by LR-PCR/NGS to be at 37% heteroplasmy in her muscle sample, but was not detected in her blood or in blood samples from her sister and daughter. These results suggest that the mutation most likely arose sporadically in the proband's muscle tissue. Therefore, the risk of passing this likely somatic mutation to her children is low.

4.3 *Low Heteroplasmy in Two Tissues May Suggest Low Heteroplasmy in the Germline*

The family in Fig. 5c has been described in Sect. 3.3.4. The mother carried a low heteroplasmy m.1630 mutation in the blood and the muscle, the affected child inherited both mutations at much higher heteroplasmy in both tissues. However, the variant was not detected by Sanger sequencing in the blood sample of her asymptomatic older sibling. This case suggested that the mother could have germline mosaicism. The affected child inherited higher heteroplasmy, which was probably present in all tissues, while the asymptomatic child inherited much lower heteroplasmic mutations.

4.4 De Novo and Near Homoplasmic Transmission in One Generation from Low Heteroplasmic Mother

The example in Fig. 5d shows a 2-year-old boy that presented with developmental delay, seizures, hypotonia, hearing loss, failure to thrive, abnormal brain MRI, and lactic acidemia. NGS detected an m.8969G>A (p.S148N, ATP6) pathogenic variant at 99.5% and 96% heteroplasmy in the muscle and blood, respectively. Since this mutation was not detected in the blood specimens from the proband's asymptomatic mother and an older sibling, it likely occurred *de novo* in the proband; however, germline mosaicism cannot be excluded. The determination of whether an mtDNA mutation is inherited or sporadic is clearly dependent on the sensitivity of the detection method used. An accurate detection of low-level heteroplasmy therefore plays an important role in the molecular diagnosis and counseling of mtDNA- related disorders. Transmission from apparently 0% to near homoplasmy in one generation has been reported [28].

These four examples illustrated the power of LR-PCR NGS-based analyses of mtDNA in detecting low-level heteroplasmy in various tissues and assessing whether a mutation is inherited, *de novo*, sporadic, somatic, or possibly germline mosaicism.

5 Notable Advances

5.1 An Innovative Comprehensive Solution for a Complex, Multi-step Disease Diagnosis

Traditionally, diagnosis of mtDNA disorders first involved the analysis of common mutations by various screening methods and Southern blot analysis for large deletions. If negative, the entire mtDNA was studied by PCR based Sanger sequencing of 24–36 overlapping amplicons. These procedures are tedious step-by-step approaches, and require additional steps for deletion detection and quantification of variant heteroplasmy, which is typically inaccurate and not reproducible. The application of LR-PCR using a pair of back-to-back primers (Fig. 1) for specific amplification of the circular double stranded mtDNA followed by deep NGS is an innovative approach that not only can simultaneously detect point mutations and large deletions, but can also reliably quantify variant heteroplasmy as low as 1%, in addition to the capability and sensitivity of detecting and distinguishing single and multiple mtDNA deletions. This LR-PCR based NGS analysis is a significant improvement over previous methodologies in terms of simplicity, speed, accuracy, and reliability, making it the new gold standard for mtDNA analysis.

5.2 *NUMT, The Old Problem Has a New Effective Solution*

Computational in silico searches for the presence of homologous mtDNA sequences in the nuclear genome (NUMT) have identified more than a thousand NUMT sites with greater than 80% homology with mtDNA sequence [7, 29]. By mapping capture-enriched mtDNA sequencing data from an mtDNA depleted *rho* zero cell line readily revealed the scope of NUMTs. It is estimated that 0.1% of sequence reads from the total genomic DNA mapped to the mtDNA reference sequence. The presence of NUMTs resulted in many false positive and numerous low heteroplasmic variant calls. Software has been developed to distinguish low-level heteroplasmy from sequencing error to allow the removal of NUMTs [30]. However, the false negative and false positive rates were still significantly above acceptable levels for the application to clinical diagnostic laboratories [29, 31]. Although reads generated from NUMT regions can be partially dissected by a stringent alignment algorithm, regions that are nearly identical to mtDNA are practically impossible to remove. Therefore, regardless of stringency, sequence enrichment using specific mtDNA probes or as non-specific by-products of exome capture will inevitably co-capture NUMTs, which will then confound mutational analyses and lead to inaccurate clinical diagnosis.

mtDNA enrichment using multiple amplicons not only fails to address NUMT problems but also introduces additional errors, including SNPs within primer binding sites, resulting in reduced amplification efficiency, masked SNP data, and possibly preferential amplification of NUMTs. The LR-PCR approach uses carefully designed primers near the origin of replication, which contains the least number of reported SNPs occurring at lowest frequencies. Moreover, the primer sites are usually Sanger sequenced separately for each sample to ensure that no SNPs are sitting at the primer sites. If SNPs are found at the primer sites, alternative LR-PCR primers without SNPs can be used to repeat the LR-PCR/NGS. The major advantage of the "Zhang-Wong method" enrichment approach over the conventional Sanger sequencing, capture based or amplicon based method is the unbiased enrichment of the entire mtDNA while avoiding the interference of NUMTs and SNPs [1, 7].

5.3 *Distinguishing Single and Multiple mtDNA Deletions*

LR-PCR amplification of the entire mtDNA as a whole allows accurate detection of mtDNA large deletions with unequivocally mapped breakpoints in one single step (Fig. 2). This approach eliminates the drawbacks of Southern blot technique in its low sensitivity detecting low heteroplasmic mtDNA deletions, and its inability to map the breakpoint. Since LR-PCR preferentially amplifies smaller circular mtDNA molecules, the ability to detect low levels of mtDNA multiple deletions is greatly enhanced, leading to an increased clinical diagnostic yield. mtDNA multiple deletions are usually observed to be associated with aging and oxidative damage. However, most importantly, genetic defects in nuclear genes responsible for mtDNA biosynthesis and integrity maintenance, such as *POLG, TWINKLE, OPA1, RRM2B, TK2* [32–36] often cause secondary mtDNA multiple deletions in muscle samples. Further studies of 15 muscle samples with mtDNA multiple deletions revealed that all patients were adults. Five of them had mutations identified in nuclear genes; *POLG, RRM2B* and *OPA1* [7]. Nine of the 15 (60%) were older than 50 years [7]. These observations suggest that accumulation of mtDNA multiple deletions may be secondary to nuclear gene defects and aging. Undoubtedly, the LR-PCR based NGS approach described here facilitates the diagnosis of mitochondrial disorders caused by nuclear genes by detecting mtDNA multiple deletions followed by the subsequent identification of the responsible nuclear gene defects.

5.4 *Accurate and Reliable Quantification of Heteroplasmy Facilitates Genetic Counseling*

The degree of heteroplasmy of a pathogenic variant and its tissue distribution contribute to disease phenotype and severity. The quantification of heteroplasmy was first developed based on restriction fragment length polymorphism (RFLP) generated by the discriminating target variant. However, agarose gel analysis is not sensitive. This was improved by labeling with radioactive P^{32} nucleotide at the last cycle of PCR to enhance the sensitivity [3, 4]. Recently, real time allele refractory mutation system (ARMS) qPCR and pyrosequencing [21, 22] were validated for a limited number of common pathogenic variants. These methods share several drawbacks. They can only be applied to known and validated target variants and have low sensitivity, high variability, and limited reproducibility. This is because these methods are all short PCR based. PCR efficiencies are greatly influenced by the discriminating nucleotide in the mutant and wild type, which also affects their ratio, thus, heteroplasmy. Subtle differences in PCR conditions will affect the amplification efficiency and ratio dramatically, thus, affecting reproducibility. Another pitfall is that due to the high frequencies of mtDNA SNPs along the entire mtDNA, it is difficult to select primers without any SNPs.

The LR-PCR NGS that unbiasedly enriches all nucleotides in mtDNA will provide accurate quantification. In addition, with 5–20,000× deep coverage, it is sensitive enough to detect heteroplasmy as low as 1% at every nucleotide position of the mtDNA without the interference of NUMTs and SNPs. Accurate and reliable quantification is critical because it gives a "yes" or "no" answer to a carrier mother that has a great impact on the risk of transmitting pathogenic variant to her offspring. For example, if a mother of an affected child with m.3243A>G, and NGS analysis showed a 0.9% heteroplasmy for m.3243A>G. Normally, a 0.9% heteroplasmy would be filtered out as negative since it is below 1% cut-off. However, in this case, since the affected is affected with m.3243A>G, it is important to check if the mother is really zero or low heteroplasmy. To do so, first, the raw NGS data of all samples from the same batch were examined for the heteroplasmy at m.3243 position, and all showed less than 0.01%, yet the mother had 0.9%, which is significant, suggesting that it is not error by chance. Second, repeat the analysis with a second DNA extraction to show the reproducible results, confirming no sample mixed up and no batch effect. Thus, the mother carried a low heteroplasmy of m.3243. The risk of transmitting this variant to next fetus is significant. On the other hand, if a mother of an affected child with a 20% novel pathogenic variant had 0.1% heteroplasmy, and all samples analyzed with the sample batch showed heteroplasmy ranging from 0.08% to 0.11% at this rare variant position, then, the mother did not carry the rare pathogenic variant, and it likely occurred *de novo*. Without reliable quantification of the heteroplasmy, it will not be possible to provide accurate genetic counseling to patients and family.

6 Conclusion

LR-PCR using one pair of back-to-back primers to unbiasedly amplify the mtDNA as a whole followed by deep NGS analysis not only eliminates potential NUMTs and SNP interference but also renders a uniform coverage of the entire circular 16.6 kb mtDNA, which is essential for reliable and accurate detection of single and multiple large mtDNA deletions. Deep coverage and stringent quality control allow the detection of low-level heteroplasmy with high confidence. With the higher sensitivity, specificity and accuracy on the detection of a much broader spectrum of mutation types, ever before achieved, the LR-PCR/NGS of "Zhang-Wong method" becomes the new gold standard for comprehensive analysis of the mitochondrial genome.

References

1. Zhang, W., Cui, H., Wong, L.J.: Comprehensive one-step molecular analyses of mitochondrial genome by massively parallel sequencing. Clin. Chem. **58**(9), 1322–1331 (2012)
2. Dimmock, D., Tang, L.Y., Schmitt, E., Wong, L.J.: A quantitative evaluation of the mitochondrial DNA depletion syndrome. Clin. Chem. **56**(7), 1119–1127 (2010)
3. Shanske, S., Wong, L.J.: Molecular analysis for mitochondrial DNA disorders. Mitochondrion. **4**(5–6), 403–415 (2004)
4. Wong, L.J., Boles, R.G.: Mitochondrial DNA analysis in clinical laboratory diagnostics. Clin. Chim. Acta. **354**(1–2), 1–20 (2005)
5. Smeitink, J., van den Heuvel, L., DiMauro, S.: The genetics and pathology of oxidative phosphorylation. Nat. Rev. Genet. **2**(5), 342–352 (2001)

6. Wong, L.-J.C., Scaglia, F., Graham, B.H., Craigen, W.J.: Current molecular diagnostic algorithm for mitochondrial disorders. Mol. Genet. Metab. **100**(2), 111–117 (2010)
7. Cui, H., Li, F., Chen, D., et al.: Comprehensive next-generation sequence analyses of the entire mitochondrial genome reveal new insights into the molecular diagnosis of mitochondrial DNA disorders. Genet. Med. **15**(5), 388–394 (2013)
8. Zaragoza, M.V., Fass, J., Diegoli, M., Lin, D., Arbustini, E.: Mitochondrial DNA variant discovery and evaluation in human cardiomyopathies through next-generation sequencing. PLoS One. **5**(8), e12295 (2010)
9. Calvo, S.E., Compton, A.G., Hershman, S.G., et al.: Molecular diagnosis of infantile mitochondrial disease with targeted next-generation sequencing. Sci. Transl. Med. **4**(118), 118ra110 (2012)
10. Tang, S., Batra, A., Zhang, Y., Ebenroth, E.S., Huang, T.: Left ventricular noncompaction is associated with mutations in the mitochondrial genome. Mitochondrion. **10**(4), 350–357 (2010)
11. Hirano, M., Shtilbans, A., Mayeux, R., et al.: Apparent mtDNA heteroplasmy in Alzheimer's disease patients and in normals due to PCR amplification of nucleus-embedded mtDNA pseudogenes. Proc. Natl. Acad. Sci. U. S. A. **94**(26), 14894–14899 (1997)
12. Parfait, B., Rustin, P., Munnich, A., Rotig, A.: Co-amplification of nuclear pseudogenes and assessment of heteroplasmy of mitochondrial DNA mutations. Biochem. Biophys. Res. Commun. **247**(1), 57–59 (1998)
13. Tsuzuki, T., Nomiyama, H., Setoyama, C., Maeda, S., Shimada, K.: Presence of mitochondrial-DNA-like sequences in the human nuclear DNA. Gene. **25**(2–3), 223–229 (1983)
14. Ware, S.M., El-Hassan, N., Kahler, S.G., et al.: Infantile cardiomyopathy caused by a mutation in the overlapping region of mitochondrial ATPase 6 and 8 genes. J. Med. Genet. **46**(5), 308–314 (2009)
15. Brautbar, A., Wang, J., Abdenur, J.E., et al.: The mitochondrial 13513G>A mutation is associated with Leigh disease phenotypes independent of complex I deficiency in muscle. Mol. Genet. Metab. **94**(4), 485–490 (2008)
16. Wang, J., Brautbar, A., Chan, A.K., et al.: Two mtDNA mutations 14487T>C (M63V, ND6) and 12297T>C (tRNA Leu) in a Leigh syndrome family. Mol. Genet. Metab. **96**(2), 59–65 (2009)
17. Kara, B., Arikan, M., Maras, H., Abaci, N., Cakiris, A., Ustek, D.: Whole mitochondrial genome analysis of a family with NARP/MILS caused by m.8993T>C mutation in the MT-ATP6 gene. Mol. Genet. Metab. **107**(3), 389–393 (2012)
18. Tang, S., Huang, T.: Characterization of mitochondrial DNA heteroplasmy using a parallel sequencing system. BioTechniques. **48**(4), 287–296 (2010)
19. Zhang, W., Cui, H., Wong, L.J.: Application of next generation sequencing to molecular diagnosis of inherited diseases. Top. Curr. Chem. **336**, 19–46 (2014)
20. Rohlin, A., Wernersson, J., Engwall, Y., Wiklund, L., Bjork, J., Nordling, M.: Parallel sequencing used in detection of mosaic mutations: comparison with four diagnostic DNA screening techniques. Hum. Mutat. **30**(6), 1012–1020 (2009)
21. Venegas, V., Halberg, M.C.: Quantification of mtDNA mutation heteroplasmy (ARMS qPCR). Methods Mol. Biol. **837**, 313–326 (2012)
22. Wang, J., Venegas, V., Li, F., Wong, LJ.: Analysis of mitochondrial DNA point mutation heteroplasmy by ARMS quantitative PCR. Curr. Protoc. Hum. Genet. (2011); Chapter19.6.1–19.6.16
23. Shanske, S., Pancrudo, J., Kaufmann, P., et al.: Varying loads of the mitochondrial DNA A3243G mutation in different tissues: implications for diagnosis. Am. J. Med. Genet. A. **130A**(2), 134–137 (2004)
24. Cox, R., Platt, J., Chen, L.C., Tang, S., Wong, L.J., Enns, G.M.: Leigh syndrome caused by a novel m.4296G>A mutation in mitochondrial tRNA isoleucine. Mitochondrion. **12**(2), 258–261 (2012)
25. Massie, R., Wang, J., Chen, L.C., et al.: Mitochondrial myopathy due to novel missense mutation in the cytochrome c oxidase 1 gene. J. Neurol. Sci. **319**(1–2), 158–163 (2012)

26. Tang, S., Halberg, M.C., Floyd, K.C., Wang, J.: Analysis of common mitochondrial DNA mutations by allele-specific oligonucleotide and Southern blot hybridization. Methods Mol. Biol. **837**, 259–279 (2012)
27. Rahman, S., Poulton, J., Marchington, D., Suomalainen, A.: Decrease of 3243 A-->G mtDNA mutation from blood in MELAS syndrome: a longitudinal study. Am. J. Hum. Genet. **68**(1), 238–240 (2001)
28. White, S.L., Collins, V.R., Wolfe, R., et al.: Genetic counseling and prenatal diagnosis for the mitochondrial DNA mutations at nucleotide 8993. Am. J. Hum. Genet. **65**(2), 474–482 (1999)
29. Li, M., Schroeder, R., Ko, A., Stoneking, M.: Fidelity of capture-enrichment for mtDNA genome sequencing: influence of NUMTs. Nucleic Acids Res. **40**(18), e137 (2012)
30. Li, M., Stoneking, M.: A new approach for detecting low-level mutations in next-generation sequence data. Genome Biol. **13**(5), R34 (2012)
31. Li, M., Schonberg, A., Schaefer, M., Schroeder, R., Nasidze, I., Stoneking, M.: Detecting heteroplasmy from high-throughput sequencing of complete human mitochondrial DNA genomes. Am. J. Hum. Genet. **87**(2), 237–249 (2010)
32. Milone, M., Brunetti-Pierri, N., Tang, L.Y., et al.: Sensory ataxic neuropathy with ophthalmoparesis caused by POLG mutations. Neuromuscul. Disord. **18**(8), 626–632 (2008)
33. Milone, M., Younge, B.R., Wang, J., Zhang, S., Wong, L.J.: Mitochondrial disorder with OPA1 mutation lacking optic atrophy. Mitochondrion. **9**(4), 279–281 (2009)
34. Van Hove, J.L., Cunningham, V., Rice, C., et al.: Finding twinkle in the eyes of a 71-year-old lady: a case report and review of the genotypic and phenotypic spectrum of TWINKLE-related dominant disease. Am. J. Med. Genet. A. **149A**(5), 861–867 (2009)
35. Shaibani, A., Shchelochkov, O.A., Zhang, S., et al.: Mitochondrial neurogastrointestinal encephalopathy due to mutations in RRM2B. Arch. Neurol. **66**(8), 1028–1032 (2009)
36. Chanprasert, S., Wang, J., Weng, S.W., et al.: Molecular and clinical characterization of the myopathic form of mitochondrial DNA depletion syndrome caused by mutations in the thymidine kinase (TK2) gene. Mol. Genet. Metab. **110**(1–2), 153–161 (2013)

Exome Sequencing in the Clinical Setting

Theodore Chiang, Magalie Leduc, Mari Tokita, Teresa Santiago-Sim, and Yaping Yang

Abstract Exome sequencing has become a powerful tool in disease gene discoveries as well as diagnostic evaluations of genetic disorders. Clinical exome sequencing can identify contributing single nucleotide variants (SNVs) and some classes of copy-number variants (CNVs) in approximately 30% of unselected cases. For clinical laboratories that offer exome sequencing, it is important to establish reliable and robust NGS analysis, interpretation and reporting pipelines and procedures in order to ensure test quality, reliability and meet required turn-around-time (TAT). The clinical report for exome sequencing usually includes primary molecular findings related to the patient's clinical phenotype, as well as opted in medically actionable secondary findings and carrier status for autosomal recessive disorders. Clinical exome sequencing, which was initially regarded as mainly the "last resort" for patients who had previously tested negative for specific genes, karyotype, and/or microarray studies, can now be utilized as an effective first tier test, usually for patients with nonspecific phenotypes or phenotypes suggesting substantial genetic heterogeneity. Additionally, the improvement of the methodologies, analysis tools, and TAT has brought new applications of exome sequencing into the genetic testing of patients, esp. prenatal patients or patients who are at high risk for potentially life-threatening conditions and are therefore considered critically ill.

Clinical NGS panel testing, exome sequencing and more rarely whole genome sequencing (WGS) currently co-exist. In the context of decreasing cost of NGS,

Theodore Chiang; Magalie Leduc; Mari Tokita; Teresa Santiago-Sim are contributed equally to this work and are co-first authors. *Correspondence: Y.Y. (yapingy@bcm.edu)

T. Chiang
Human Genome Sequencing Center, Baylor College of Medicine, Houston, TX 77030, USA

M. Leduc • Y. Yang (✉)
Department of Molecular and Human Genetics, Baylor College of Medicine, Houston, TX 77030, USA

Baylor Genetics Laboratories, Houston, TX 77030, USA
e-mail: yapingy@bcm.edu

M. Tokita • T. Santiago-Sim
Department of Molecular and Human Genetics, Baylor College of Medicine, Houston, TX 77030, USA

L.-J.C. Wong (ed.), *Next Generation Sequencing Based Clinical Molecular Diagnosis of Human Genetic Disorders*, DOI 10.1007/978-3-319-56418-0_14

rigorous comparisons of different clinical NGS platforms are needed in order to better understand their clinical utilities and harness the power of whole-wide NGS sequencing in the clinical arena.

Keywords Exome • NGS • Genetic testing • Prenatal testing

1 Introduction to the NGS/Exome Sequencing Process

Next-generation sequencing (NGS) allows for simultaneous interrogation of multiple genes with dramatically reduced cost of sequencing per base. As a result, NGS-based analysis has changed the landscape of research and clinical diagnostic testing. Many clinical laboratories are now able to offer a wide array of disease-targeted NGS panels and exome Sequencing. Whole genome sequencing (WGS) has also been available in a few clinical laboratories.

In general, the process of NGS involves breaking down patient DNA into short fragments, followed by adaptor ligation during library construction, target region enrichment (capture), and sequencing. The capture step is needed for targeted analysis of a subset of genomic regions (exome or panel sequencing, but not WGS); the process involves an additional step prior to sequencing. In this step, the targeted regions are enriched from the total genomic DNA by means of probe hybridization. Enriched libraries are then loaded on NGS sequencers to undergo sequencing in a massively parallel fashion. With NGS becoming routine practice and the availability of many open source software for the analysis of such data, certain tools are becoming standards in common bioinformatic pipelines.

2 Variant Calls and Annotations in the Bioinformatics Pipeline

Whether the sequencing occurs in the clinical or research laboratory, there are generally three broad steps to go from raw sequencing data to annotated variants (Fig. 1). The first step consists of processing and mapping reads to a reference genome. This would include trimming and removing duplicate reads, as well as local realignment and readjusting base quality scores around problematic regions. The second step would involve calling Single Nucleotide Variants (SNVs) and sort insertion-deletions (INDELs), as well as, in some laboratories, detecting copy number variations (CNVs). The third step is annotating and classifying the variation type and assigning candidate consequences to the variants. It is important to note that the choice, design and implementation of the wide assortment of available tools into a bioinformatics pipeline can significantly affect the accuracy and identification of variants. With the increasing number of possible combination of various tools,

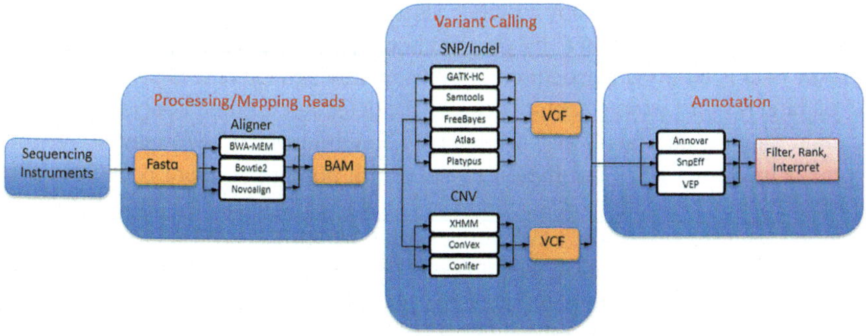

Fig. 1 Common bioinformatics pipelines generally consists of three broad steps between raw sequencing data to a filtered list of candidate causal variants. White boxes lists some common options for tools used in each of these processes. Orange boxes are standard file formats in each step

one can build many different variant-calling pipelines. Recent reports of systematic comparison of variant callers provide valuable insight and guidance to choosing variant callers when building a pipeline [1].

Although the list of tools may be long, there are invariably predominant tools that have become standards and widely used in the community. For read aligners, the major three are BWA-MEM (http://arxiv.org/abs/1303.3997), Bowtie [2] and Novoalign (http://www.novocraft.com). For SNVs and INDELs variant callers, the major ones are: GATK-HC (Genome Analysis Tool Kit HaplotypeCaller) [3], Samtools [4], Freebayes (http://arxiv.org/abs/1207.3907), Atlas [5], Platypus [6]. For classifying and annotating variants, the major three tools are: Annovar [7], SnpEff [8] and Variant Effect Predictor (VEP) [9].

Since there are obviously differences in the bioinformatics assumptions and rules that are codified into a given tool, it is not surprising that pipelines with different components do not always produce identical results when compared to each other. Nevertheless, since the process of aligning and calling variants are usually well integrated with variant annotation in any given complete pipeline, the real challenge is trying to understand the behavior and expected outcomes of such a pipeline in relation to the biological consequence of variants. This complexity arises, not only because of the chosen tool (static factors), but is also affected by dynamic factors, such as the transcript set used and the constantly evolving database resources in the public domain. One example is a recent report by McCarthy's et al., which concluded that both the selection of annotation software and transcript set can affect variant classification and concordance when comparing two different tools using the same transcript set [10]. In other words, variant calling and annotation is a tight intertwinement of several major components, namely, pipeline tools, public data resources, the actual single sample of interest, and the aggregate information of all previously analyzed clinical samples.

The ability to detect copy number variants (CNVs) in NGS is becoming an important component of any analysis pipeline. Estimating copy number in from the exome data is highly dependent on both the quality and the coverage read depth of exon targets in the exome capture design. The detection is limited only to genic and exonic regions, and therefore pure intergenic and intronic CNVs may be missed. Similar to calling SNP and INDELs, it is worth noting that there is considerable variability in CNV calling algorithms, and so it's recommended to try or combine multiple approaches. Tools like CoNVex [11], CoNIFER [12] and XHMM [13], among others, are commonly used to call CNVs from exome data. Some of these tools use normalization methods that can also take into account batch effects and other background noises that should be removed to improve signal detection of copy losses or gains. Other considerations include whether the samples of interest are part of a large (cohort case control), small (tumor/normal comparison) study, or a growing collection of samples from a clinical laboratory with varying types of phenotypes.

3 Variant Classification and Clinical Reporting for Exome Cases

Exome sequencing is a highly complex test in which a large number of rare variants are detected. Bioinformatic pipelines are built to filter variants based on frequencies and additional criteria to help reduce the number of variants to assess. As a basic criteria, a 1% minimum allele frequency (MAF) is commonly used to filter for rare variants. This wide-ranging cut off value allows for the detection of any potentially disease-causing variants. This cut off value is also relaxed for variants that have previously been reported as disease-causing in the literature. Furthermore, additional internal information can be added to reduce the number of variants to assess. This filtering process helps in reducing the number of variants to review and classify. Once filtered, the list of variants needs to be assessed based on the newly updated ACMG/AMP guidelines for variant interpretation [14]. Studies have shown that intra- and inter-laboratory discrepancies exist regarding variant classification [15], indicating the urgent need of optimizing and expanding the current framework ACMG/AMP guidelines and improving communications among laboratories.

One additional uniqueness and challenge of genome wide test such as exome sequencing as opposed to targeted testing is the unfamiliarity and ambiguity of the patient's disorder. The phenotype of the patient is often unclear and broad. Therefore, while the variants are classified based on the ACMG guidelines, extensive genotype/phenotype correlation is also necessary in the analysis of variants from clinical exome sequencing. Thus, the challenge lies in the number of variants to assess, the assessment of the classification based on the ACMG guidelines and the uncertain phenotype/genotype correlation.

As with any other clinical genetic testing, the content of clinical exome sequencing reports follows the current CLIA regulations and requirements (42 CFR §493.1291) and CAP recommendations [16]. However, the major challenges and differences with single genes and panels reports lies in (1) the complexity of the test and its interpretation and (2) the possible discovery of secondary variants unrelated to the patient's phenotype but that could affect the patient's health (incidental findings) and reproductive risk (carrier status). Therefore, the development of a clear and concise report is essential to effectively communicate the clinically relevant findings back to the referring clinician and patient [17].

3.1 *Key Elements of Clinical Exome Report Content and Organization*

The exact content and organization of exome reports is dependent on the type of exome testing requested and the clinical laboratory issuing the report (See Table 1). Nevertheless, common key elements are essential and are as follow:

3.1.1 A clear and concise title indicating the most relevant molecular finding.

3.1.2 The clinical indication/referral for exome testing. This clinical information is essential for the clinical laboratory to provide an accurate correlation between the molecular findings and the clinical presentation of the patient during the review analysis of the exome data.

3.1.3 The primary molecular findings including a list of variants detected and related to the patient's clinical phenotype, followed by an interpretation of the results. For clarity, exome reports usually contain a table with the variants' information followed by the interpretation of the results. These variants are interpreted and categorized following the modified ACMG guidelines [14] and include pathogenic, likely pathogenic variants and VUS related to the patient's clinical phenotype. Relevant information on the report usually include the disease associated with defects in the gene and a description of the disorder, inheritance pattern(s) of the disorder, the gene symbol, the classification, genomic coordinate, nucleotide and amino acid positions, and zygosity of the variant, if the parents are available, the inheritance from each parent is indicated and de novo and compound heterozygous variants are also indicated. If the variant has previously been reported in public databases or in the literature, references such as dbSNP, ClinVar, or PMID numbers and population frequencies are specified to help the clinician understanding the meaning of the findings. Additional information may also include (1) whether the variant has been confirmed by a second methodology such as Sanger sequencing; (2) the predicted pathogenecity of the variant based on bioinformatic algorithms such as SIFT, Polyphen2 [18, 19] and (3) the coverage depth of the gene and whether all exons of the gene were entirely covered by NGS: this information is relevant if a single heterozygous variant is detected in an autosomal recessive gene disorder, to insure that no additional variant in trans configuration may be missed by exome sequencing.

3.1.4 Reports of medically actionable secondary findings and carrier status. Exome sequencing may detect secondary findings such as pathogenic variants known to affect the health of an individual and potentially medically actionable and pathogenic variants in autosomal recessive disorder predicting the carrier status of the individual. Among clinical exome cohorts, medically actionable variants are detected in about 3–4% of patients referred for exome [20, 21]. ACMG recommends the report of known pathogenic variants (and expected pathogenic for some genes) from 56 (version 1) and 59 (version 2) genes mostly related to cardiovascular disorders and cancer for which treatment or medical recommendations are available [22, 23]. The guidelines are of tremendous help for clinical laboratories. In the meantime, several differences in reporting variants in genes outside the "ACMG59" and approaching of returning results among clinical laboratories still exist. First, regarding the definition of medically actionable variants, while pathogenic variants in the 59 ACMG recommended genes are analyzed and returned, additional pathogenic variants in specific genes considered medically actionable by the clinical laboratory may be returned when encountered during data analysis [20, 21]. For instance, Tarailo-Graovac et al. detected an individual who was compound heterozygosity for two pathogenic variants in CFTR but had no reported clinical phenotype of the disorder [21]. While this gene may be on the return carrier list by some clinical laboratories, this type of findings was considered as medically actionable. Second, regarding the reporting, while the ACMG guidelines originally recommended the mandatory return of the actionable variants, the guideline has been modified and laboratories have chosen over time to give an option to opt in or out to fulfill the patient's request. Overall, it was estimated that over 90% of the patients choose to receive medically actionable variants [24]. Finally, regarding the mechanism of return, some laboratories report medically actionable variants for the proband only and request additional consent and counseling for the parents to obtain knowledge of their status. An alternative approach is to report these variants for proband and parents at the same time, for which opt-in authorization and counselling is necessary prior to testing to address the concerns of the family. Other laboratories report these findings on a separate report, available to the patient and the parents. In all cases, these medically actionable variants warrantee genetics counseling and recommendations in addition to the primary findings related to the clinical phenotype in patients.

Carrier status was not included in the ACMG recommendations for reporting of incidental findings. Currently there is no official guideline for carrier status reporting specifically for exome sequencing. Different clinical laboratories choose to report a different set of gene for carrier status, based on disease severity, frequency and recommendations from professional societies such as ACMG and ACOG. For instance, pathogenic variants in CFTR may be reported because of the high prevalence of the disorder in population of European descent. Similarly, hemoglobin S may also be considered for return due to the high frequency in populations at risk for sickle cell disease. Other laboratories choose not to report any carrier findings and recommend a separate universal carrier screening if needed by the patient. Thus the report of carrier status is not consistent among clinical laboratories in the exome sequencing context.

3.1.5 Methodologies and recommendations based on the molecular findings. Because of the constant improvement of technologies and variations in variant annotations and variant calling algorithms between clinical laboratories, methodologies are necessary to fully comprehend the test. Limitations of the exome test should be also mentioned such as triplet nucleotide disorders, large deletion and duplications and further testing should be recommended if indicated.

3.2 Additional Optional Report Categories

In addition to the primary and secondary findings, some clinical laboratories have chosen to report on pharmacogenomic variants, including known pathogenic variants VKORC1/CYP2C9 for instance that can alter warfarin metabolism and known pathogenic variants in CYP2C19, that can alter Plavix metabolism. Moreover, mitochondrial sequencing may be available as part of exome testing. Thus, clinicians need to be aware of the variations between the type of exome test and variation between laboratories, when ordering a whole exome sequencing test.

In the case of trio exome analysis, de novo variants and compound heterozygous variants in genes unknown to cause a disorder or in genes unrelated to the patient's clinical presentation are reported in an additional table. If the exome does not provide a molecular diagnosis, this additional information may become relevant in the future as more genes causing disease are discovered.

3.3 Special Cases

The organization and content of exome reports will vary based on the type of exomes ordered. Examples of specific exome test reports are mentioned below and compared in Table 1.

3.3.1 Prenatal Trio Exome

This report includes all variants related to the prenatal indications as well as variants in disease genes unrelated to the prenatal indications but likely to cause significant disorders during childhood. Because of the nature of this test, the incidental findings are reported after birth as requested. Although this test is a trio test, de novo and compound heterozygous variants in non-disease causing genes are not reported as these will not facilitate a clinical diagnosis.

Table 1 Exome report content

CategorieCategories	Proband exome	Trio exome	Adult exome	Prenatal exome
Primary findings				
Variants related to the patient's phenotype				
Pathogenic/likely pathogenic/VUS	Yes	Yes	Yes[a]	Yes
Benign/likely benign	No	No	No	No
Secondary findings	Yes[b]	Yes[b]	Yes[b]	No[b]
Carrier	Yes	Yes	Yes	No[b]
Adult onset disorders	No	No	Yes[a]	No
Childhood onset disorders	Yes	Yes	Yes	Yes
Additional variants				
Non related to the phenotype of the patients	No[c]	No[c]	No[c]	No[c]
De novo and compound heterozygous in non-disease causing genes	No[c]	Yes	No[c]	No[c]
Optional variant categories depending on the laboratory Pharmacogenomics Mitochondrial DNA				

[a]patient present with a phenotype
[b]IF and carrier variants may be reported in the same report as the primary findings or on a separate report. Options to opt-out is available
[c]Available upon request

3.3.2 Adult Screening Exome

This test is offered by several laboratories to individuals, usually in good health, with no significant abnormal clinical presentation. Reports include the IF, carrier status and pathogenic findings in adult conditions.

3.4 Variants Usually Not Included in Exome Reporting

Variants not reported in exome reports consist of clinically irrelevant variants including: (1) Variants in disease genes not related to the patient's clinical phenotype; (2) Benign and likely benign variants; and (3) Variants in genes unknown to cause Mendelian disorders including susceptibility genes. These variants may be available in secondary reports available from the clinical laboratories and may help in future diagnosis. For instance, if an exome does not provide a molecular diagnosis, variants located in genes unknown to cause disease at the time of the report may become clinically relevant later as new genes causing disorders are discovered.

Additionally, pathological findings in adult neurological degenerative disorders including Huntington, Alzheimer and Parkinson's diseases, are usually not reported unless the patient is an adult referred for testing with one of these specific clinical phenotype, although specific gene testing is recommended in this particular case.

3.5 *Additional Considerations and Challenges*

Regarding the delivery and communication of clinical exome reports, clinical laboratories have primarily issued PDF reports, easily printable and delivered to the referring clinic. Due to the complex nature of the exome sequencing, the return of results needs more dynamic interactions between the referring physician and the clinical laboratory. An interactive web-based reporting portal with hyperlinks to the relevant web-based clinical and genomic information will help the physicians better understand the information provided in the exome reports. In addition, the interactive web-based reporting system should also enable better communication and interaction between the physician and clinical laboratory.

4 Clinical Utility of Whole Exome Sequencing

In an analysis of 500 patients evaluated in a medical genetics clinic setting, Shashi et al. (2014) reported that conventional diagnostic evaluation (i.e. clinical exam, biochemical testing, CGH array, and phenotype-directed sequencing) failed to establish a specific etiology in approximately 50% of patients with suspected genetic disorders [25]. This statistic is a testament to the great challenge of genetic diagnosis which is complicated by the rarity of many genetic syndromes and by the potential for phenotype and locus heterogeneity to obscure the causative gene. Without a diagnosis, patients are left with uncertainties about disease progression and long-term prognosis, may be ineligible for medically-indicated social services, and are often subject to additional and potentially invasive diagnostic testing (e.g. muscle biopsy).

Recent studies have demonstrated an inarguable role for exome sequencing in the diagnostic assessment of such unsolved cases. Using a proband-only approach, Yang et al. (2014) reported a diagnosis rate of 25% among 2000 consecutively tested patients for whom traditional approaches failed to elucidate a genetic etiology [20]. A combination of trio-based and proband-only testing yielded a diagnosis in 26% of 814 patients evaluated by exome sequencing and reported by Lee et al. (2014) [26]. In a smaller cohort of Canadian patients, Sawyer et al. (2016) identified a pathogenic variant in a known disease gene in 29% of cases, and this specifically in patients who were previously extensively evaluated and nearing the end of a protracted diagnostic odyssey [27]. In an unselected cohort of 500 patients, exome sequencing detected a positive or likely positive result in a recognized disease gene in 30% of patients (Farwell et al. 2015) [28]. Thus, although the process of exome sequencing data acquisition and variant analysis may differ between clinical laboratories, the general approach of sequencing the exome consistently yields a diagnosis in at least one out of every four patients tested.

Several groups have examined the effect of genetic diagnosis by exome sequencing on subsequent patient management. Valencia et al. (2015) reviewed in detail the first 40 pediatric exome cases performed at a single institution [29]. Consistent with other reports, the overall diagnostic rate was 30%. All patients who received a molecular diagnosis were considered meaningfully impacted by the result in that exome sequencing brought an end to the diagnostic odyssey and enabled disorder-specific genetic counseling. In addition, variants detected by exome sequencing resulted in a targeted treatment plan in three patients, an altered approach to clinical management in one patient, and disorder-specific surveillance in four patients [29]. Thevenon et al. (2016) similarly studied 43 patients with intellectual disability or epileptic encephalopathy at a single institution who underwent exome sequencing analysis [30]. Fourteen patients received a molecular diagnosis; in two cases this enabled prenatal testing and in two cases disease management was altered by the exome findings. In a third study, six of 105 patients diagnosed by exome sequencing were reported to have had a dramatic change in management as a consequence of the exome result (Sawyer et al. 2016) including, for example, a patient whose diagnosis was modified from infantile myofibromatosis to fibrodysplasia ossificans progressiva by exome sequencing resulting in discontinuation of chemotherapy [27].

The diagnostic yield of exome sequencing may be further increased if the testing context permits careful patient selection, rigorous phenotyping, and functional analyses. In a recent study of 41 deeply-phenotyped patients with intellectual disability and suspected metabolic disease evaluated with proband or trio-based exome, Tarailo-Graovac et al. (2016) found a molecular diagnosis in a remarkable 68% of patients [21]. However, this high diagnostic rate was contingent upon the establishment of two novel disease genes and the recognition of phenotypic expansion associated with 22 known disease genes; functional studies were performed to provide evidence of pathogenicity for variants in a subset of these genes. These exome sequencing results were reported to have altered or influenced subsequent clinical management in 44% of patients. While not feasible in a high throughput clinical setting, this work demonstrates that comprehensive phenotypic assessment together with the time allowance to pursue new gene discovery and the availability of resources to functionally address questions of phenotypic expansion may greatly augment the solve rate achievable by exome sequencing.

Although in most cases a clinical approach based on syndrome recognition abets the diagnostic process, it can also be a source of bias as the true phenotypic spectrum of many genetic disorders is not known. In addition, clinicians often operate under the assumption of Occam's razor – that the simplest explanation is the most likely. Studies of clinical exome cases have demonstrated the power of comprehensive sequencing to address diagnostic holes that may result from unavoidable clinician bias. For example, Farwell et al. (2015) specifically describe several cases in which autosomal recessive inheritance was suspected on the basis of a family history of consanguinity however *de novo* dominant events were ultimately detected by exome sequencing [28]. In 362 families tested by exome sequencing, Sawyer et al. (2016) found causative variants in established disease genes in 26 patients who escaped diagnosis because of atypical disease presentation [27]. Yang et al. (2014) found *bona fide* pathogenic variants in two disease-associated genes in 23 patients in their cohort resulting in blended and likely convoluted phenotypes, and also reported somewhat counterintuitively that X-linked disorders were found in equal numbers of male and female patients [20]. These scenarios underscore the value of unbiased genetic analysis in the diagnostic evaluation of unsolved cases.

The benefit of exome sequencing in augmenting diagnostic yield and providing medically-actionable information should be weighed against the potential cost to the individual and effect on societal healthcare expenditures. Individual costs may be financial or may come in the form of increased anxiety and/or additional medical surveillance following detection of a variant of uncertain clinical significance. To minimize personal cost and justify exome sequencing in a resource-limited context, judicious application of the test to those patients most likely to benefit is essential. The American College of Medical Genetics and Genomics policy statement on the Clinical Application of Genomic Sequencing (2012) suggests consideration of exome sequencing in affected patients with (1) non-diagnostic clinical features for whom a genetic etiology is likely (i.e. positive family history), (2) a disorder characterized by substantial locus heterogeneity, (3) a defined genetic disorder for which a molecular diagnosis has not been established by existing assays; and (4) for prenatal evaluation in cases where a clear diagnosis remains elusive after conventional genetic testing [https://www.acmg.net/staticcontent/ppg/clinical_application_of_genomic_sequencing.pdf]. The application of exome sequencing may also be appropriate for patients suspected of having a genetic condition for which no clinically-validated assay exists. The appropriate use of exome sequencing also requires recognition that exome sequencing as a methodology does not detect all forms of genetic variation. For example, single nucleotide variants and small insertion/deletion events (<10 bp) are reliably identified on exome sequencing whereas trinucleotide repeats, copy number variants, large insertion/deletion events, structural variants, aneuploidy, and epigenetic changes are not (Biesecker et al. 2014) [31]. In addition, technical limitations hinder complete coverage of the exome and in most cases, a small subset of genes lack the depth of coverage required for rigorous diagnostic assessment of that region. As such, ensuring coverage of key genes through online tools and acquiring a basic knowledge of the mechanism of gene disruption for disorders high on the differential diagnosis are important considerations prior to exome testing.

The sensitivity of exome sequencing may be further enhanced when a parent-child trio-based approach is used. Trio exome readily identifies *de novo* events and provides upfront phase data for variants found in autosomal recessive genes. Lee et al. (2014) compared 410 trio exome cases with 338 proband-only cases. The diagnostic rate was significantly higher in the trio exome cohort (31% vs. 22%, $p = .003$) although this was not a randomized comparison [26]. In addition to improved diagnostic yield, a second benefit of the trio-based approach is reduced turnaround time, which permits exome reporting in a time frame suitable for prenatal diagnosis and testing of critically-ill patients. Carss et al. (2014) performed a proof of principle study in which trio exome sequencing was performed on 30 fetuses and neonates with structural anomalies detected on prenatal ultrasound [32]. Three *de novo* likely causative variants were detected yielding a diagnostic rate of 10%. Drury et al. (2015) examined the utility of proband-only and trio-based exome sequencing for diagnosis of fetuses with abnormal ultrasound findings [33]. Return of results did not occur during pregnancy but pertinent findings were shared with families afterward. A definitive diagnosis was established by proband exome sequencing in 2 of 14 cases (14%) and by trio exome in 3 of 10 cases (30%). In the largest study to date, Normand et al. (unpublished data) reviewed 92 cases of prenatal WES performed on fetal samples obtained by amniocentesis/chorionic villus sampling or on products of conception. In 15 of 42 probandonly cases and 21 of 50 trio cases (~39% overall) a molecular diagnosis was ascertained by WES, suggesting a promising role for WES in improving prenatal diagnosis.

For critically-ill patients, studies have also evaluated the utility of whole genome sequencing (WGS) as WGS does not require a capture step and can be performed expeditiously. Fifty-seven percent of 35 acutely ill patients with heterogeneous clinical phenotypes reported by Willig et al. (2015) were found to have a causative variant on rapid trio-based WGS [34]. Soden et al. (2014) employed a rapid WGS protocol in 15 patients with primarily neurological phenotypes from neonatal or pediatric intensive care units; a molecular diagnostic rate of 73% was achieved in this cohort [35]. Notably, the fastest time to final report for rapid WGS in this study was 6–10 days, suggesting that trio-based exome sequencing (which can be performed clinically with a turnaround time of 2–3 weeks) is a reasonable alternative to WGS in critically-ill patients. Meng et al. (unpublished data) reviewed 40 patients tested clinically with critical or time-sensitive exome sequencing. The median turnaround time was 12.8 days and a potential or partial diagnosis was established in 52.5% of patients. In at least 14 cases, the results of exome sequencing influenced subsequent patient care decisions, demonstrating the utility and feasibility of exome sequencing in the critical care setting.

5 Exome Sequencing Versus NGS Panel Versus WGS

In order to select the most appropriate test for each patient, from single-gene to whole genome sequencing, it is essential for clinicians to understand the strengths, limitations, and diagnostic indications for each test. It is important to note that, even in the era of NGS technology, the traditional approach of single-gene testing still holds great utility for many disorders. Single-gene testing is preferred for patients who present with distinctive clinical findings that point to a particular Mendelian genetic disorder, for which the causative gene has been established. On the other hand, for disorders associated with wide clinical variability and genetic locus heterogeneity, a multigene panel approach or the whole exome or genome sequencing approach may provide greater benefit over single-gene tests.

Gene panel testing is preferred for patients who present with disorders associated with multiple causative genes, and/or present with a phenotype that cannot clearly point to one disorder. An advantage of targeted gene panels over whole exome and whole genome sequencing is comprehensive sequence coverage because these panels are often combined with complementary technologies such as Sanger sequencing or long range PCR to fill gaps that NGS fails to cover (due to high GC content, sequence homology, repetitive sequences, etc.). Some panels are also complemented with aCGH to simultaneously detect exon-level copy-number changes in targeted genes. Another advantage of targeted gene panels is better depth of coverage that provides greater confidence in variants detected, and shorter turnaround time.

Selecting the most appropriate gene panel can be a challenge for ordering physicians. Because clinical laboratories may use different stringencies for gene inclusion, the number of genes incorporated into a panel may vary significantly among laboratories even for the same clinical indications (Xue Y et al. Genet Med. 2015) [36]. Therefore, it is essential to know which genes show strong disease association and are therefore more relevant to the patient phenotype versus those that were linked with the disease based only on association studies or single studies. It is also important to note that addition of newly-identified disease genes may take time before they are added to existing panels. Many laboratories may also decide against adding new genes if not cost effective. Therefore, some laboratories have shifted to performing whole exome sequencing and limiting the analysis to genes associated with a particular phenotype and filling up the gaps with Sanger sequencing.

Clinical whole exome sequencing is currently indicated for patients who have either remained undiagnosed after single- or multi-gene panel testing, or for disorders with extreme heterogeneity and clinical variability that multigene testing is deemed less cost effective. Although exomes are intended to cover all protein coding regions of the genome, certain genomic regions (e.g. repetitive regions, high GC regions) decrease the performance of assay. Clinical exome sequencing usually has slightly lower coverage (usually up to 95–98%) than clinical NGS panels.

Whole exome sequencing typically uncovers approximately 20,000–50,000 variants per exome (Gilissen C et al. Eur J Hum Genet 2012) [37], and identifying the causal variant(s) thus can be a challenge. Computational tools have been developed that aid in the automation of variant prioritization, however, up to hundreds of variants still require careful manual inspection and curation. In addition, there is a growing concern about the potential of this test to identify incidental findings and the how to appropriately communicate them to patients (Kiltzman R, et al. *JAMA* 2013) [38]. As an important side note, when ordering exome sequencing, it is very useful to provide all clinical findings to the clinical molecular geneticist to help with variant interpretation.

Despite these limitations, exome sequencing has demonstrated great success as both a gene discovery and diagnostic tool. Large studies on the clinical utility of whole exome sequencing on a range of disorders have reported an overall molecular diagnostic rate of approximately 25–28% [20, 26, 39], with the yield higher for trio exomes than proband exomes (Lee H et al. *JAMA* 2014) [26]. Patients who have had whole exome sequencing are commonly children, since many genetic conditions present during childhood. In a report by Yang et al., the highest rate of a positive diagnosis was in a group of patients with a nonspecific neurological disorder (Yang Y, et al. *JAMA* 2014) [20].

Whole genome sequencing is considered to be the most comprehensive genetic test to date, covering approximately 98% of the genome [40, 41]. Because whole genome sequencing does not require an enrichment step, it generates a more uniform coverage of the genome over exome sequencing. Also, longer reads available for whole genome sequencing allows for better calling of copy number variations, rearrangements and other structural variations. In a report by Gilissen et al., whole genome sequencing was applied to patients with severe intellectual disability and their unaffected parents and reached a diagnostic yield of 42% (Gilissen C et al. *Nature* 2014) [41].

Despite the rapidly falling costs of sequencing, widespread application of whole genome sequencing to clinical diagnostics has been hampered by challenges in data analysis and relatively high costs of infrastructure needed to store, manage and analyze whole genome data. With the majority of causative variants identified so far in Mendelian disease occurring in coding regions, whole exome sequencing currently appears to be a more cost-effective approach and more practical alternative to whole genome sequencing (Teer JK, Mullikin JC. Hum Mol Genet.2010) [42]. Additionally, because variation in noncoding regions is less well understood than variation in the coding region, it is more difficult to predict which variants might be relevant to a trait of interest in whole genome datasets.

References

1. Hwang, S., Kim, E., Lee, I., Marcotte, E.M.: Systematic comparison of variant calling pipelines using gold standard personal exome variants. Sci. Rep. **5**, 17875 (2015)
2. Langmead, B., Salzberg, S.L.: Fast gapped-read alignment with Bowtie 2. Nat. Methods. **9**(4), 357–359 (2012)
3. Van der Auwera, G.A., Carneiro, M.O., Hartl, C., et al.: From FastQ data to high confidence variant calls: the Genome Analysis Toolkit best practices pipeline. Curr. Protoc. Bioinformatics. **43**, 11.10.11–33 (2013)
4. Li, H.: A statistical framework for SNP calling, mutation discovery, association mapping and population genetical parameter estimation from sequencing data. Bioinformatics. **27**(21), 2987–2993 (2011)
5. Challis, D., Yu, J., Evani, U.S., et al.: An integrative variant analysis suite for whole exome next-generation sequencing data. BMC Bioinf. **13**, 8 (2012)
6. Rimmer, A., Phan, H., Mathieson, I., et al.: Integrating mapping-, assembly- and haplotype-based approaches for calling variants in clinical sequencing applications. Nat. Genet. **46**(8), 912–918 (2014)
7. Yang, H., Wang, K.: Genomic variant annotation and prioritization with ANNOVAR and wANNOVAR. Nat. Protoc. **10**(10), 1556–1566 (2015)
8. Cingolani, P., Platts, A., Wang le, L., et al.: A program for annotating and predicting the effects of single nucleotide polymorphisms, SnpEff: SNPs in the genome of Drosophila melanogaster strain w1118; iso-2; iso-3. Fly (Austin). **6**(2), 80–92 (2012)
9. McLaren, W., Pritchard, B., Rios, D., Chen, Y., Flicek, P., Cunningham, F.: Deriving the consequences of genomic variants with the Ensembl API and SNP effect predictor. Bioinformatics. **26**(16), 2069–2070 (2010)
10. McCarthy, D.J., Humburg, P., Kanapin, A., et al.: Choice of transcripts and software has a large effect on variant annotation. Genome Med. **6**(3), 26 (2014)
11. Amarasinghe, K.C., Li, J., Halgamuge, S.K.: CoNVEX: copy number variation estimation in exome sequencing data using HMM. BMC Bioinf. **14**(Suppl 2), S2 (2013)

12. Krumm, N., Sudmant, P.H., Ko, A., et al.: Copy number variation detection and genotyping from exome sequence data. Genome Res. **22**(8), 1525–1532 (2012)
13. Fromer, M., Moran, J.L., Chambert, K., et al.: Discovery and statistical genotyping of copy-number variation from whole-exome sequencing depth. Am. J. Hum. Genet. **91**(4), 597–607 (2012)
14. Richards, S., Aziz, N., Bale, S., et al.: Standards and guidelines for the interpretation of sequence variants: a joint consensus recommendation of the American College of Medical Genetics and Genomics and the Association for Molecular Pathology. Genet. Med. **17**(5), 405–424 (2015)
15. Amendola, L.M., Jarvik, G.P., Leo, M.C., et al.: Performance of ACMG-AMP variant-interpretation guidelines among nine Laboratories in the Clinical Sequencing Exploratory Research Consortium. Am. J. Hum. Genet. **98**(6), 1067–1076 (2016)
16. Gulley, M.L., Braziel, R.M., Halling, K.C., et al.: Clinical laboratory reports in molecular pathology. Arch. Pathol. Lab. Med. **131**(6), 852–863 (2007)
17. Dorschner, M.O., Amendola, L.M., Shirts, B.H., et al.: Refining the structure and content of clinical genomic reports. Am. J. Med. Genet. C Semin. Med. Genet. **166C**(1), 85–92 (2014)
18. Kumar, P., Henikoff, S., Ng, P.C.: Predicting the effects of coding non-synonymous variants on protein function using the SIFT algorithm. Nat. Protoc. **4**(7), 1073–1081 (2009)
19. Adzhubei, I.A., Schmidt, S., Peshkin, L., et al.: A method and server for predicting damaging missense mutations. Nat. Methods. **7**(4), 248–249 (2010)
20. Yang, Y., Muzny, D.M., Xia, F., et al.: Molecular findings among patients referred for clinical whole-exome sequencing. JAMA. **312**(18), 1870–1879 (2014)
21. Tarailo-Graovac, M., Shyr, C., Ross, C.J., et al.: Exome sequencing and the Management of Neurometabolic Disorders. N. Engl. J. Med. **374**(23), 2246–2255 (2016)
22. Green, R.C., Berg, J.S., Grody, W.W., et al.: ACMG recommendations for reporting of incidental findings in clinical exome and genome sequencing. Genet. Med. **15**(7), 565–574 (2013)
23. Kalia, S.S., Adelman, K., Bale, S.J., et al.: Recommendations for reporting of secondary findings in clinical exome and genome sequencing, 2016 update (ACMG SF v2.0): a policy statement of the American College of Medical Genetics and Genomics. Genet. Med. **19**(2), 249–255 (2016)
24. Smith, L.A., Douglas, J., Braxton, A.A., Kramer, K.: Reporting incidental findings in clinical whole exome sequencing: incorporation of the 2013 ACMG recommendations into current practices of genetic counseling. J. Genet. Couns. **24**(4), 654–662 (2015)
25. Shashi, V., McConkie-Rosell, A., Rosell, B., et al.: The utility of the traditional medical genetics diagnostic evaluation in the context of next-generation sequencing for undiagnosed genetic disorders. Genet. Med. **16**(2), 176–182 (2014)
26. Lee, H., Deignan, J.L., Dorrani, N., et al.: Clinical exome sequencing for genetic identification of rare mendelian disorders. JAMA. **312**(18), 1880–1887 (2014)
27. Sawyer, S.L., Hartley, T., Dyment, D.A., et al.: Utility of whole-exome sequencing for those near the end of the diagnostic odyssey: time to address gaps in care. Clin. Genet. **89**(3), 275–284 (2016)
28. Farwell, K.D., Shahmirzadi, L., El-Khechen, D., et al.: Enhanced utility of family-centered diagnostic exome sequencing with inheritance model-based analysis: results from 500 unselected families with undiagnosed genetic conditions. Genet. Med. **17**(7), 578–586 (2015)
29. Valencia, C.A., Husami, A., Holle, J., et al.: Clinical impact and cost-effectiveness of whole exome sequencing as a diagnostic tool: a pediatric center's experience. Front Pediatr. **3**, 67 (2015)
30. Thevenon, J., Duffourd, Y., Masurel-Paulet, A., et al.: Diagnostic odyssey in severe neurodevelopmental disorders: toward clinical whole-exome sequencing as a first-line diagnostic test. Clin. Genet. **89**(6), 700–707 (2016)
31. Biesecker, L.G., Green, R.C.: Diagnostic clinical genome and exome sequencing. N. Engl. J. Med. **370**(25), 2418–2425 (2014)
32. Carss, K.J., Hillman, S.C., Parthiban, V., et al.: Exome sequencing improves genetic diagnosis of structural fetal abnormalities revealed by ultrasound. Hum. Mol. Genet. **23**(12), 3269–3277 (2014)

33. Drury, S., Williams, H., Trump, N., et al.: Exome sequencing for prenatal diagnosis of fetuses with sonographic abnormalities. Prenat. Diagn. **35**(10), 1010–1017 (2015)
34. Willig, L.K., Petrikin, J.E., Smith, L.D., et al.: Whole-genome sequencing for identification of mendelian disorders in critically ill infants: a retrospective analysis of diagnostic and clinical findings. Lancet Respir. Med. **3**(5), 377–387 (2015)
35. Soden, S.E., Saunders, C.J., Willig, L.K., et al.: Effectiveness of exome and genome sequencing guided by acuity of illness for diagnosis of neurodevelopmental disorders. Sci. Transl. Med. **6**(265), 265ra168 (2014)
36. Xue, Y., Ankala, A., Wilcox, W.R., Hegde, M.R.: Solving the molecular diagnostic testing conundrum for Mendelian disorders in the era of next-generation sequencing: single-gene, gene panel, or exome/genome sequencing. Genet. Med. **17**(6), 444–451 (2015)
37. Gilissen, C., Hoischen, A., Brunner, H.G., Veltman, J.A.: Disease gene identification strategies for exome sequencing. Eur. J. Hum. Genet. **20**(5), 490–497 (2012)
38. Klitzman, R., Appelbaum, P.S., Chung, W.: Return of secondary genomic findings vs patient autonomy: implications for medical care. JAMA. **310**(4), 369–370 (2013)
39. Yang, Y., Muzny, D.M., Reid, J.G., et al.: Clinical whole-exome sequencing for the diagnosis of mendelian disorders. N. Engl. J. Med. **369**(16), 1502–1511 (2013)
40. Lupski, J.R., Reid, J.G., Gonzaga-Jauregui, C., et al.: Whole-genome sequencing in a patient with Charcot-Marie-tooth neuropathy. N. Engl. J. Med. **362**(13), 1181–1191 (2010)
41. Gilissen, C., Hehir-Kwa, J.Y., Thung, D.T., et al.: Genome sequencing identifies major causes of severe intellectual disability. Nature. **511**(7509), 344–347 (2014)
42. Teer, J.K., Mullikin, J.C.: Exome sequencing: the sweet spot before whole genomes. Hum. Mol. Genet. **19**(R2), R145–R151 (2010)

Family-Based Next-Generation Sequencing Analysis

Xia Wang, Linyan Meng, and Magalie S. Leduc

Abstract Next-generation sequencing (NGS) has been extensively used in genomic study and clinical diagnostic arena in the past decade. With the decreasing cost of NGS, family-based sequencing analysis has been increasingly used to identify causal genes for Mendelian disorders and to aid the rare variants association analysis for common complex traits. By incorporating relatedness among family members, several family-based variant calling algorithms have been developed to increase variant calling accuracy, and showed superior performance comparing with naïve calling algorithms. Using parent-offspring trio sequencing, numerous causal genes and risk alleles for neuropsychiatric and other disorders have been successfully identified. In addition, family-based NGS can help to increase the statistic power for rare variants association analysis for common complex traits. Family-based NGS is also found to be particularly useful in clinical diagnostic laboratories. Comparing with proband-only sequencing, family-based NGS helps to quickly narrow down variants inheritance, increasing the diagnostic yield and decreasing the turnaround time of clinical tests. In this review, we will first discuss the methodology for family-based NGS analysis, focusing on variant calling, and then review the application of such design in research and clinical settings.

Keywords Next-generation sequencing • Family-based sequencing • Trio • Variant calling • Disease-gene identification • *de novo*

1 Introduction

Next-generation sequencing (NGS) has revolutionized the genomic study in both research and clinical settings in the past decade [1, 2]. Aided by NGS, numerous causal genes for Mendenlian disorders and risk alleles for common disorders have been discovered, paving the way for mechanistic study and target treatment [3, 4]. In clinically diagnostic laboratories, NGS has been strictly validated and applied as

X. Wang, Ph.D. (✉) • L. Meng, Ph.D. (✉) • M.S. Leduc, Ph.D. (✉)
Department of Molecular and Human Genetics, Baylor College of Medicine, Houston, TX, USA
e-mail: xiaw@bcm.edu; lmeng@bcm.edu; Magalie.Leduc@bcm.edu

L.-J.C. Wong (ed.), *Next Generation Sequencing Based Clinical Molecular Diagnosis of Human Genetic Disorders*, DOI 10.1007/978-3-319-56418-0_15

diagnostic testing for inherited disorders that could not have been easily diagnosed by other conventional sequencing methods, confirming clinical diagnosis, leading to more accurate prognosis, and allowing target treatment. By the rapidly decreasing cost of NGS technology, family-based NGS experiment design has been used more and more in both research and clinical settings [5, 6]. Several advantages arise from family-based NGS analysis. First, genotyping all family members helps to determine the inheritance pattern of target variants and prioritize candidate variants related to the phenotypes. For example, when only parent-offspring trio is available to study Mendelian disorders, trio sequencing quickly identifies *de novo* variants (DNV) in autosomal dominant disease genes or two *in trans* variants in autosomal recessive genes that related to clinical phenotypes, and excludes variants that do not fit the Mendelian inheritance pattern, resulting in very few candidate genes. Second, by joint variant calling across all family members, modeling Mendelian inheritance, and incorporating shared haplotypes, family-based NGS greatly increases the accuracy of variant calling. Third, family-based design helps to identify rare variants underlying common complex traits. Rare variants play important roles in common complex disorders [4, 7]. Families including multiple affected members are enriched for rare causal variants, thus increase the statistic power for rare variants association study [8]. Numerous examples have been published of identifying variants in known disease genes using trio or family-based exome sequencing since 2010. It has also been very successfully used in discovering novel disease genes, especially for those with *de novo* mutations. Because of these advantages and the continuing decrease of the cost for NGS technology, trio whole exome sequencing has been widely used for research of rare Mendelian disorders and common complex genetic disorders. Several global initiatives to explore Mendelian conditions, including the Centers for Mendelian Genomics (CMGs) in the US, the Finding of Rare Disease Genes (FORGE) in Canada, the Deciphering Development Disease study group (DDD) in UK, and centers in other countries, has also employed the family-based sequencing strategy when possible. In addition, it was quickly adapted into clinical services as well with great success. In this chapter, we first discuss the methodology for family-based NGS analysis, focusing on variant calling (Sect. 2), and then review the application of such design in research (Sect. 3) and clinical settings (Sect. 4).

2 Family-Based NGS Methodology: Variant Calling

Family based NGS has been widely used to study the genetic basis of common and rare human traits. Variant calling is a crucial analytic step to convert raw NGS data into human readable genetic variations. One big advantage of family study design is that it enables us to leverage the information of all family members to improve the accuracy of variant calling, which in turn fundamentally enhances our ability to narrow genetic loci associated with traits of interests. In this section, we will review various variant calling algorithms that specifically designed for family-based sequencing analysis.

Variant calling is an important step in the analysis of NGS data. Variant calling identifies variant sites in sequencing samples and estimates the genotype of those variants. Due to the high throughput and short-reads of NGS, variant calling suffers from several sources of errors happened during the generation and processing of NGS data, such as artifacts introduced by library preparation and PCR, reads alignment errors, inflated or nonlinear base quality values, and errors caused by low coverage. To tackle these challenges and call variants as accurately as possible, several variant calling algorithms have been developed to model these possible errors and to estimate the likelihood of observed sequencing variants to be true variants in the genome, such as GATK, SAMtools, and SOAP2 [9]. It should be noted that the concordance among various individual variant callers are low, indicating the complexity of NGS data and rooms for algorithm improvements [10].

One emerging algorithmic improvement of variant calling is to call variants simultaneously across all samples, namely joint or multi-sample calling. The idea is to utilize prior information across all different samples, such as patterns of allele frequency distribution and population frequency, to improve the variant calling accuracy in individual sample. For example, low quality variants that are missed by individual calling may be called by joint calling if accumulative evidences across all samples support those variants to be true. For another example, if the prior population frequency of a certain variant is known, we can use this prior information to adjust the likelihood of the variant's genotype. Indeed, by genotyping variants on chromosome 20 of 62 CEU individuals using GATK, preliminary data from Nielsen et al. (2011) showed that joint calling is overall more accurate than individual calling [9]. By genotyping variants in 641 low-coverage whole genome sequencing (WGS) samples using GATK, Bizon et al. (2014) showed that joint calling is more sensitive (even for singletons only appear once in multiple samples being analyzed) than individual calling, although it also causes more false positives [11]. In addition, Liu et al. (2013) evaluated four different variant calling algorithms on 20 WGS samples and found that join calling is more sensitive than individual calling [12]. In most situations, joint calling is more accurate than individual calling. Table 1 summarizes available algorithms for joint calling.

Family-based sequencing analysis aims to identify DNV or inherited variants in the family that are related with phenotypes or clinical conditions. In this scenario, joint calling algorithms that call variants across all family members can be used to improve the variant calling accuracy. More importantly, relatedness between family members can provide additional information to improve the variant calling. First, modeling the Mendelian inheritance of variants within the family greatly reduces sequencing errors and false calls [5, 16]. Second, shared haplotypes among family members can be used to infer variant genotypes [17]. By taking into account the relatedness of family members, family-based joint calling greatly increases the calling accuracy. For example, low quality variant calls in the parents that are missed by individual calling may be picked up by family-based joint calling, leading to "correction" of the false positive DNV calls in the proband. Similarly, low quality variant calls in the proband that are missed by individual calling may be picked up by family-based joint calling, resulting in "rescue" of the real DNV in the proband [18]. Conrad et al. (2011) used methods including two family-based joint calling

Table 1 Available algorithms for joint calling.

Algorithms	Inputs	Comments	References
GATK	Raw reads	A package for reads mapping, alignment, processing, and SNV calling	[13]
glfTools	Raw candidate variants	SNV calling	(http://csg.sph.umich.edu//abecasis/glfTools/)
SAMtools	Raw reads	A package for reads manipulation and SNV calling	[14]
VarScan	Aligned reads	Germline/somatic SNV/CNV calling	[15]

algorithms, FIGL and FPIR, to identify DNV in the genomes of two trios and estimated the genome-wide mutation rate. [19] The FPIR method was later formalized by Cartwright et al. (2012) and was thought to be the first model-based approach to family-based variant calling [18]. Li et al. (2012) developed a probabilistic method for DNV detention (PolyMutt) in families and showed superior performance over standard joint calling algorithm GATK that ignores the family relatedness information [20]. Chen et al. (2013) incorporates both relatedness and linkage disequilibrium (LD) information within the family and developed the TrioCaller algorithm for variant calling and haplotype inference [21]. Similarly, Kojima et al. (2013) utilized both pedigree and haplotype information for variant calling and showed their algorithm, PedigreeCaller, is more accurate than other family-based variant caller, such as TrioCaller and PolyMutt, on real parent-offspring NGS data set from HapMap project [22]. Cleary et al. (2014) developed the RTG Variant algorithm using both relatedness and haplotype information and showed that their method is especially fast and scalable to large pedigrees [17]. It should be noted that LD information is less useful for the calling/inferring of rare or private variants because there is not enough LD information associated with these variants. Santoni et al. (2014) developed the VariantMaster algorithm for both family-based and tumor/normal pair variant calling and showed better accuracy and precision than other algorithms [23]. Because DNV calling suffers from various NGS false positive errors, Liu et al. (2014) developed a machine learning based algorithm DNMFilter to filter candidate DNV calls from family sequencing data [24]. The DNMFilter greatly reduces the false positive DNV and can be used as a complement module to any other DNV callers mentioned above. To be more user-friendly and streamline the process of family-based variant calling, Li et al. (2015) developed a web tool mirTrios for family-based variant calling, annotation, and candidate gene prioritization [25]. Table 2 summarizes the available algorithms for family-based variant calling.

In summary, joint calling across all family members, modeling Medenlian inheritance, and incorporating LD information can greatly improve the family-based variant calling. When design family-based sequencing experiments, one should almost always utilize joint calling and family-based variant calling algorithms to increase calling accuracy and identify candidate disease-causing variants.

Table 2 Available algorithms for family-based variant calling.

Algorithms	Input	Comments	References
FPIR		The first model-based approach to family-based variant calling	[18, 19]
PolyMutt	Raw candidate variants	Family-based variant calling	[20]
Famseq	Variants	Family-based variant calling	[26]
TrioCaller	Variants	Family-based variant calling and haplotype inference	[21]
PedigreeCaller	Aligned reads	Family-based variant calling and haplotype inference	[22]
Denovogear	Variants	Family-based variant calling and haplotype inference	[27]
RTG Variant	Aligned reads	Family-based variant calling and haplotype inference, fast and scalable to large pedigrees	[17]
VariantMaster	Aligned reads, variants	Family-based and tumor/normal pair variant calling	[23]
DNMFilter	Variants	A standalone algorithm to filter and reduce false positive DNV calls, complementary to other DNV callers.	[24]
mirTrios	Variants	A web tool for family-based variant calling, annotation, and candidate gene prioritization	[25]

3 Family-Based NGS in Research Setting

3.1 Family-Based NGS in Healthy Populations

A basic understanding of germline mutation rate in the human genome in healthy individuals is necessary to understand and interpret DNV in patient cohorts. Family based NGS gives access for the first time to an estimation of the mutation rate in human populations by detecting all *de novo* variants from one generation to the next. The development of whole genome sequencing technology helped estimate the germline mutation rate in healthy individuals with the use of trio analysis. First, Roach et al. re-sequenced a family of four individuals (parents and two siblings) and estimated the mutation rate at 1.1×10^{-8} per base per haploid genome [5]. Subsequently, Conrad et al. performed whole genome sequencing in two trios (proband-parents) and found 35 and 49 *de novo* germline variants in each trio respectively [19]. This mutation rate is consistent with other reports from Iceland with a mutation rate of 1.2×10^{-8} per base per generation [28]. Recently, Francoli et al. re-sequenced 250 Dutch families (trio parent-offsprings) allowing for a better understanding of the mutation rate and mechanism that yield to DNV in a larger population [29]. The team uncovered 11,020 DNV. Contributing factor to the mutation rate included paternal age: offsprings of older father carried more DNV and 95% of the global mutation rate variation in humans could be explained by paternal age only [29]. Interestingly, functional regions of the genome were enriched for DNV: about 1.2% of DNV were located in exons. Additionally, regulatory regions were also enriched for DNV, likely due to the presence of CpG dinucleotides. As a consequence, it was hypothesized that DNV in offsprings of older father are likely to be functional. This phenomenon is already observed in well-known Mendelian disorders such as achondroplasia [OMIM: 100,800]. The majority of the patients with asymptomatic parents (99%) carry one of two mutations in the *FGFR3* gene [30]. The incidence of these DNV is correlated with the father's age [31]. Thus, estimating the number of germline DNV in healthy individuals and trios and understanding the contributing factors that can alter the mutation rate in humans, helps interpreting DNV in our genome.

3.2 Studies for Rare Mendelian Disorders

The pioneer work for family-based genome sequencing was first done by two independent groups in 2010 [32, 33]. In one study, whole-genome sequencing was performed in a family with two siblings affected with two autosomal recessive diseases, Miller syndrome and ciliary dyskinesia, and their unaffected parents. Based on the principle that adjacent variant in the family genomes have the same inheritance state unless recombination has occurred between the bases during meiosis, the authors were able to filter out ~70% of the sequencing errors, delineate inheritance blocks, and narrowed down the candidate genes from 34, if parent sequence was unavailable, to only four genes. In another study, ten undiagnosed individuals with moderate to severe mental retardation and negative family history were analyzed by whole exome sequencing, together with their unaffected parents. Nine *de novo* non-synonymous mutations from seven families were identified and confirmed by Sanger, of which six were concluded as strong candidates based on gene function, evolutionary conservation, and mutation impact. An additional pathogenic mutation in an X-linked gene was found in one family, which was later confirmed to occur in the mother of this proband. These two pieces of work beautifully demonstrated the use of family sequencing analysis under the two most common paradigm, compound heterozygous inheritance and DNV. Since then, family based WES analysis, or more commonly used, trio WES, has been widely applied in research to help identify molecular causes for monogeneic as well as complex genetic disorders.

Many genetic disorders occur sporadically, i.e. the proband is the only affected individual with no other similarly affected family members. In these cases, it is highly likely that the genetic disturbance arose during meiosis and account for the clinical phenotype. Indeed for people who received definitive diagnosis with clinical WES, about 40–50% of cases are attributed to DNV, no matter through proband only analysis or trio-based analysis [6, 34, 35]. Identifying such variants through proband only WES requires prior knowledge about the gene-phenotype association and careful correlation of the sequencing data with the clinical notes. Even this, while the disease gene is unknown or the clinical presentation is atypical, it is very difficult to identify candidate genes when analyzing proband only. In contrast, such information is readily available when both parents are concurrently analyzed and compared. Taking advantage of trio-based whole exome sequencing, many new disease genes were discovered. Examples include Weaver syndrome (*EZH2*) [36], Baraitser-Winter syndrome (*ACTB* and *ACTG1*) [37], Cantu syndrome (*ABCC9*) [38], mental retardation type 21 (*CTCF*) [39], SHORT syndrome (*PIK3R1*) [40], Harstfield syndrome (*FGFR1*) [41], and epileptic encephalopathy (*SCN8A*) [42] to name a few.

Parental mosacism is frequently reported in families with apparent DNV in the affected child, including true germline mosacism, which cannot be easily detected by current methodologies, and somatic mosacism, which is present in one or more somatic tissues like blood. It is estimated that the recurrent risk will be substantially increased when mutations were found in parental blood comparing to those confined to the germline [43]. However, detection of parental mosaic mutations, especially those at a low level, is often limited by the testing sensitivity. For example the detection limit of Sanger sequencing is around 5–10% [44]. In contrast, with the increase of sequencing depth, targeted massive parallel sequencing has the ability to detect low level mosaicism. In a recent study of 174 Dravet syndrome patients with *SCN1A* variants, Sanger sequencing revealed parental mosaicism in only five cases. However, when targeted resequencing by NGS was performed, disease-causing mutations from an additional 15 cases (8.6%) were found to be inherited from a mosaic parent, and the level of mosaicism ranges 1.1–32.6% [45], indicating that parental mosaicism is possibly more common than expected and it can be missed by proband-only WES analysis, which only uses Sanger sequencing in parental samples. Trio exome sequencing, on the other hand, offers an advantage on this aspect. This is demonstrated in a family with two affected siblings of hypomyelination with atrophy of the basal ganglia and cerebellum. Quartet WES analysis revealed that the pathogenic c.745G>A variant in the *TUBB4A* gene is inherited from a mosaic mother, with 25% mutant allele frequency [46]. Although this mosaic level is within the Sanger detection range, as a proof-of-principle, it showed that family-based exome sequencing is able to detect parental mosaicism. With the increased application of trio exome sequencing, we will be seeing similar cases with low level parental mosaicism being reported more often.

Autosomal recessive inheritance is another mechanism for sporadic genetic disorders. With the availability of the both parental sequence, family-based WES analysis can quickly identify those homozygous or compound heterozygous variants, as demonstrated in one of the earliest family exome sequencing cases by Roach et al. mentioned above [5]. Additional examples include skeletal dysplasia (*POP1*) [47], non-syndromic mental retardation (*TECR*) [48], opsismodysplasia (*INPP1*) [49], Alstrom syndrome (*ALMS1*) [50], and others. More interestingly, in a patient affected with congenital disorder of glycosylation, trio WES analysis identified a *de novo* frame-shift mutation [c.701dup (p.Tyr234*)] and a missense mutation [c.1900T>G (p.Trp634Gly)] inherited from the mother in the COG2 gene. Further RNA study suggested a trans configuration of the two variants in the proband, confirming the diagnosis [51].

3.3 Large Cohort Studies for Complex Diseases

While some diseases have distinct clinical features and highly penetrant phenotype, many are difficult to be diagnosed even by the most experienced genetic specialist. As a result, patients with rare diseases either went through several years of "diagnostic odyssey" to receive a definitive answer, or in most cases, remain undiagnosed. The complication and difficulty of genetic diagnosis, on one hand, can be attributed to the similar and non-specific clinical presentations of many genetic diseases, which is usually further complicated by incomplete penetrance and atypical presentation of the disorders. On the other hand, this comes from our little understanding of the underlying genetic basis of many diseases. While targeted gene analysis (gene panels) and proband only WES has profoundly accelerated the diagnostic timeframe, family-based exome or genome sequencing remain to be the most powerful tool in the search for new disease genes. This is especially true when the disease or phenotype is highly locus heterogeneous, i.e. the disease causing gene may be distinct in each family, although the clinical phenotypes may appear similar as other families. With the advance of next generation sequencing technology and the ever decreasing sequencing cost, unprecedentedly large scale studies with hundreds of parent-proband trios were continuously reported for autism, schizophrenia, intellectual disability, developmental delay, cerebral palsy, etc. Here we are going to take the extremely complicated neurodevelopmental disorders as examples to illustrate trio exome sequencing analysis as a robust tool in overcoming such broad genetic heterogeneity in this patient group.

Developmental delay and/or intellectual disability are estimated to be the major complaint of families seeking for genetic help. However, more than half of patients cannot receive a definitive diagnosis despite extensive workup, especially those with non-syndromic intellectual disability [52, 53]. Several studies on this patient population by trio exome sequencing have generated a consistent mutation profile [54–56]. From 151 trio families with non-syndromic severe intellectual disability, DNV were found in 96 affected probands, 26 of them received definitive diagnosis (17.1%) and 28 genes deemed as disease-causing candidates. In contrast, only three cases were considered diagnosed with X-linked genes (0.02%) and two with autosomal recessive genes (one case has one variant inherited and one variant *de novo*, 0.01%). In a more recent large cohort of 1133 patients with developmental disorders, mainly intellectual disability or developmental delay, similar diagnostic profile was obtained with 22.5% of cases diagnosed by DNV (autosomal and X-linked), and only 3.6% with autosomal recessive diseases and 0.2% with X-linked diseases [56]. Such mutation profile was even replicated in a whole genome sequencing study for 50 ID trios previously undiagnosed by chromosome microarray and whole exome sequencing. 60% of patients from this prescreened cohort received a conclusive genetic diagnosis by *de novo* events (39% SNV and 21% CNV) and only 2% by recessive inheritance [57]. Based on these findings, it is concluded that DNV is an important cause of intellectual disability. This conclusion is also supported by a study of schizophrenia patients, which specifically focused on recessive genotypes [58].

Autism spectrum disorders (ASDs) is another relatively common but even more complicated and less understood neurodevelopmental disorder. It is estimated to affect up to 1% of human population [59]. A strong genetic component is known to exist. However, despite the tremendous research effort, no major gene has been identified in this patient population. A two-class risk model was proposed [60, 61]. The low-risk families have only one affected proband (sporadic simplex cases), comprising 99% of the disease population. The genetic cause in this category is more likely to be *de novo* that occur in a parental germline. The high-risk families, on the contrary, have two or more affected individuals (multiplex cases), and carry a causative mutation that transmitted in the family in an autosomal dominant fashion. This model was soon echoed by several studies revealing that large *de novo* copy number variants (CNVs) were significantly enriched among probands (8% with a CNV) when compared with unaffected siblings or controls (2%) [62–65]. Therefore, when next generation sequencing became readily available, parent-child trio sequencing was quickly applied to those simplex autism families to study the contribution of *de novo* sequencing mutations and to discover new disease genes. Similar to *de novo* CNVs, four large cohort studies with 948 trio exomes found a significant increase of *de novo* loss-of-function (LOF) sequence variants in probands comparing with unaffected siblings [66–69], with a collective odds ratio (OR) of 2.41 ($P < 1 \times 10^{-4}$) [70]. Interestingly, when DNV result in missense or synonymouse changes, they appear to be less enriched in the affected probands, probably because that these variants are less likely to be pathogenic and disease-causing (for missense, OR = 1.29, and for synonymous, OR = 0.98) [70]. Overall, it is estimated that *de novo* LOF sequencing variants contribute to ASD risk for about 10–15% of probands [68, 69]. Similar DNV burden has also been observed in the previously mentioned patient cohorts of intellectual disability or developmental disorders, consistent with the mutation profile that majority of definitive genetic diagnosis are due to DNV [55, 56].

Genes with DNV identified by a trio exome sequencing study are often flagged as candidate disease causing genes. Further supporting evidence comes from bioinformatics prediction based on the mutation and protein structure, animal models, protein networking analysis, correlation with CNV findings, discovery of additional unrelated patients with mutations in the same gene, etc. For example, with trio exome sequencing from four independent studies and further effort of targeted resequencing of candidate genes in additional ASD families, a number of candidate genes were discovered with *de novo* hits in at least two unrelated ASD probands. These genes include *CHD8*, *DYRK1A*, *GRIN2B*, *KATNAL2*, *RIMS1*, *SCN2A*, *POGZ*, *ADNP*, *ARID1B*, and *TBR1* [61]. A novel approach was taken in the recent large scale gene discovery study by Deciphering Developmental Disorders group [56]. With meta-analysis of 3477 developmental disorder trios from several published cohorts of intellectual disability, autism, epilepsy encephalopathy, and schizophrenia, a gene-specific over-abundance analysis of either *de novo* LOF mutations or clustered functional mutations was performed (as compared to the rest of exome). This strategy successfully re-discovered 20 known disease genes linked to developmental disorders, and 12 novel genes solely based on statistical grounds. It is noteworthy that many from the 12 candidate genes have already been demonstrated to be bona-fide developmental-disorder-linked genes [71, 72].

Family based whole exome and genome sequencing has revolutionized our understanding and diagnostic strategy for neurodevelopmental disorders in the past several years. However, even with the most commonly identified genes from these large cohort studies, each single gene accounts for only 0.5–1% of total patients. It is estimated that there will be approximately 300–1000 ASD genes and more than 1000 different genes in neurodevelopmental disorders [61, 70, 73]. This will continue being one of the most challenging obstacles to the field in the future and the search for new disease genes will never end.

4 Family-Based NGS in Clinical Settings

The development and the use of family-based NGS in research has helped understanding the genetic basis of Mendelian, complex and sporadic disorders. For patients with undiagnosed disorders, family-based NGS can also help to prioritize the disease-causing variants. As research laboratories streamlined the technical and analytical aspect of whole exome sequencing (WES), in 2011, several clinical CLIA /CAP accredited laboratories started developing a clinical test offering clinical whole exome sequencing to patients with undiagnosed disorders [6, 34, 35, 74]. As of August 2015, over 15 clinical laboratories are now offering clinical whole exome sequencing (Gene registry, http://www.ncbi.nlm.nih.gov/gtr/).

4.1 Trio-WES Approach in the Clinical Laboratories

Clinical genetic laboratories offer two major strategic approaches regarding clinical WES: (1) Proband exome sequencing with or without subsequent targeted Sanger sequencing of relevant variants in the parents (P-WES); and (2) Trio exome sequencing (Trio-WES) where proband and parents are simultaneously sequenced and analyzed. While the first approach is best suitable when one parent is unavailable for instance or for cost purposes, the second approach (Trio-WES) has been shown to provide advantages from a patient and laboratory perspective as explained below.

First, Trio-WES offers a better diagnosis rate. Two clinical laboratories have compared their diagnosis rate using either Trio-WES or P-WES. In 2015, Lee et al. from the clinical laboratory at UCLA reported on 814 clinical patients who underwent WES testing; 410 (about 50%) were Trio-WES. A molecular diagnostic was provided for 127 of those (31% success rate) as opposed to 74/338 (22%) patients who underwent P-WES [6]. Similarly, Farwell et al. reported a success rate of 37.3% in Trio-WES as opposed to 20.6% for P-WES [35]. In both studies, it is important to note that about half of the diagnosed patients had a DNV in a gene related to the patient's phenotype, hence providing a molecular diagnosis to the patient.

Second, trio-WES often offers a better turnaround time (TAT). Trio WES allows for the direct determination of variant inheritance from each parent. Any DNV, compound heterozygous variants and X-linked variants are directly identified without the need for Sanger confirmation in the parents, simplifying overall the WES analysis and bioinformatic process. As a consequence, the TAT is shorten, and a possible molecular diagnosis can be provided faster. At the same time, because both - proband and parents - are sequenced simultaneously, the laboratory cost of Trio-WES is higher than P-WES: it has been estimated that the cost is only about 50% higher compared to proband only [6].

Overall, the decision of ordering a Trio-WES versus a P-WES will need to be balanced between cost efficiency, TAT, success rate and clinical presentation of the patient. For instance, trio WES has been applied in perinatal patients in critical conditions [75]. A molecular diagnosis for these patients was provided within 50 h. While this timeline may not be feasible on a clinical basis, a few clinical laboratories started offering Trio-WES for critical patients with a TAT of 2–3 weeks, allowing for a fast molecular diagnosis. On the other hand, proband only WES may be a better option for families where cost and parent availability is an issue.

From a laboratory perspective, Trio-WES provides an unbiased analysis of exonic information and an invaluable genomic resource of exonic variants. Trio WES detects all DNV in disease-causing genes as well as in genes with unknown function and unlinked to genetic disorders. As much as 30% of pathogenic variants in clinical WES are located in genes discovered in the last 3 years [34]. Therefore, internal clinical databases of DNV in genes unknown to cause a disorder become a valuable resource for previously undiagnosed patients. As research continues to identify new disease-causing genes, internal database are queried and additional patients receive a molecular diagnosis. Furthermore, as the number of trio WES increases, genotype/phenotype correlation can be performed and through subsequent collaboration with research laboratories, additional disease-causing genes are discovered.

Moreover, the sequencing of asymptomatic parents offers advantages to the clinical laboratory. At this time, the ExAC database (Exome Aggregation Consortium (ExAC), Cambridge, MA, http://exac.broadinstitute.org/) is the largest and most complete resource to estimate the frequency of variants in the general population although phenotypic information is not available for these individuals. The genomic information of asymptomatic parents can help in the construction of an internal database of healthy individuals. Hence, this database is a very useful additional resource to help interpret genomic variants in the clinical laboratory in subsequent WES patients.

Finally, from a laboratory practical perspective, performing parental WES as part of the Trio WES provides a quality control to the clinical test. Sample swaps and non-paternity/maternity are controlled as Trio-WES determines whether both parental samples are the biological parents of the proband. Additionally, high throughput sequencing of the parents also allows the possible detection and estimation of low level of mosaicism in one of the parents, which may be undetectable by Sanger sequencing.

4.2 *De Novo* Findings in the Clinical Laboratories

Using both approaches (P-WES and Trio-WES), several clinical laboratories have reported on their findings and diagnostics yields [6, 34, 35, 74]. Clinical laboratories report an average of 25% molecular diagnosis rate using WES. Among these diagnosed patients, DNV represent the major molecular diagnosis in the clinical laboratory: about 42–50% of the reported pathogenic variants are *de novo*. This result and rate is however dependent on the phenotype of the patient, the methodology (P-WES and Trio-WES, explained previously) and the population background.

Neurologic presentation such as developmental delay and intellectual disability represents the major referrals for clinical WES. For instance, neurological disorders represents 87% of referral for WES [34, 74]. At the same time, DNV play a prominent part in the etiology of rare neurodevelopmental disorders, including intellectual disability and autism [76]. Therefore, trio WES is the most attractive approach to provide a molecular diagnosis in sporadic cases when an autosomal dominant disorder with a DNV is suspected in patients with neurological disorders. DNV can also occur in X-linked disorders and recessive disorders: Yang et al. showed that 40% of X-linked pathogenic variants identified by WES were *de novo* [34, 74]. In addition, DNV have been observed in patients with autosomal recessive conditions where the patient inherit one pathogenic variant from one parent while the other variant is *de novo*.

The clinical diagnostic rates reported here originated from a few laboratories from the USA and Europe, a non-ascertained diversified patient population [6, 34, 35, 74]. In consanguineous populations, homozygous pathogenic variants are expected to explain the majority of the patients' clinical presentation. Yavarna et al. have reported on the use of clinical Trio-WES in 149 probands from a consanguineous population of Qatar [77]. The diagnostic rate in this population was higher (60%), but 25% of the molecular diagnosis (20/80) were explained by a DNV in a gene related to the patient's phenotype [77]. This result demonstrates that DNV still play an important role in the molecular diagnosis of patients in all populations, including consanguineous populations.

4.3 Variant Interpretation in Family Based NGS Setting

Clinical Trio WES is a powerful, unbiased approach to not only detect all exonic DNV in a patient but also determine the inheritance of each variant from each parent. Patients referred for Trio-WES usually present with a broad and nonspecific clinical phenotype. Based on the published mutation rate and published trio-WES, about one to two *de novo* exonic variant are detected in each individual on average. Recurrent pathogenic DNV have long been observed for specific Mendelian disorders. However, most DNV identified in Trio WES are novel, previously unreported, and the interpretation of these variants is necessary to provide an accurate diagnosis.

ACMGG has recently instituted new guidelines to help categorize and interpret variants in the clinical laboratory [78]. Under these new guidelines, variant inheritance adds value and evidence to the interpretation of the variant. For instance, the finding of DNV in an autosomal dominant or X-linked disorder gene related to the patient's phenotype, is considered a "strong evidence of pathogenicity, PS2" under the condition that both parents are confirmed biological parents. Similarly, finding a pathogenic variant *in trans* configuration with a variant of interest in a recessive disorder gene is considered a "moderate evidence of pathogenicity, PM3". Both PS2 and PM3 evidences require testing of the parents and family-based NGS provides this information at the time of clinical testing. During the interpretation of the results, without parental information, the pathogenicity of novel variants may be under estimated.

Finding a novel LOF DNV in an autosomal dominant gene where haploinsufficiency is a disease mechanism is likely to be interpreted as pathogenic. For instance, the finding of a *de novo* loss of function variant in *ARID1B* most likely provides a diagnosis for mental retardation, autosomal dominant 12 [OMIM 614562]. However, caution is necessary regarding the exonic location of the variant. For instance, loss of function variants located in the last exon of the gene may escape non sense mediated decay and may not explain the phenotype of the patient. Also, the distribution of mutations may be relevant. For instance, mutations in *KAT6B* are known to cause Genitopatellar syndrome [OMIM 606170] and a SBBYSS syndrome [OMIM 603736]. Most mutations are *de novo* and located specifically in the C terminal region of the protein [79, 80]. Therefore, the interpretation of a DNV, previously unreported, in the early N terminal part of the protein for instance would be challenging as those have never been reported in any patients. Correlation with the phenotype of the patient is crucial. Finally, the interpretation of novel *de novo* missense variants as opposed to loss of function variants in a disease-causing gene is also challenging as the functionality of missense variants is often unknown. In conclusion, while the detection of a *de novo* variant may be diagnostics in a patient, careful examination and interpretation of the variant, application the ABMGG interpretation guidelines and phenotypic correlation is crucial to properly provide a diagnosis to the patient.

5 Summary

In this chapter we first described the variant calling algorithms designed for family-based NGS analysis, then reviewed the application of family-based NGS in both research and clinical settings Comparing with proband-only sequencing, family-based NGS has several advantages in the genetic study of human diseases. First, several family-based variant calling algorithms have been designed to increase the accuracy of variant calling by joint calling across all family members, modeling Mendelian inheritance within the family, and incorporating shared haplotype information. When design family-based NGS experiments, we should almost always use these family-based variant callers to increase the sensitivity and specificity of variant calling. Second, family-based NGS facilitates the prioritization of candidate disease genes for Mendelian disorders, as well as enriches signals for rare variants association study underlying complex disorders. Third, family-based NGS, such as trio-WES, allows quickly pinpointing disease-causing variants in clinical diagnostic laboratories, reducing the turnaround time, and increasing accuracy and the chance to find causal genes that may not have been associated with known diseases. Because of all these advantages, family-based NGS is becoming the standard sequencing approach in the genetic study of human disorders.

References

1. Metzker, M.L.: Sequencing technologies – the next generation. Nat. Rev. Genet. **11**, 31–46 (2010)
2. Davey, J.W., Hohenlohe, P.A., Etter, P.D., Boone, J.Q., Catchen, J.M., Blaxter, M.L.: Genome-wide genetic marker discovery and genotyping using next-generation sequencing. Nat. Rev. Genet. **12**, 499–510 (2011)
3. Bamshad, M.J., Ng, S.B., Bigham, A.W., Tabor, H.K., Emond, M.J., Nickerson, D.A., et al.: Exome sequencing as a tool for Mendelian disease gene discovery. Nat. Rev. Genet. **12**, 745–755 (2011)
4. Cirulli, E.T., Goldstein, D.B.: Uncovering the roles of rare variants in common disease through whole-genome sequencing. Nat. Rev. Genet. **11**, 415–425 (2010)
5. Roach, J.C., Glusman, G., Smit, A.F.A., Huff, C.D., Hubley, R., Shannon, P.T., et al.: Analysis of genetic inheritance in a family quartet by whole-genome sequencing. Science. **328**, 636–639 (2010)
6. Lee, H., Deignan, J.L., Dorrani, N., Strom, S.P., Kantarci, S., Quintero-Rivera, F., et al.: Clinical exome sequencing for genetic identification of rare Mendelian disorders. JAMA. **312**, 1880–1887 (2014)
7. Manolio, T.A., Collins, F.S., Cox, N.J., Goldstein, D.B., Hindorff, L.A., Hunter, D.J., et al.: Finding the missing heritability of complex diseases. Nature. **461**, 747–753 (2009)
8. Ott, J., Wang, J., Leal, S.M.: Genetic linkage analysis in the age of whole-genome sequencing. Nat. Rev. Genet. **16**, 275–284 (2015)
9. Nielsen, R., Paul, J.S., Albrechtsen, A., Song, Y.S.: Genotype and SNP calling from next-generation sequencing data. Nat. Rev. Genet. **12**, 443–451 (2011)
10. O'Rawe, J., Jiang, T., Sun, G., Wu, Y., Wang, W., Hu, J., et al.: Low concordance of multiple variant-calling pipelines: practical implications for exome and genome sequencing. Genome Med. **5**, 28 (2013)
11. Bizon, C., Spiegel, M., Chasse, S.A., Gizer, I.R., Li, Y., Malc, E.P., et al.: Variant calling in low-coverage whole genome sequencing of a native American population sample. BMC Genomics. **15**, 85 (2014)
12. Liu, X., Han, S., Wang, Z., Gelernter, J., Yang, B.-Z.: Variant callers for next-generation sequencing data: a comparison study. PLoS One. **8**, e75619 (2013)
13. McKenna, A., Hanna, M., Banks, E., Sivachenko, A., Cibulskis, K., Kernytsky, A., et al.: The genome analysis toolkit: a MapReduce framework for analyzing next-generation DNA sequencing data. Genome Res. **20**, 1297–1303 (2010)
14. Li, H., Handsaker, B., Wysoker, A., Fennell, T., Ruan, J., Homer, N., et al.: The sequence alignment/map format and SAMtools. Bioinforma. Oxf. Engl. **25**, 2078–2079 (2009)
15. Koboldt, D.C., Chen, K., Wylie, T., Larson, D.E., McLellan, M.D., Mardis, E.R., et al.: VarScan: variant detection in massively parallel sequencing of individual and pooled samples. Bioinforma. Oxf. Engl. **25**, 2283–2285 (2009)

16. Taylor, J.C., Martin, H.C., Lise, S., Broxholme, J., Cazier, J.-B., Rimmer, A., et al.: Factors influencing success of clinical genome sequencing across a broad spectrum of disorders. Nat. Genet. **47**, 717–726 (2015)
17. Cleary, J.G., Braithwaite, R., Gaastra, K., Hilbush, B.S., Inglis, S., Irvine, S.A., et al.: Joint variant and de novo mutation identification on pedigrees from high-throughput sequencing data. J. Comput. Biol. J. Comput. Mol. Cell Biol. **21**, 405–419 (2014)
18. Cartwright, R.A., Hussin, J., Keebler, J.E.M., Stone, E.A., Awadalla, P.: A family-based probabilistic method for capturing de novo mutations from high-throughput short-read sequencing data. Stat. Appl. Genet. Mol. Biol. **11**, (2012). doi:10.2202/1544-6115.1713
19. Conrad, D.F., Keebler, J.E.M., DePristo, M.A., Lindsay, S.J., Zhang, Y., Casals, F., et al.: Variation in genome-wide mutation rates within and between human families. Nat. Genet. **43**, 712–714 (2011)
20. Li, B., Chen, W., Zhan, X., Busonero, F., Sanna, S., Sidore, C., et al.: A likelihood-based framework for variant calling and de novo mutation detection in families. PLoS Genet. **8**, e1002944 (2012)
21. Chen, W., Li, B., Zeng, Z., Sanna, S., Sidore, C., Busonero, F., et al.: Genotype calling and haplotyping in parent-offspring trios. Genome Res. **23**, 142–151 (2013)
22. Kojima, K., Nariai, N., Mimori, T., Takahashi, M., Yamaguchi-Kabata, Y., Sato, Y., et al.: A statistical variant calling approach from pedigree information and local haplotyping with phase informative reads. Bioinforma. Oxf. Engl. **29**, 2835–2843 (2013)
23. Santoni, F.A., Makrythanasis, P., Nikolaev, S., Guipponi, M., Robyr, D., Bottani, A., et al.: Simultaneous identification and prioritization of variants in familial, de novo, and somatic genetic disorders with VariantMaster. Genome Res. **24**, 349–355 (2014)
24. Liu, Y., Li, B., Tan, R., Zhu, X., Wang, Y.: A gradient-boosting approach for filtering de novo mutations in parent-offspring trios. Bioinforma. Oxf. Engl. **30**, 1830–1836 (2014)
25. Li, J., Jiang, Y., Wang, T., Chen, H., Xie, Q., Shao, Q., et al.: mirTrios: an integrated pipeline for detection of de novo and rare inherited mutations from trios-based next-generation sequencing. J. Med. Genet. **52**, 275–281 (2015)
26. Peng, G., Fan, Y., Palculict, T.B., Shen, P., Ruteshouser, E.C., Chi, A.-K., et al.: Rare variant detection using family-based sequencing analysis. Proc. Natl. Acad. Sci. U. S. A. **110**, 3985–3990 (2013)
27. Ramu, A., Noordam, M.J., Schwartz, R.S., Wuster, A., Hurles, M.E., Cartwright, R.A., et al.: DeNovoGear: de novo indel and point mutation discovery and phasing. Nat. Methods. **10**, 985–987 (2013)
28. Kong, A., Frigge, M.L., Masson, G., Besenbacher, S., Sulem, P., Magnusson, G., et al.: Rate of de novo mutations and the importance of father's age to disease risk. Nature. **488**, 471–475 (2012)
29. Francioli, L.C., Polak, P.P., Koren, A., Menelaou, A., Chun, S., Renkens, I., et al.: Genome-wide patterns and properties of de novo mutations in humans. Nat. Genet. **47**, 822–826 (2015)
30. Shiang, R., Thompson, L.M., Zhu, Y.Z., Church, D.M., Fielder, T.J., Bocian, M., et al.: Mutations in the transmembrane domain of FGFR3 cause the most common genetic form of dwarfism, achondroplasia. Cell. **78**, 335–342 (1994)
31. Goriely, A., Hansen, R.M., Taylor, I.B., Olesen, I.A., Jacobsen, G.K., McGowan, S.J., et al.: Activating mutations in FGFR3 and HRAS reveal a shared genetic origin for congenital disorders and testicular tumors. Nat. Genet. **41**, 1247–1252 (2009)
32. Rusk, N.: Keeping it in the family. Nat. Methods. **7**, 350 (2010)
33. Vissers, L.E.L.M., de Ligt, J., Gilissen, C., Janssen, I., Steehouwer, M., de Vries, P., et al.: A de novo paradigm for mental retardation. Nat. Genet. **42**, 1109–1112 (2010)
34. Yang, Y., Muzny, D.M., Xia, F., Niu, Z., Person, R., Ding, Y., et al.: Molecular findings among patients referred for clinical whole-exome sequencing. JAMA. **312**, 1870–1879 (2014)
35. Farwell, K.D., Shahmirzadi, L., El-Khechen, D., Powis, Z., Chao, E.C., Tippin Davis, B., et al.: Enhanced utility of family-centered diagnostic exome sequencing with inheritance model-based analysis: results from 500 unselected families with undiagnosed genetic conditions. Genet. Med. Off. J. Am. Coll. Med. Genet. **17**, 578–586 (2015)

36. Gibson, W.T., Hood, R.L., Zhan, S.H., Bulman, D.E., Fejes, A.P., Moore, R., et al.: Mutations in EZH2 cause weaver syndrome. Am. J. Hum. Genet. **90**, 110–118 (2012)
37. Rivière, J-B, van Bon, B.W.M., Hoischen, A., Kholmanskikh, S.S., O'Roak, B.J., Gilissen, C., et al.: De novo mutations in the actin genes ACTB and ACTG1 cause Baraitser-winter syndrome. Nat. Genet. **44**, 440–4, S1–2 (2012)
38. Yukinaga, H., Iwabuchi, H., Okazaki, O., Izumi, T.: Glutathione S-transferase pi trapping method for generation and characterization of drug-protein adducts in human liver microsomes using liquid chromatography-tandem mass spectrometry. J. Pharm. Biomed. Anal. **67–68**, 186–192 (2012). doi:10.1016/j.jpba.2012.04.035
39. Gregor, A., Oti, M., Kouwenhoven, E.N., Hoyer, J., Sticht, H., Ekici, A.B., et al.: De novo mutations in the genome organizer CTCF cause intellectual disability. Am. J. Hum. Genet. **93**, 124–131 (2013)
40. Dyment, D.A., Smith, A.C., Alcantara, D., Schwartzentruber, J.A., Basel-Vanagaite, L., Curry, C.J., et al.: Mutations in PIK3R1 cause SHORT syndrome. Am. J. Hum. Genet. **93**, 158–166 (2013)
41. Simonis, N., Migeotte, I., Lambert, N., Perazzolo, C., de Silva, D.C., Dimitrov, B., et al.: FGFR1 mutations cause Hartsfield syndrome, the unique association of holoprosencephaly and ectrodactyly. J. Med. Genet. **50**, 585–592 (2013)
42. Vaher, U., Nõukas, M., Nikopensius, T., Kals, M., Annilo, T., Nelis, M., et al.: De novo SCN8A mutation identified by whole-exome sequencing in a boy with neonatal epileptic encephalopathy, multiple congenital anomalies, and movement disorders. J. Child Neurol. **29**, NP202–NP206 (2014)
43. Campbell, I.M., Yuan, B., Robberecht, C., Pfundt, R., Szafranski, P., McEntagart, M.E., et al.: Parental somatic mosaicism is underrecognized and influences recurrence risk of genomic disorders. Am. J. Hum. Genet. **95**, 173–182 (2014)
44. Chen, Z., Moran, K., Richards-Yutz, J., Toorens, E., Gerhart, D., Ganguly, T., et al.: Enhanced sensitivity for detection of low-level germline mosaic RB1 mutations in sporadic retinoblastoma cases using deep semiconductor sequencing. Hum. Mutat. **35**, 384–391 (2014)
45. Xu, X., Yang, X., Wu, Q., Liu, A., Yang, X., Ye, A.Y., et al.: Amplicon resequencing identified parental mosaicism for approximately 10% of "de novo" SCN1A mutations in children with Dravet syndrome. Hum. Mutat. **36**, 861–872 (2015)
46. Simons, C., Wolf, N.I., McNeil, N., Caldovic, L., Devaney, J.M., Takanohashi, A., et al.: A de novo mutation in the β-tubulin gene TUBB4A results in the leukoencephalopathy hypomyelination with atrophy of the basal ganglia and cerebellum. Am. J. Hum. Genet. **92**, 767–773 (2013)
47. Glazov, E.A., Zankl, A., Donskoi, M., Kenna, T.J., Thomas, G.P., Clark, G.R., et al.: Whole-exome re-sequencing in a family quartet identifies POP1 mutations as the cause of a novel skeletal dysplasia. PLoS Genet. **7**, e1002027 (2011)
48. Çalışkan, M., Chong, J.X., Uricchio, L., Anderson, R., Chen, P., Sougnez, C., et al.: Exome sequencing reveals a novel mutation for autosomal recessive non-syndromic mental retardation in the TECR gene on chromosome 19p13. Hum. Mol. Genet. **20**, 1285–1289 (2011)
49. Below, J.E., Earl, D.L., Shively, K.M., McMillin, M.J., Smith, J.D., Turner, E.H., et al.: Whole-genome analysis reveals that mutations in inositol polyphosphate phosphatase-like 1 cause opsismodysplasia. Am. J. Hum. Genet. **92**, 137–143 (2013)
50. Long, P.A., Evans, J.M., Olson, T.M.: Exome sequencing establishes diagnosis of Alström syndrome in an infant presenting with non-syndromic dilated cardiomyopathy. Am. J. Med. Genet. A. **167A**, 886–890 (2015)
51. Kodera, H., Ando, N., Yuasa, I., Wada, Y., Tsurusaki, Y., Nakashima, M., et al.: Mutations in COG2 encoding a subunit of the conserved oligomeric golgi complex cause a congenital disorder of glycosylation. Clin. Genet. **87**, 455–460 (2015)
52. Shashi, V., McConkie-Rosell, A., Rosell, B., Schoch, K., Vellore, K., McDonald, M., et al.: The utility of the traditional medical genetics diagnostic evaluation in the context of next-generation sequencing for undiagnosed genetic disorders. Genet. Med. Off. J. Am. Coll. Med. Genet. **16**, 176–182 (2014)

53. Rauch, A., Hoyer, J., Guth, S., Zweier, C., Kraus, C., Becker, C., et al.: Diagnostic yield of various genetic approaches in patients with unexplained developmental delay or mental retardation. Am. J. Med. Genet. A. **140**, 2063–2074 (2006)
54. de Ligt, J., Willemsen, M.H., van Bon, B.W.M., Kleefstra, T., Yntema, H.G., Kroes, T., et al.: Diagnostic exome sequencing in persons with severe intellectual disability. N. Engl. J. Med. **367**, 1921–1929 (2012)
55. Rauch, A., Wieczorek, D., Graf, E., Wieland, T., Endele, S., Schwarzmayr, T., et al.: Range of genetic mutations associated with severe non-syndromic sporadic intellectual disability: an exome sequencing study. Lancet Lond. Engl. **380**, 1674–1682 (2012)
56. Deciphering Developmental Disorders Study: Large-scale discovery of novel genetic causes of developmental disorders. Nature. **519**, 223–228 (2015)
57. Gilissen, C., Hehir-Kwa, J.Y., Thung, D.T., van de Vorst, M., van Bon, B.W.M., Willemsen, M.H., et al.: Genome sequencing identifies major causes of severe intellectual disability. Nature. **511**, 344–347 (2014)
58. Rees, E., Kirov, G., Walters, J.T., Richards, A.L., Howrigan, D., Kavanagh, D.H., et al.: Analysis of exome sequence in 604 trios for recessive genotypes in schizophrenia. Transl. Psychiatry. **5**, e607 (2015)
59. Fombonne, E.: Epidemiology of pervasive developmental disorders. Pediatr. Res. **65**, 591–598 (2009)
60. Zhao, X., Leotta, A., Kustanovich, V., Lajonchere, C., Geschwind, D.H., Law, K., et al.: A unified genetic theory for sporadic and inherited autism. Proc. Natl. Acad. Sci. U. S. A. **104**, 12831–12836 (2007)
61. Ronemus, M., Iossifov, I., Levy, D., Wigler, M.: The role of de novo mutations in the genetics of autism spectrum disorders. Nat. Rev. Genet. **15**, 133–141 (2014)
62. Sebat, J., Lakshmi, B., Malhotra, D., Troge, J., Lese-Martin, C., Walsh, T., et al.: Strong association of de novo copy number mutations with autism. Science. **316**, 445–449 (2007)
63. Marshall, C.R., Noor, A., Vincent, J.B., Lionel, A.C., Feuk, L., Skaug, J., et al.: Structural variation of chromosomes in autism spectrum disorder. Am. J. Hum. Genet. **82**, 477–488 (2008)
64. Levy, D., Ronemus, M., Yamrom, B., Lee, Y., Leotta, A., Kendall, J., et al.: Rare de novo and transmitted copy-number variation in autistic spectrum disorders. Neuron. **70**, 886–897 (2011)
65. Sanders, S.J., Ercan-Sencicek, A.G., Hus, V., Luo, R., Murtha, M.T., Moreno-De-Luca, D., et al.: Multiple recurrent de novo CNVs, including duplications of the 7q11.23 Williams syndrome region, are strongly associated with autism. Neuron. **70**, 863–885 (2011)
66. O'Roak, B.J., Vives, L., Girirajan, S., Karakoc, E., Krumm, N., Coe, B.P., et al.: Sporadic autism exomes reveal a highly interconnected protein network of de novo mutations. Nature. **485**, 246–250 (2012)
67. Neale, B.M., Kou, Y., Liu, L., Ma'ayan, A., Samocha, K.E., Sabo, A., et al.: Patterns and rates of exonic de novo mutations in autism spectrum disorders. Nature. **485**, 242–245 (2012)
68. Iossifov, I., Ronemus, M., Levy, D., Wang, Z., Hakker, I., Rosenbaum, J., et al.: De novo gene disruptions in children on the autistic spectrum. Neuron. **74**, 285–299 (2012)
69. Sanders, S.J., Murtha, M.T., Gupta, A.R., Murdoch, J.D., Raubeson, M.J., Willsey, A.J., et al.: De novo mutations revealed by whole-exome sequencing are strongly associated with autism. Nature. **485**, 237–241 (2012)
70. Krumm, N., O'Roak, B.J., Shendure, J., Eichler, E.E.: A de novo convergence of autism genetics and molecular neuroscience. Trends Neurosci. **37**, 95–105 (2014)
71. Lalani, S.R., Zhang, J., Schaaf, C.P., Brown, C.W., Magoulas, P., Tsai, A.C.-H., et al.: Mutations in PURA cause profound neonatal hypotonia, seizures, and encephalopathy in 5q31.3 microdeletion syndrome. Am. J. Hum. Genet. **95**, 579–583 (2014)
72. EuroEPINOMICS-RES Consortium: Epilepsy Phenome/genome project, Epi4K consortium. De novo mutations in synaptic transmission genes including DNM1 cause epileptic encephalopathies. Am. J. Hum. Genet. **95**, 360–370 (2014)
73. van Bokhoven, H.: Genetic and epigenetic networks in intellectual disabilities. Annu. Rev. Genet. **45**, 81–104 (2011)

74. Yang, Y., Muzny, D.M., Reid, J.G., Bainbridge, M.N., Willis, A., Ward, P.A., et al.: Clinical whole-exome sequencing for the diagnosis of mendelian disorders. N. Engl. J. Med. **369**, 1502–1511 (2013)
75. Saunders, C.J., Miller, N.A., Soden, S.E., Dinwiddie, D.L., Noll, A., Alnadi, N.A., et al.: Rapid whole-genome sequencing for genetic disease diagnosis in neonatal intensive care units. Sci. Transl. Med. **4**, 154ra135 (2012)
76. Veltman, J.A., Brunner, H.G.: De novo mutations in human genetic disease. Nat. Rev. **13**, 565–575 (2012)
77. Yavarna, T., Al-Dewik, N., Al-Mureikhi, M., Ali, R., Al-Mesaifri, F., Mahmoud, L., et al.: High diagnostic yield of clinical exome sequencing in middle eastern patients with Mendelian disorders. Hum. Genet. **134**, 967–980 (2015)
78. Richards, S., Aziz, N., Bale, S., Bick, D., Das, S., Gastier-Foster, J., et al.: Standards and guidelines for the interpretation of sequence variants: a joint consensus recommendation of the American College of Medical Genetics and Genomics and the Association for Molecular Pathology. Genet. Med. Off. J. Am. Coll. Med. Genet. **17**, 405–424 (2015)
79. Campeau, P.M., Kim, J.C., Lu, J.T., Schwartzentruber, J.A., Abdul-Rahman, O.A., Schlaubitz, S., et al.: Mutations in KAT6B, encoding a histone acetyltransferase, cause Genitopatellar syndrome. Am. J. Hum. Genet. **90**, 282–289 (2012)
80. Clayton-Smith, J., O'Sullivan, J., Daly, S., Bhaskar, S., Day, R., Anderson, B., et al.: Whole-exome-sequencing identifies mutations in histone acetyltransferase gene KAT6B in individuals with the say-barber-Biesecker variant of Ohdo syndrome. Am. J. Hum. Genet. **89**, 675–681 (2011)

Next Generation of Carrier Screening

Anastasia Fedick and Jinglan Zhang

Abstract Preconception or prenatal carrier screening is an effective approach for preventing devastating recessive diseases. Traditionally, DNA-based carrier testing used targeted genotyping panels to detect common mutations among specific ethnic groups. While the sensitivity of this approach is generally acceptable, private or ultra-rare mutations will be missed. Next generation sequencing (NGS) has been adopted in recent years for carrier screening in order to increase test sensitivity for pan-ethnic individuals. Important issues regarding NGS-based carrier screening panels' workflow design, variant interpretation, reporting, and genetic counseling follow-ups are reviewed and discussed in this chapter.

Keywords Next generation sequencing • Carrier screening

1 Introduction

Carrier screening was first introduced for Tay-Sachs disease (TSD) in the Ashkenazi Jewish (AJ) population in the early 1970s. Due to its tremendous success in reducing the incidence of TSD in AJ by over 90%, screening for cystic fibrosis and Gaucher disease began in 1993 and was offered to other ethnic groups [1, 2]. Since then, the American College of Medical Genetics (ACMG) and the American College of Obstetricians and Gynecologists (ACOG) have added several additional diseases, including Bloom syndrome, Canavan disease, familial dysautonomia, Fanconi anemia group C, mucolipidosis IV, and Niemann-Pick disease type A to the list of diseases for which individuals with Ashkenazi Jewish ancestry should be screened [3, 4]. Carrier screening has also been recommended for diseases prevalent in different ethnic groups, such as the hemoglobinopathies in individuals of African and Mediterranean origin [5]. These professional recommendations and increased

A. Fedick (✉)
Icahn School of Medicine at Mount Sinai, New York, NY 10029, USA

J. Zhang (✉)
Baylor College of Medicine, Houston, TX 77030, USA
e-mail: Jinglan.Zhang@bcm.edu

L.-J.C. Wong (ed.), *Next Generation Sequencing Based Clinical Molecular Diagnosis of Human Genetic Disorders*, DOI 10.1007/978-3-319-56418-0_16

public awareness have led to an increasing demand for efficient screening of a broad panel of mutations.

Targeted genotyping for AJ diseases on high-throughput platforms offered improvements over Sanger sequencing for its low cost, rapid interpretation, and high throughput [6–8]. As more genes and mutations were identified at increased prevalence in different populations, many carrier screening tests were expanded to include diseases that were not specifically recommended by ACOG and ACMG, which further underscored the necessity for high-throughput screening [9, 10]. It has been recommended that screening for cystic fibrosis (CF), spinal muscular atrophy (SMA) and fragile-X syndrome be offered pan-ethnically as they have comparable carrier frequencies in many ethnic groups [11, 12]. The idea of universal carrier screening led to the development of new panels to include many mutations regardless of their ethnic prevalence [13]. These expanded carrier screening genotyping panels aimed to be more comprehensive but were still forced to only focus on the most prevalent mutations in various ethnic groups [14–16]. While performing carrier screening in this manner was very efficient, the analysis was limited to the specific and pre-selected mutations based on prior studies.

The advent of NGS has allowed carrier screening tests to advance from targeted mutation testing of specific genes to comprehensive (whole-gene) multi-gene testing. Therefore, NGS-based carrier screening is currently primarily limited by the information available about the clinical impact of a detected variant rather than whether the variant is included in a pre-defined list of variants that can be detected [17].

2 Advantages and Challenges of NGS-Based Carrier Screening

The major advantage of NGS-based carrier screening panels is obvious: comprehensiveness. Essentially every nucleotide in a gene can be assessed by NGS rather than limited to the known, pre-selected mutations. In large genes, where many mutations have been reported, NGS allows for a complete evaluation in a cost-effective fashion. For example, the detection rate of the 23-mutation CF panel recommended by ACOG and ACMG differs significantly based on different geographic regions, ranging from 94% in the AJ population to only 64% and 49% in the African American and Asian American populations, respectively [18]. A study comparing mutation frequency distributions for a 32-CF mutation targeted panel and a 69-mutation targeted panel found that, not surprisingly, the larger panel detected around 20% more mutations than the smaller panel for both African-American and Hispanic-American individuals [19]. These uneven detection rates stem from the prevalence of the targeted mutations included in the genotyping panels in each population. Sequencing the entire *CFTR* gene, which has at least 1700 known mutations, doubled the carrier detection rate in patients from 1.06% to 2.13% compared to a focused genotype approach, with four patients harboring mutations not included in the ACOG panel [20]. Similarly, it has been shown that using NGS to test for

TSD is more effective than using traditional enzyme testing or targeted genotyping panels, since enzyme testing can generate false-positive and false-negative results, due to the presence of pseudodeficiency or B1 alleles, respectively. Allelic-specific targeted panels are usually ethnicity-dependent, making their utility limited and not useful for pan-ethnic screening [21]. NGS is also the test of choice for diseases such as very long-chain acyl-CoA dehydrogenase (VLCAD) deficiency, for which there is no particular prevalent ethnic mutation, making targeted genotyping much less efficient.

The comprehensiveness of NGS testing also makes receiving a negative result much more reassuring when compared to a genotyping test in terms of residual risk of variants that are undetected. While variant interpretation is more challenging in NGS, the sequencing information is retained, making it possible to reanalyze data periodically as new information becomes available, thereby allowing patients to receive updated results without the need for sample retesting. In genotyping panels, only the target mutation information is obtained, so if new mutations are discovered later, complete retesting on an updated panel needs to be performed. This renders NGS a practical advantage in terms of reduced cost and decreased patient sampling.

Using NGS for carrier screening does have its challenges. NGS is technically demanding due to its complexity in library construction, instrument operation, the alignment of DNA sequences, variant calling, and variant interpretation. In addition, the test needs to be conducted in a high-throughput manner in order to provide timely results.

3 The Technology and Informatics Solution for Carrier Screening

Library preparation for NGS can involve target capture or amplification-based enrichment technologies. A capture-based method is preferred in large carrier screening gene panels for its higher capacity for target enrichment. Using either method, unique sample indexing barcodes are linked to individual samples through short nucleotide sequences known as sequencing adapters so that samples can be combined together before being pooled for capture and sequencing. Following sequencing, samples are demultiplexed based on the capture oligonucleotide and aligned to a reference sequence before variant calling and annotation can be performed. Quality metrics are particularly important for filtering out noise and can be applied at the assay level to determine the overall quality of a particular sequencing run, such as monitoring the depth and uniformity of coverage, the portion of reads that map to a non-targeted region, etc. In addition, it is also necessary at the sample level to assess the quality of any given sample's specific data, and finally at the variant level to assess sample-specific variant calls [17].

Bioinformatic pipelines can be created and used for filtering out common polymorphisms, identifying known deleterious mutations, providing functional annotations and pathogenicity predictions for novel variants, and for prioritizing variants that need to be further evaluated [22]. In order to determine which nucleotide variants are pathogenic, likely pathogenic, variants of uncertain significance, or benign, several integrated software tools are available to allow for automated variant calling, including Genome Analysis Toolkit (GATK) [23], ANNOVAR [24], the Ensembl Variant Effect Predictor [25], and Sequence Variant Analyzer (SVA) [26], among others. All of these programs attempt to make corrections in systemic errors and thereby accurately call and identify disease-related variants. There are, however, as yet unresolved flaws in the software programs that limit their clinical utility. After the initial variant assessment, additional information, including known variation sites, family history, and population information, are integrated into the original variant calling algorithms to increase the variant identification sensitivity and specificity. GATK uses the adaptive error modeling system to identify false positive variants based on their dissimilarity to known variants, which allows for the generation of variant calls that have varying degrees of confidence, and can be included or excluded in downstream analysis [23]. For carrier screening of Mendelian disorders, this strategy can be used to increase the overall variant call sensitivity for recurrent mutations.

However, when assessing large panels, not all variants can be validated due to both cost issues and the lack of publically available positive control samples, especially for rare diseases. Typically, newly developed panels that are used in clinical testing undergo analytical validation by the Clinical Laboratory Improvement Amendment (CLIA), where the accuracy and precision, analytical sensitivity and specificity, reportable range, reference intervals, and other test performance characteristics are proven and documented to be reliable and robust. Traditionally such tests would be compared to the original gold standard, which for NGS is Sanger sequencing, however this is not feasible for all NGS tests. Additionally, there is a lack of appropriate control samples that can be used to detect errors due to the quantity and variation of data generated by these tests. Therefore, methods-based proficiency testing that focuses on the kinds of variants that are intended to be detected by a given test should be used to determine test quality, as well as sample exchanges with other laboratories or blinded retesting of previously tested samples [27]. The methods-based tests are able to address some of the technical aspects of the NGS workflow such as sequence read generation, sequence mapping, alignment, variant calling, and annotation of pathogenicity, while the blinded retesting of samples can assess a laboratory's ability to accurately interpret variant calls in terms of pathogenicity [28].

Table 1 The common diseases and mutations included in expanded carrier screening panels

Disorder and gene	Gene(s)	Approximate carrier frequencies	Common pathogenic variants
Alpha-thalassemia	HBA1 and HBA2	1 in 3 to 1 in 50	p.D75H, p.L126P, p. Ter143Q,c.94_95delAG
Bloom syndrome	BLM	1 in 100 for Ashkenazi Jewish	c.2207_2212delinsTAGTTC
Becker muscular dystrophy	DMD	1 in 5600–7250 males	N/A
Beta-thalassemia	HBB	1 in 12.5 African Americans	p.Q40X, p.K18X
Canavan disease	ASPA	1 in 40 for Ashkenazi Jewish	p. Glu285Ala, p.Tyr231Ter, p. Ala305Glu
Congenital adrenal hyperplasia	CYP21A2	1 in 300 to 1 in 23,000	p.E319X, p.R357W, p.I173N, p. P31L, p.V282 L, p.P454S
Cystic fibrosis	CFTR	1 in 29 to 1 in 61	c.1521_1523delCTT
Duchenne muscular dystrophy	DMD	1 in 5600–7250 males	N/A
Dihydrolipoamide dehydrogenase deficiency	DLD	1 in 94 for Ashkenazi Jewish	p.G229C
Familial dysautonomia	IKBKAP	1 in 36 for Ashkenazi Jewish	c.2204+6T>C, p.R696P
Familial hyperinsulinism	ABCC8	1 in 68 for Ashkenazi Jewish	c.3989-9G>A, p.F1387del, p. V187D, p.E1506K
Fanconi anemia group C	FANCC	1 in 300	c.456+4A>T, c.67delG
Fragile X	FMR1	1 in 113 to 1 in 259	N/A
Glycogen storage disease, type 1A	G6PC	1 in 72 for Ashkenazi Jewish	p.R83C, p.Q347X
Maple syrup urine disease, type A	BCKDHA	1 in 150 for Mennonites	p.Y438N
Maple syrup urine disease, type B	BCKDHB	1 in 113 for Ashkenazi Jewish	p.R183P, p.G278S, p.E372X
Mucolipidosis, type IV	MCOLN1	1 in 100 for Ashkenazi Jewish	c.406-2A>G, 6.4 kb del

(continued)

Table 1 (continued)

Disorder and gene	Gene(s)	Approximate carrier frequencies	Common pathogenic variants
Niemann-Pick disease, type AB	SMPD1	1 in 100 for Ashkenazi Jewish	p.L304P, p.H423Y, p.R498L, c.1829_1831delGCC
Tay-Sachs disease	HEXA	1 in 30 for Ashkenazi Jewish	c.1274_1277dupTATC, c.1421+1G>C, c.1073+1G>A, p. G269S
Usher syndrome, type 1F	PCDH15	1 in 100 for Ashkenazi Jewish	p.R245X, p.T1867del
Usher syndrome, type III	CLRN1	1 in 100 for Ashkenazi Jewish	p.N48K, p.Y176X

4 Pan-Ethnicity Carrier Screening

Carrier screening has traditionally been based on specific ethnicities, however, given the population structure of many countries, pan-ethnic carrier screening is beneficial for individuals of mixed or uncertain ethnicities and individuals from ethnic backgrounds that were not largely represented in targeted-genotyping tests (Table 1). There are often inaccuracies when obtaining pedigrees in terms of ethnicity, with a study in Colorado reporting incomplete or inaccurate ethnic information for 30% of dried blood spots obtained for newborn screening from at least one parent [29].

Since race and ethnicity are self-perceived concepts to most individuals, they are not always reliable indicators of geographic ancestry, which is necessary for determining genetic risk in specific populations. Ethnic terms can also be too broad to be useful. For example, while many individuals from Africa are at an increased risk for sickle cell disease due to balanced selection by malaria resistance, there is a low prevalence of the disease in southern Africa, so identifying a patient simply as African does not accurately infer their genetic risk for this disease [30]. There is also the potential for stigmatization, discrimination, and notions of privilege if offering carrier screening to only certain ethnic or population groups.

In order to determine the cost savings for a pan-ethnic carrier screening program, the frequency of each disease being tested along with its cost of treatment and cost savings of prevention, as well as its genotype-phenotype correlation, the accuracy and comprehensiveness of the overall screening test, and the population's acceptance of the testing, must be taken into account. Overall, a pan-ethnic carrier screening test can be considered cost-effective if it reduces the frequency of high-risk pregnancies and the prevalence of diseases in the next generation. Interestingly, newborn screening panels, which are downstream of carrier screening programs, are by definition pan-ethnic since they apply to all pregnancies, indicating an established benefit over subpopulation screening for this type of testing. In addition to the cost, pan-ethnic screening panels can be advantageous since they are more robust due to the prevalence of admixture. Hence, there is no need to obtain race and ethnicity information for the purpose of pre-test approvals. As a result, the stigmatization of genetic diseases within certain ethnic populations is minimized [16].

5 NGS Based Carrier Screening Test Sensitivity and Specificity

To calculate a meaningful test or clinical sensitivity for NGS-based carrier screening tests, the analytical sensitivity of different mutation types (SNVs, small insertions/deletions, indels, and large insertions or deletions), the prevalence of these mutations in different ethnic groups, and the disease incidence in specific populations are required. However, much of this knowledge is scarce in the literature, thus making an accurate estimate of the test sensitivity often not feasible.

Additional problems for current NGS technology include capturing, sequencing, and/or calling variants that occur in difficult regions such as: sequences with high GC content, genes with pseduogenes, triplet expansions, highly homologous regions, complex rearrangements, repeat elements including LINES and SINES, etc. If a complex mutation that is not amenable for NGS variant discovery represents the predominating mutation associated with a disease, then the overall test sensitivity will be greatly reduced. For example, there is a large deletion in the *GJB6* gene that causes hearing loss that is a founder mutation in the AJ population and unless parallel testing is performed using an alternative method to detect the deletion, the sensitivity provided by NGS alone will be low. That being said, it is possible to detect single or multiple exon deletions using capture NGS [31, 32], but since less stringent filters and alignment conditions are used, the number of false positives due to misalignments will also increase. Similarly, the *FKTN* c.1167insA mutation is another founder mutation in the AJ population that involves an insertion in a homopolymer stretch causing Walker Walburg syndrome, which can sometimes have problematic detection based on the chemistry of the NGS platform. If a patient has a known familial risk of inheriting that specific mutation, using an NGS panel would not necessarily be the best method, especially if special variant calling algorithms for complex mutations were not previously optimized. Some genes may also need to be carefully considered when included in NGS panels. For example, the *GBA* gene which causes Gaucher's Disease has a known pseudogene with 96% homology [33]. Since the L444P mutation is a known founder mutation in the AJ population, it would need to be clearly stated if Gaucher disease was not included in an NGS Ashkenazi carrier screening panel, as well as if targeted mutation analysis for such mutations were being performed concurrently using a different technology. Similarly, some panels will also target deep intronic sequences or UTRs if mutations have been previously reported in those areas to increase the overall test sensitivity.

NGS has also improved the detection rate for diseases caused by exonic CNVs. Duchenne and Becker muscular dystrophies (DMD/BMD) for example, are primarily caused by large deletions (~60%) or duplications (~7%) spanning one or more exons, as well as small insertions or deletions, SNVs, and splice site or intronic mutations. The dystrophin gene responsible for DMD/BMD is very large at 2.5 Mb and consists of 79 exons, 78 introns, and 8 promoters. Prior to NGS, a combination of multiplex ligation-dependent probe amplification and Sanger sequencing was necessary to test for DMD/BMD. Targeted NGS, however, can now be used to diagnose DMD patients and female carriers, with one study citing 99.99% specificity and 98.96% sensitivity for CNVs compared to traditional methods and 100% accuracy for SNVs [34]. This method was also able to detect partial deletions and duplications in patients, allowing for precise breakpoints to be reported, which is important for gene therapy. The referenced study used a three-step computational framework to detect exonic CNVs. First, the sequencing depth across all the exons in *DMD* from a reference group of healthy individuals was normalized and used to establish a Gaussian distribution, where a mutated exon would be seen as an outlier. Inter-batch differences and NGS method variability were then corrected by using intra-batch ratios to further distinguish true CNVs from false positives. This methodology was more difficult to apply to females, however, due to the diploid X chromosome, and four false-negative results involving female carriers occurred. Additionally, the described method was not able to detect deep intronic mutations and complex rearrangements.

Similarly, NGS techniques can now be used to determine the carrier status for spinal muscular atrophy (SMA). Autosomal recessive forms of SMA are caused by variant forms of the SMN locus. There are two nearly identical copies of the genes *SMN1* and *SMN2* that are only distinguishable by a few SNVs, one of which (c.840C>T) causes exon-skipping and reduced gene functionality. Current difficulties in determining *SMN1* carrier status in addition to the having *SMN1* and *SMN2* paralogs is that the human population has varying copy numbers of *SMN1* and *SMN2*. Prior to NGS, traditional SMA screening involved quantitative polymerase chain reactions (qPCR) where specific primers amplified only the *SMN1*-specific region of exon 7, and copy number was determined by comparing the *SMN1* cycle threshold to a control gene or genes. Now, Larson et al., has developed a method that utilizes a Bayesian hierarchical model to determine an individual's carrier probability by examining reads at six loci (three nucleotide positions that are unique to each gene) in both *SMN1* and *SMN2* from NGS data [35]. Although the exact test sensitivity and specificity have not been reported, the carrier probability is measured on a continuous scale, and complete concordance was reported when compared to current qPCR testing methods in a limited number of samples studied [35]. Limitations of this NGS method are the same as for the standard qPCR methods, which include not being able to take haplotype phase into account, or being able to identify silent carriers, which are individuals who carry two copies of *SMN1* in cis, or on one allele, but none on the other (2 + 0). However, in instances such as the Ashkenazi Jewish population where a specific polymorphism (g.27134T>G) is associated with the silent carrier genotype [36], testing for that polymorphism could easily be incorporated into the NGS protocol.

When using NGS, one wants to make sure that the overall sensitivity and specificity are at least as good as the traditional single gene tests that they are replacing. Since NGS involves the sequencing of many small DNA fragments, adequate depth of coverage is warranted for acceptable specificity and sensitivity. Especially since target enrichment is often uneven across various gene regions, it is often necessary to sequence to high depths to get the necessary coverage across all of the target nucleotides within a given gene panel. When considering NGS for carrier screening, false-negative results are usually more concerning than false positive results, as positive results are normally confirmed by a second method, which essentially eliminates the false-positive calls before the results are reported. This can be challenging, however, if an "unknown" or unreported missense change is the major carrier allele, since even though it will be detected by NGS, it will probably be called as a variant of uncertain significance (VUS) until new evidence regarding its pathogenicity becomes available, and then has the potential to be reported. As previously mentioned, however, NGS information is retained, making reanalysis at a later time a possibility.

6 Variant Interpretation for Carrier Screening

The results interpretation is much more complex in NGS carrier screening than for a genotyping panel consisting of only targeted and well characterized pathogenic mutations. Variant interpretation for carrier screening is different from that of diagnostic testing because in the latter, the patient's phenotype and clinical data aids in determining a variant's pathogenicity. Novel variants are identified when using NGS for carrier testing and need to be categorized for the first time by the laboratory. Unfortunately, the same variant can be interpreted differently by different laboratories, causing confusion for the clinician and patient [37]. This is in part due to a lack of standard approaches in terms of sequencing quality control matrices, variant prioritization rules, access to private variant databases and the contribution of different clinical and laboratory expertise in the analysis. In one study, nine patients were analyzed at eight different genetic centers using different methods (including arrays, targeted gene panels, whole exome, and whole genome sequencing), with none of the labs issuing the same report in regards to variants of uncertain significance (VUS), and with one lab identifying a different causal mutation and therefore an alternative diagnosis for one patient [38]. This study emphasized the inherent inconsistency in variant interpretation, and the need for standardization in the field.

While the ACMG has provided guidelines to help with categorizing variants [39], it remains difficult to determine from the biomedical literature if a variant should be classified as a VUS or likely pathogenic. Since variant interpretation is often based on published results, there can be ambiguity in establishing the genotype to phenotype correlation, with a representative example being the eventual removal of the I148T and 1078delT mutations from the ACMG-recommended *CFTR* screening panel after extensive studies suggested that they were ultimately not disease related [3]. Similarly, a pathogenic mutation can be found in a gene that is not well-characterized for a certain disease, or not consistently implicated in the particular phenotype, making the interpretation difficult [40]. While reputable database such as the Human Gene Mutation Database (HGMD) [41] or Online Mendelian Inheritance of Man (OMIM) [42] can be used as starting points, many variants detected by NGS are novel and will therefore not be found in these databases. Additionally, since healthy individuals also have rare variants, benign variants have mistakenly been reported as disease alleles in these databases based on their low frequency and apparent segregation in families. This raises the concern that some reported pathogenic variants are merely the result of multiple testing bias, which suggests that if enough genes are tested, such as in whole exome sequencing, an association is more likely to be made between a candidate gene and a phenotype. When reviewing literature, multiple families or multiple affected family members, as well as the variant's frequency in control samples and the availability of functional studies such as quantitative measurements of mRNA, protein levels, or enzymatic activity can aide in making a more accurate variant classification [43].

For novel changes, the likelihood of a variant being pathogenic can increase based on the nature of the variant, such as if it occurs at the invariant splice donor or acceptor site (ie positions 1 or 2 of the intron), is a nonsense mutation (ie. generates a premature stop codon), or if it is a frameshift mutation [17]. For truncating variants, it is important to document where the change occurs in the gene. For example, if the variant is located at the extreme 3' end of the gene, the likelihood of it being pathogenic will decrease if nonsense mediated decay is no longer expected to occur. Determining if nearby truncating variants have been previously reported to be disease-causing can also aide in determining a variant's pathogenicity.

In order to quickly and accurately interpret variants, many labs have developed their own variant curation system, where variants are deposited from each case into a laboratory-specific database so that interpretation of the clinical impact of the sequence change is consistent. When a variant is seen for the first time a comprehensive analysis of the variant's potential pathogenicity must be performed. One of the first criteria usually includes the frequency of the variant in the general population. For missense, nonsense and short insertions and/or deletions, the National Center for Biotechnology Information (NCBI) dbSNP database [44] is the largest source of variation data, and it reports on the clinical significance when available. dbSNP works with the 1000 Genomes Project Consortium that expanded on the International HapMap Project [45] to catalog genetic variation shared within and between members of various populations. There are also databases for large structural rearrangements, copy number variants and large indels, such as NCBI's database of genomic structural variation (dbVar) [46] and the Database of Genomic Variants (DGV) [47]. Databases such as ExAC (a compilation of approximately 65,000 exomes), 1000 genomes, or 6500 ESP exomes are also crucial for determining allele frequencies in various populations, since most pathogenic variants will not have frequencies over

5%. In addition to allele frequency, determining pathogenicity often involves a literature and database search. There are several databases that contain variants with disease associations that are useful in identifying pathogenic variants. The OMIM database [42] and the SwissVar database [48] are catalogues of human genes and their associated diseases, including associated phenotypes and, when known, specific variants, while HGMD [41] features disease variants along with their respective publications. NCBI's ClinVar database includes variants associated with phenotypes, along with links to evidence behind the variant interpretation as well as the source of the information. Locus specific databases also exist that report variants in a single gene, often related to a single disease database, e.g., bioPKU for phenylketonuria. Unfortunately, not all of these databases are accurate, with 27% of HMGD entries found to be common polymorphisms, mis-annotated variants, sequencing errors, or to lack significant evidence for pathogenicity [49]. Similarly, 17% of variants in ClinVar have been found to be interpreted differently when submitted by more than one laboratory [37]. Therefore, using these reference databases alone is not sufficient to determine if variants are truly pathogenic.

Some bioinformatics tools use gene-to-gene or protein-to-protein interactions, as well as pathway information to identify potentially pathogenic genes or mutations. These programs focus on a variety of factors, including functional similarities to known mutations, the localization of a variant in the gene, sequence divergence from orthologs, cross-species associations, evolutionary information such as homology, possible functional, folding, or aggregation effects, or structural instability [22, 50–52]. In silico models are also available to help interpret novel variants, including MutPred [53], nsSNPAnalyzer [54], Panther [55], PhD-SNP SVMProfile [56], PolyPhen [57], PolyPhen2 [58], SIFT [59], SNAP [60], SNPs&GO [61], and Mutation Taster [62]. When using these tools, it is important to understand what information the various models use to classify pathogenicity (e.g., Bayesian, mathematical, or empirically derived rules), and the weight given to different attributes. Use of conservation data and in silico prediction models can be useful, however, since they were not developed and validated for clinical use, they should not be used as strong evidence when making clinical interpretations. These programs tend to have higher specificity as opposed to sensitivity, meaning that they are better at predicting variants which are not pathogenic than ones that are pathogenic [40]. A recent comparison of the programs found that results correlated poorly when compared to each other, suggesting that the properties of the variants taken into account needs to be considered when analyzing each variant [50].

Since carrier screening panels have expanded to include diseases that have significant variation in their presentation, including variable age of onset or low penetrance (e.g., Gaucher disease and cystic fibrosis), it is possible for an individual to learn that they potentially have a particular disease when their intention was only to be screened for reproductive purposes [63]. Since such results would be generally unexpected, it is important to annotate variants so that as new information is found, variant information in the curated database can be updated and patients informed of changes in the perceived pathogenicity.

7 Reporting and Test Limitations

NGS reports need to include all the exons that are adequately covered and the sensitivity for important mutations such as founder mutations, complex mutations, or known mutations that are located in homologous regions where detection rates may be low. Information on how post-residual risk is calculated when population prevalence is not known, follow-up testing for partners, and implications for variants that are associated with reduced penetrance, or mild, or late-onset diseases, also needs to be included. The variant reporting criteria (e.g., if only pathogenic mutations will be reported), test limitations, and situations where complex mutation testing was performed by an alternative testing strategy (e.g., for repetitive regions such as exon 10 for *CFTR*), should also be reported in order to inform the referring clinician of the scope and limitations of the testing.

When reporting variants, it is extremely important to include the transcript ID, which is the reference mRNA that is basis of the report. Any coding variants identified should have the gene name, cDNA change, protein change, and any other useful identifiers (e.g., rs numbers or genomic coordinates) included in the report in order to identify the mutation as clearly as possible. A common problem is inconsistent assignment of the amino acid position, causing considerable confusion as to which amino acid is mutated. Many laboratories will also choose to divide their reports into sections, where pathogenic variants and VUSes are listed in separate sections, in an attempt to be clear on what is important for the clinical interpretation. Reports should also list any genomic regions that had low sequence coverage or that were excluded from analysis because of technical problems, affecting the ability to accurately infer the patient's residual risk. In addition, reports should include a summary of all findings, along with an interpretation explaining how the variant(s) was categorized, in addition to citing specific publications appropriately. A general disclosure stating that variant categorization is not always clear should be applied when necessary.

Limitations in terms of reporting are often indicated in the consent forms, and should be explicitly reviewed with the clinician or patient ordering the test. These caveats can include the reporting of susceptibility genes and risk alleles. Limitations in terms of clinical NGS tests include a lack of specific national standards for quality assurance, quality control, test accessioning and reporting, and proficiency evaluations [49]. Panel specific limitations in terms of probe placement, mapping artifacts, and depth of coverage should also be disclosed.

8 Genetic Counseling, Partner Follow-Up and Prenatal Testing

When using a large NGS panel for carrier screening, the likelihood of identifying a positive result is high [17]. Currently, most individuals being screened are women who are already pregnant. Therefore, it is important for future parents to receive proper genetic counseling in order to be informed about their increased risk of having an affected child. The immediate option is to test the carrier's partner if this has not already been performed. Since most laboratories will not report a VUS when performing general population carrier screening (only pathogenic or likely pathogenic variants are typically included), it is necessary to request a full report from the laboratory to ensure all variants, including VUSes, are discussed. This will allow the couple to consider all reproductive options, including pre-implantation genetic diagnosis for individuals who are not yet pregnant, and/or the use of donor gametes, prenatal testing, and adoption.

In the event that both individuals are carriers for a recessive disease but have not yet conceived a child, pre-implantation genetic diagnosis (PGD) can be performed where the presence or absence of parental mutations can be determined in embryos prior to implantation [64]. If conception has taken place naturally, prenatal testing for the parental mutations can be performed. Guidelines exist for prenatal carrier screening that generally state that the conditions screened for should be severe (cause cognitive disabilities, need surgical or medical intervention, and/or affect the quality of life). Additionally, instances where a prenatal diagnosis may result in prenatal intervention to improve perinatal outcome, delivery management, or allow for the necessary prenatal education of parents are also approved [63].

Genetic counseling is particularly important for carrier couples in order to ensure that the risk of having an affected fetus is made clear. It is particularly important for cases where the disease phenotype is not fully penetrant, especially when considering terminating a pregnancy or partaking in a fetal intervention. Discussions about the risks and benefits in identifying variants associated with adult-onset phenotypes should also be offered during pre- and post-testing counseling sessions. Ethical concerns arise when the benefit of doing the analysis is not always clear, especially in instances of late-onset conditions or incomplete penetrance, and may warrant involvement of medical ethicists in the decision making [65].

9 Summary Paragraph

In the short amount of time since its conception, NGS has managed to revolutionize the field of genetics. From the high-throughput capabilities of the technology to the comprehensiveness of the provided results, the benefits of NGS can easily be seen when compared to alternative techniques. Just as the technology has continued to improve, so has the ability to manage and interpret the resulting data, not only in terms of classifying variants but also in terms of counseling patients. NGS has already become the method of choice for many diagnostic and carrier screening tests, and as the financial costs of the test declines while the methodology continues to advance, it will soon become the new gold standard.

References

1. Kronn, D., Jansen, V., Ostrer, H.: Carrier screening for cystic fibrosis, Gaucher disease, and Tay-Sachs disease in the Ashkenazi Jewish population: the first 1000 cases at New York University Medical Center, New York, NY. Arch. Intern. Med. **158**(7), 777–781 (1998)
2. Zuckerman, S., et al.: Carrier screening for Gaucher disease: lessons for lxow-penetrance, treatable diseases. JAMA. **298**(11), 1281–1290 (2007)
3. Watson, M.S., et al.: Cystic fibrosis population carrier screening: 2004 revision of American College of Medical Genetics mutation panel. Genet. Med. **6**(5), 387–391 (2004)
4. Gross, S.J., et al.: Carrier screening in individuals of Ashkenazi Jewish descent. Genet. Med. **10**(1), 54–56 (2008)
5. American College of, O. and C.o.G. Gynecologists: ACOG (American College of Obstetricians and Gynecologists) committee opinion. Genetic screening for hemoglobinopathies, number 238, July 2000 (replaces number 168, February 1996). Committee on genetics. Int. J. Gynaecol. Obstet. **74**(3), 309–310 (2001)
6. Klugman, S., et al.: Detection of carriers in the Ashkenazi Jewish population: an objective comparison of high-throughput genotyping versus gene-by-gene testing. Genet. Test. Mol. Biomarkers. **17**(10), 763–767 (2013)
7. Fedick, A., et al.: High-throughput carrier screening using TaqMan allelic discrimination. PLoS One. **8**(3), e59722 (2013)
8. Schrijver, I., et al.: Comprehensive arrayed primer extension array for the detection of 59 sequence variants in 15 conditions prevalent among the (Ashkenazi) Jewish population. J. Mol. Diagn. **9**(2), 228–236 (2007)
9. Haque, I.S., et al.: Modeled fetal risk of genetic diseases identified by expanded carrier screening. JAMA. **316**(7), 734–742 (2016)
10. Baskovich, B., et al.: Expanded genetic screening panel for the Ashkenazi Jewish population. Genet. Med. **18**(5), 522–528 (2016)
11. Grody, W.W., et al.: Laboratory standards and guidelines for population-based cystic fibrosis carrier screening. Genet. Med. **3**(2), 149–154 (2001)
12. Prior, T.W., Professional, P., Guidelines, C.: Carrier screening for spinal muscular atrophy. Genet. Med. **10**(11), 840–842 (2008)
13. Farkas, D.H., et al.: The suitability of matrix assisted laser desorption/ionization time of flight mass spectrometry in a laboratory developed test using cystic fibrosis carrier screening as a model. J. Mol. Diagn. **12**(5), 611–619 (2010)
14. Lazarin, G.A., et al.: An empirical estimate of carrier frequencies for 400+ causal Mendelian variants: results from an ethnically diverse clinical sample of 23,453 individuals. Genet. Med. **15**(3), 178–186 (2013)
15. Tanner, A.K., et al.: Development and performance of a comprehensive targeted sequencing assay for pan-ethnic screening of carrier status. J. Mol. Diagn. **16**(3), 350–360 (2014)
16. Srinivasan, B.S., et al.: A universal carrier test for the long tail of Mendelian disease. Reprod. Biomed. Online. **21**(4), 537–551 (2010)
17. Hallam, S., et al.: Validation for clinical use of, and initial clinical experience with, a novel approach to population-based carrier screening using high-throughput, next-generation DNA sequencing. J. Mol. Diagn. **16**(2), 180–189 (2014)
18. American College of, Obstetricians and Gynecologists Committee on Genetics: ACOG Committee opinion no. 486: update on carrier screening for cystic fibrosis. Obstet. Gynecol. **117**(4), 1028–1031 (2011)
19. Zvereff, V.V., et al.: Cystic fibrosis carrier screening in a North American population. Genet. Med. **16**(7), 539–546 (2014)
20. Loukas, Y.L., et al.: Clinical diagnostic next-generation sequencing: the case of CFTR carrier screening. Scand. J. Clin. Lab. Invest. **75**(5), 374–381 (2015)
21. Hoffman, J.D., et al.: Next-generation DNA sequencing of HEXA: a step in the right direction for carrier screening. Mol. Genet. Genomic. Med. **1**(4), 260–268 (2013)

22. Capriotti, E., et al.: Bioinformatics for personal genome interpretation. Brief. Bioinform. **13**(4), 495–512 (2012)
23. DePristo, M.A., et al.: A framework for variation discovery and genotyping using next-generation DNA sequencing data. Nat. Genet. **43**(5), 491–498 (2011)
24. Wang, K., Li, M., Hakonarson, H.: ANNOVAR: functional annotation of genetic variants from high-throughput sequencing data. Nucleic Acids Res. **38**(16), e164 (2010)
25. McLaren, W., et al.: Deriving the consequences of genomic variants with the Ensembl API and SNP effect predictor. Bioinformatics. **26**(16), 2069–2070 (2010)
26. Ge, D., et al.: SVA: software for annotating and visualizing sequenced human genomes. Bioinformatics. **27**(14), 1998–2000 (2011)
27. Schrijver, I., et al.: Opportunities and challenges associated with clinical diagnostic genome sequencing: a report of the Association for Molecular Pathology. J. Mol. Diagn. **14**(6), 525–540 (2012)
28. Schrijver, I., et al.: Methods-based proficiency testing in molecular genetic pathology. J. Mol. Diagn. **16**(3), 283–287 (2014)
29. Githens, J.H., et al.: Newborn screening for hemoglobinopathies in Colorado. The first 10 years. Am. J. Dis. Child. **144**(4), 466–470 (1990)
30. Hinton, C.F., Grant, A.M., Grosse, S.D.: Ethical implications and practical considerations of ethnically targeted screening for genetic disorders: the case of hemoglobinopathy screening. Ethn. Health. **16**(4-5), 377–388 (2011)
31. Wang, J., et al.: Capture-based high-coverage NGS: a powerful tool to uncover a wide spectrum of mutation types. Genet. Med. **18**(5), 513–521 (2016)
32. Feng, Y., et al.: Improved molecular diagnosis by the detection of exonic deletions with target gene capture and deep sequencing. Genet. Med. **17**(2), 99–107 (2015)
33. Brown, J.T., et al.: Polymorphisms in the glucocerebrosidase gene and pseudogene urge caution in clinical analysis of Gaucher disease allele c.1448T>C (L444P). BMC Med. Genet. **7**, 69 (2006)
34. Wei, X., et al.: Targeted next-generation sequencing as a comprehensive test for patients with and female carriers of DMD/BMD: a multi-population diagnostic study. Eur. J. Hum. Genet. **22**(1), 110–118 (2014)
35. Larson, J.L., et al.: Validation of a high resolution NGS method for detecting spinal muscular atrophy carriers among phase 3 participants in the 1000 genomes project. BMC Med. Genet. **16**, 100 (2015)
36. Luo, M., et al.: An Ashkenazi Jewish SMN1 haplotype specific to duplication alleles improves pan-ethnic carrier screening for spinal muscular atrophy. Genet. Med. **16**(2), 149–156 (2014)
37. Rehm, H.L., et al.: ClinGen–the clinical genome resource. N. Engl. J. Med. **372**(23), 2235–2242 (2015)
38. Vrijenhoek, T., et al.: Next-generation sequencing-based genome diagnostics across clinical genetics centers: implementation choices and their effects. Eur. J. Hum. Genet. **23**(9), 1270 (2015)
39. Richards, S., et al.: Standards and guidelines for the interpretation of sequence variants: a joint consensus recommendation of the American College of Medical Genetics and Genomics and the Association for Molecular Pathology. Genet. Med. **17**(5), 405–424 (2015)
40. Facio, F.M., Lee, K., O'Daniel, J.M.: A genetic counselor's guide to using next-generation sequencing in clinical practice. J. Genet. Couns. **23**(4), 455–462 (2014)
41. Stenson, P.D., et al.: Human Gene mutation database: towards a comprehensive central mutation database. J. Med. Genet. **45**(2), 124–126 (2008)
42. Amberger, J., et al.: McKusick's Online Mendelian Inheritance in Man (OMIM). Nucleic Acids Res. **37**(Database issue), D793–D796 (2009)
43. Raffan, E., Semple, R.K.: Next generation sequencing–implications for clinical practice. Br. Med. Bull. **99**, 53–71 (2011)
44. Sherry, S.T., et al.: dbSNP: the NCBI database of genetic variation. Nucleic Acids Res. **29**(1), 308–311 (2001)

45. International HapMap Consortium: The international HapMap project. Nature. **426**(6968), 789–796 (2003)
46. Church, D.M., et al.: Public data archives for genomic structural variation. Nat. Genet. **42**(10), 813–814 (2010)
47. Zhang, J., et al.: Development of bioinformatics resources for display and analysis of copy number and other structural variants in the human genome. Cytogenet. Genome Res. **115**(3-4), 205–214 (2006)
48. Mottaz, A., et al.: Easy retrieval of single amino-acid polymorphisms and phenotype information using SwissVar. Bioinformatics. **26**(6), 851–852 (2010)
49. Bell, C.J., et al.: Carrier testing for severe childhood recessive diseases by next-generation sequencing. Sci. Transl. Med. **3**(65), 65ra4 (2011)
50. Thusberg, J., Olatubosun, A., Vihinen, M.: Performance of mutation pathogenicity prediction methods on missense variants. Hum. Mutat. **32**(4), 358–368 (2011)
51. Thusberg, J., Vihinen, M.: Pathogenic or not? And if so, then how? Studying the effects of missense mutations using bioinformatics methods. Hum. Mutat. **30**(5), 703–714 (2009)
52. Niroula, A., Vihinen, M.: Variation interpretation predictors: principles, types, performance, and choice. Hum. Mutat. **37**(6), 579–597 (2016)
53. Li, B., et al.: Automated inference of molecular mechanisms of disease from amino acid substitutions. Bioinformatics. **25**(21), 2744–2750 (2009)
54. Bao, L., Zhou, M., Cui, Y.: nsSNPAnalyzer: identifying disease-associated nonsynonymous single nucleotide polymorphisms. Nucleic Acids Res. **33**(Web Server issue), W480–W482 (2005)
55. Thomas, P.D., et al.: PANTHER: a library of protein families and subfamilies indexed by function. Genome Res. **13**(9), 2129–2141 (2003)
56. Capriotti, E., Calabrese, R., Casadio, R.: Predicting the insurgence of human genetic diseases associated to single point protein mutations with support vector machines and evolutionary information. Bioinformatics. **22**(22), 2729–2734 (2006)
57. Ramensky, V., Bork, P., Sunyaev, S.: Human non-synonymous SNPs: server and survey. Nucleic Acids Res. **30**(17), 3894–3900 (2002)
58. Adzhubei, I.A., et al.: A method and server for predicting damaging missense mutations. Nat. Methods. **7**(4), 248–249 (2010)
59. Ng, P.C., Henikoff, S.: Predicting deleterious amino acid substitutions. Genome Res. **11**(5), 863–874 (2001)
60. Bromberg, Y., Rost, B.: SNAP: predict effect of non-synonymous polymorphisms on function. Nucleic Acids Res. **35**(11), 3823–3835 (2007)
61. Calabrese, R., et al.: Functional annotations improve the predictive score of human disease-related mutations in proteins. Hum. Mutat. **30**(8), 1237–1244 (2009)
62. Schwarz, J.M., et al.: MutationTaster evaluates disease-causing potential of sequence alterations. Nat. Methods. **7**(8), 575–576 (2010)
63. Edwards, J.G., et al.: Expanded carrier screening in reproductive medicine-points to consider: a joint statement of the American College of Medical Genetics and Genomics, American College of Obstetricians and Gynecologists, National Society of Genetic Counselors, Perinatal Quality Foundation, and Society for Maternal-Fetal Medicine. Obstet. Gynecol. **125**(3), 653–662 (2015)
64. Treff, N.R., et al.: Evaluation of targeted next-generation sequencing-based preimplantation genetic diagnosis of monogenic disease. Fertil. Steril. **99**(5), 1377–1384 e6 (2013)
65. Golden-Grant, K., Merritt 2nd, J.L., Scott, C.R.: Ethical considerations of population screening for late-onset genetic disease. Clin. Genet. **88**(6), 589–592 (2015)

Erratum to: Application of Next-Generation Sequencing to Hearing Loss

Lisa Dyer, Xinjian Wang, Kejian Zhang, John Greinwald Jr., and C. Alexander Valencia

Erratum to:
Chapter 5 in: L.-J.C. Wong (ed.), ***Next Generation Sequencing Based Clinical Molecular Diagnosis of Human Genetic Disorders*****,**
DOI 10.1007/978-3-319-56418-0_5

In chapter titled "**Application of Next-Generation Sequencing to Hearing Loss**", the authors realized that the first author of the chapter should be Lisa Dyer not Xinjian Wang. Xinjian Wang should be second author. The correction has been updated in the chapter now.

The updated online version of this chapter can be found at
http://dx.doi.org/10.1007/978-3-319-56418-0_5

L.-J.C. Wong (ed.), *Next Generation Sequencing Based Clinical Molecular Diagnosis of Human Genetic Disorders*, DOI 10.1007/978-3-319-56418-0_17

Index

L.-J.C. Wong (ed.), *Next Generation Sequencing Based Clinical Molecular Diagnosis of Human Genetic Disorders*, DOI 10.1007/978-3-319-56418-0

The manufacturer's authorised representative in the EU is Springer Nature Customer Service Centre GmbH, Europaplatz 3, 69115 Heidelberg, Germany. If you have any concerns regarding our products, please contact ProductSafety@springernature.com

Printed and bound by CPI Group (UK) Ltd, Croydon, CR0 4YY
01/10/2026
02225870-0012